# 计算机英语

## 第5版

刘艺 王春生 等编著

机械工业出版社
CHINA MACHINE PRESS

**图书在版编目（CIP）数据**

计算机英语 / 刘艺等编著．—5 版．—北京：机械工业出版社，2020.9（2024.11 重印）

ISBN 978-7-111-66502-1

I. 计…　II. 刘…　III. 电子计算机 – 英语　IV. TP3

中国版本图书馆 CIP 数据核字（2020）第 170128 号

本书涉及计算机与计算机科学、计算机体系结构、计算机语言与编程、软件开发、软件工程、数据库、计算机网络、因特网、移动与云计算、计算机安全、计算机文化、智能世界等深刻影响我们生活的信息技术。本书以计算机领域英语时文和经典原版教材为基础，通过大量精心挑选的阅读材料，配以相应的注释和练习，使读者能够快速掌握计算机领域的大量专业词汇以及相关的语法等，并提高阅读和检索计算机原版文献资料的能力。

本书选材广泛、图文并茂，采用双色印刷，极大地方便了读者的学习和查阅。书后还附有词汇表和缩略语表。本书可作为高等院校计算机及相关专业“计算机英语”课程的教材，也可供参加计算机水平考试的考生、IT 行业的工程技术人员以及其他有需要的读者学习和参考。

出版发行：机械工业出版社（北京市西城区百万庄大街 22 号　邮政编码：100037）

责任编辑：游　静　　　　责任校对：殷　虹

印　　刷：北京捷迅佳彩印刷有限公司　　　　版　　次：2024 年 11 月第 5 版第 8 次印刷

开　　本：185mm×260mm　1/16　　　　印　　张：22.5

书　　号：ISBN 978-7-111-66502-1　　　　定　　价：79.00 元

客服电话：（010）88361066　68326294

# 前　言

20 年前，新千禧年伊始，温莉芳老师问我能不能编写一本全新的计算机英语教材。对这个想法我们当时讨论得很兴奋，一拍即合。因为我们要编写的计算机英语教材是和国际 IT 教学接轨的易学易用的教材，能改变国内计算机英语教材沉闷落后的状态。那时国内的计算机英语教材大多选材不系统，内容枯燥，语言不地道。而温莉芳老师作为 IT 出版界的急先锋，已经开始大量引进国外优秀的计算机原版教材，对计算机英语教材的出版有着国际眼光。她要求计算机英语教材不仅仅是为了专业英语教学，使学生提高阅读计算机原版资料的能力，更是为了培养学生的计算机文化素养。国内学生没有机会上国外的"计算机科学导论"或"计算机科学概论"这样的公共课，但是通过计算机英语的学习可以掌握这方面的知识和专业词汇。我当时参与了温莉芳老师引进的计算机原版教材的翻译，包括"计算机科学丛书"中的大师力作《计算机科学导论》。"计算机科学丛书"是温莉芳老师从 1998 年开始策划引进的，以每年 25 种左右的规模持续出版了 20 多年，至今共出版 600 余种，总发行量近千万册。

在温莉芳老师的支持下，我邀请王春生老师一起编写《计算机英语》一书。王春生老师是一个在学术和工作中一丝不苟的人。一开始我按照 ACM 推荐的 CS0、CS1、CS2 课程体系编排好了课文，但王春生老师不满意，他坚持从语言地道、行文严谨的教材或正式出版物中取材，反对我使用网上的文章和非英语母语作者（主要是印度裔和华裔）的文章。他的这种"偏执"有时让我为难，因为在美国 IT 行业，是大量非英语国家的技术移民在做贡献，实际交流中，看中的是技术而不是语言。但是王春生老师坚持认为教材中不能出现不地道、不正确的英语，这和实际交流不一样。在他的坚持下，我们这几版的《计算机英语》在选材上一直保持了语言的地道、精准和优雅，虽然增加了编写难度，但是也造就了这部教材与众不同的严谨风格。

在版式设计上，我们也做了创新，《计算机英语》分左右两栏排版，右栏为课文内容，左栏标注课文对应行的生词音标和含义。课文中的难句讲解、缩略词和专有名词注释都放在脚注中，极大地方便了读者的阅读和理解。2000 年我们采用的新颖版式给读者提供了一流的阅读体验，因而沿用至今。最近几年某些出版社的计算机英语教材也悄悄模仿这一版式。

值得一提的是，我们编写的《计算机英语》自世纪之交第 1 版问世以来，历经数次修订再版，20 年来始终深受读者的喜爱，销量在同类教材中一直领先。

在 20 年的时光中，我们紧跟 IT 时代的步伐，不停地对《计算机英语》进行修订，推出新版，直到如今的第 5 版。这 20 年计算机和因特网以惊人的速度发展着，各种技术和应用竞相推出，令人目不暇接。以本教材中相关课文的选材为例：第 1 版中我们的文章还在谈论信息高速公路、电子邮件；第 2 版开始讨论电子商务、GIS；第 3 版涉及 Web 2.0、网购；第 4 版转向关注移动应用、云计算、物联网；第 5 版则变成人工智能、增强现实、雾计算和边缘计算。每次版本的更新，我们都引入了论述计算机和 IT 领域重大技术发展和应用的前瞻文章，以保持与时俱进。

第 5 版同样紧跟计算机日新月异的发展趋势，在保留第 4 版中计算机和网络原理、编程开发的经典核心内容的基础上，新增和改写了计算机硬件及 IT 应用、互联网和人工智能等新技术内容。在保持全书 36 篇课文不变的体量下，更换了 8 篇课文，修改了 8 篇课文，更新内容超过 40%。可以说第 5 版是历次版本中改动最大的一版。

考虑到长期使用本书的教师和读者的用书习惯，本书的编写格式与第 4 版保持一致，即课文中有难点和知识点注释、生词标注，课后有练习题，书后还附有练习参考答案、词汇表和缩略语表。我们还在线提供每个单元课文 A 的参考译文㊀。

英语在计算机及 IT 行业有着其他语言所不能替代的功能。无论是学习最新的计算机技术，还是使用最新的计算机软硬件产品，都离不开对计算机英语的熟练掌握。而且，对于生活在信息时代的人们，计算机已成为工作、学习、生活中的必备工具，掌握一定程度的计算机英语大有裨益。正是为了适应这种要求，不少院校纷纷开设了计算机英语课程。有些院校不仅把它作为计算机专业的必修课，还将其作为一门实用的选修课推广到其他专业。

本书是按照教育部《大学英语教学指南》对专门用途英语的要求，为计算机英语课程而编写的教材。在满足计算机专业英语教学的同时，我们并没有过分沉溺于晦涩抽象的理论和专业术语之中，而是注重实际应用与调动学习兴趣。本书选材广泛，内容丰富，涉及计算机基础知识、硬件结构、程序设计、软件工程、应用开发、网络通信、信息安全等深刻影响我们生活的信息技术。这次经过全面修订的第 5 版更加适应深化计算机英语教学改革的需要。

这次修订主要完成以下工作：

- **课文更新**　第 5 版替换了内容过时或不太合适的课文，同时替换了相应的练习题，新课文占全书总量的近 1/4。新增的课文紧跟技术潮流，把握主流趋势。既有涉及人工智能、增强现实、雾计算与边缘计算、面向服务的软件工程、计算机专业人员的道德准则等方面的最新时文，也有对计算机类别、数据挖掘、防病毒软件等原有课文的全面内容更新。借此也引入了大量新的技术术语和缩略语，为读者尽快独立阅读英文资料，融入英文技术社区提供帮助。
- **对保留的课文进行修改**　在本次修订中，我们对近 1/4 的课文及练习进行了较大幅度的内容更新，并对其他一些保留课文中的疏漏进行了改正，调整了部分注解，使之更加贴切。

㊀ 在线资源请访问机工网站 www.cmpedu.com 下载。——编辑注

- **保持全书的体系结构** 本书参考 ACM 推荐的 CS0、CS1、CS2 课程体系，并保持第 4 版的 12 个教学单元不变。新版中替换了过时的文章，并调整和优化了内容结构，新增了“智能世界”单元。

本书在出版过程中得到了机械工业出版社的鼎力支持，同时洪蕾、王珊珊、刘哲雨、王子凡等也为本书的编写和顺利出版付出了心血，在此一并表示感谢！

尽管本书在编写过程中，在资料的查核、术语的汉译、生词的注音以及文字的规范等方面都做了大量工作，但由于计算机领域的发展日新月异，许多新术语尚无确定的规范译法，加上编者水平有限，书中难免有不尽如人意之处，恳请广大读者不吝赐教。

刘 艺
2020 年 7 月于南京
E-mail：anygooddream@163.com

# 使用说明

1．本教材共 12 单元，每单元包括三篇文章。课文 A 为精读材料，课文 B 和课文 C 为泛读材料。三篇课文均围绕同一主题。课文 A 一般为该主题的概述，课文 B 和课文 C 多就该主题的某一具体方面或具体例子展开讨论。课文长度一般为 1200～1800 词。对于课文 A，要求能正确理解和熟练掌握其内容。对于课文 B 和课文 C，要求能掌握中心大意，抓住主要事实。

2．课文 A 配有四项练习，即“课文理解填空”“词组中英文互译”“完形填空”和“段落翻译”；课文 B 和课文 C 配有两项练习，即“课文理解填空”和“词组中英文互译”。各项练习均与课文内容和计算机专业紧密结合，旨在巩固和拓展所学内容。

3．生词均用彩色粗体在课文中标出，并在课文旁边的文本框中进行注释，以便于阅读和记忆。相同生词原则上只在首次出现之处进行注释，但书后附有词汇表备查。每个生词一般标注一个发音，但有的常用异读音也标了出来，中间用逗号分隔；如系英美发音差异，英国发音在前，美国发音在后，中间用分号分隔。生词注音中的斜体音标表示该音可读可不读；短划线（-）用于截同示异，代表与前面所注发音相同的部分。

4．计算机英语的特点之一是大量使用缩略语。本教材对缩略语的处理方式是，在首次出现之处加脚注说明，并将所有缩略语收入书后的缩略语表，以备查阅和方便记忆。另外，在计算机英语中，缩略语所代表的词组或术语在大小写上有比较随意和不一致的现象，本教材原则上尊重原文所采用的形式。

5．教育部 2017 年印发的《大学英语教学指南》（以下简称《指南》）将大学英语教学目标分为基础、提高、发展三个等级。《指南》规定：基础目标是针对大多数非英语专业学生的英语学习基本需求确定的，是大部分学生本科毕业时应达到的基本要求。《指南》说明：大学英语教学的主要内容可分为通用英语、专门用途英语和跨文化交际三个部分，专门用途英语课程以增强学生运用英语进行专业和学术交流、从事工作的能力，提升学生学术和职业素养为目的。《指南》还指出：各高校可根据人才培养规格和学生需要开设专门用途英语课程，也可在通用英语体系内，纳入通用学术英语和职业英语等内容。根据《指南》的精神，以及教学上的易操作性，本教材主要用于专门开设的计算机专业英语课，建议本科生安排在第五学期至第七学期，研究生安排在第一学年。生词选注参照“基础目标”应掌握的词汇，所注生词有两类：一类是通用词，即超出“基础目标”的词汇；另一类是计算机及相关专业词汇。

编　者

2020 年 7 月

# 目　录

# Unit 1 Computer and Computer Science

# （计算机与计算机科学）

## Section A

## Computer Overview

**overview**
/ ˈəuvəvjuː /
n. 概述；概观

**calculation**
/ ˌkælkjuˈleiʃ ən /
n. 计算

**numeric(al)**
/ njuːˈmerik(əl) /
a. 数字的；数值的

**banking** / ˈbæŋkiŋ /
n. 银行业（务）

### I. Introduction

A computer is an electronic device that can receive a set of instructions, or program, and then carry out this program by performing **calculations** on **numerical** data or by manipulating other forms of information.

The modern world of high technology could not have come about[1] except for the development of the computer. Different types and sizes of computers find uses throughout society in the storage and handling of data, from secret governmental files to **banking** transactions to private household

1 *come about*：发生，产生。

accounts. Computers have opened up a new era in manufacturing through the techniques of **automation**, and they have enhanced modern communication systems. They are essential tools in almost every field of research and applied technology, from constructing models of the universe to producing tomorrow's weather reports, and their use has in itself opened up new areas of **conjecture**. Database services and computer networks make available a great variety of information sources[1]. The same advanced techniques also make possible invasions of personal and business **privacy**. Computer crime has become one of the many risks that are part of the price of modern technology.

## II. History

The first adding machine, a **precursor** of the digital computer, was devised in 1642 by the French scientist, mathematician, and philosopher Blaise Pascal[2]. This device employed a series of ten-toothed wheels, each tooth representing a **digit** from 0 to 9. The wheels were connected so that numbers could be added to each other by advancing the wheels by a correct number of teeth. In the 1670s the German philosopher and mathematician Gottfried Wilhelm Leibniz[3] improved on this machine by devising one that could also multiply.

The French inventor Joseph-Marie Jacquard[4], in designing an automatic **loom**, used thin, **perforated** wooden boards to control the weaving of complicated designs. During the 1880s the American **statistician** Herman Hollerith[5] **conceived** the idea of using perforated cards, similar to Jacquard's boards, for processing data. Employing a system that passed **punched cards** over **electrical contacts**, he was able to **compile** statistical information for the 1890 United States **census**.

### 1. The Analytical Engine

Also in the 19th century, the British mathematician and inventor

**automation** /ˌɔːtəˈmeiʃən/
n. 自动化
**conjecture** /kənˈdʒektʃə/
n. 推测，猜想
**privacy** /ˈprivəsi, ˈprai-; ˈprai-/
n. 隐私；秘密
**precursor** /ˌpriːˈkəːsə, pri-/
n. 先驱
**digit** /ˈdidʒit/
n. 数字
**loom** /luːm/
n. 织机
**perforated** /ˈpəːfəreitid/
a. 穿孔的
**statistician** /ˌstætiˈstiʃən/
n. 统计学家；统计员
**conceive** /kənˈsiːv/
v.（构）想出
**punched card**
穿孔卡片
**electrical contact**
电触点
**compile** /kəmˈpail/
v. 汇编；编译
**census** /ˈsensəs/
n. 人口普查
**analytic(al)** /ˌænəˈlitik(əl)/
a. 分析的
**Analytical Engine**
分析机，解析机

[1] *Database services and computer networks make available a great variety of information sources.*：数据库服务和计算机网络使各种各样的信息源可供使用。这句话的宾语（a great variety of information sources）较长，因此将宾语的补语（available）放到了宾语的前面。文中下一句话也属于同样情况。

[2] *Blaise Pascal*：布莱斯·帕斯卡（1623—1662），法国数学家、物理学家、哲学家，概率论创立者之一。

[3] *Gottfried Wilhelm Leibniz*：戈特弗里德·威廉·莱布尼兹（1646—1716），德国自然科学家、哲学家，微积分、数理逻辑的先驱，提出了二进制。

[4] *Joseph-Marie Jacquard*：约瑟夫-玛丽·雅卡尔（1752—1834），法国著名的织机工匠，纹板提花机的主要改革家。

[5] *Herman Hollerith*：赫尔曼·何勒里斯（1860—1929），美国发明家和统计学家。

Charles Babbage[1] worked out the principles of the modern digital computer. He conceived a number of machines, such as the **Difference Engine**, that were designed to handle complicated mathematical problems. Many historians consider Babbage and his associate, the mathematician Augusta Ada Byron[2], the true pioneers of the modern digital computer. One of Babbage's designs, the Analytical Engine, had many features of a modern computer. It had an **input stream** in the form of a **deck** of punched cards, a "store" for saving data, a "mill" for arithmetic operations, and a printer that made a permanent record[3]. Babbage failed to put this idea into practice, though it may well have been technically possible at that date.

### 2. Early Computers

**Analogue** computers began to be built in the late 19th century. Early models calculated by means of rotating **shafts** and gears. Numerical **approximations** of equations too difficult to solve in any other way were evaluated with such machines. Lord Kelvin[4] built a mechanical tide predictor that was a specialized analogue computer. During World Wars I and II, mechanical and, later, electrical analogue computing systems were used as **torpedo** course predictors in **submarines** and as **bombsight** controllers in aircraft. Another system was designed to predict spring floods in the Mississippi River[5] basin.

### 3. Electronic Computers

During World War II, a team of scientists and mathematicians, working at Bletchley Park, north of London, created one of the first all-electronic digital computers: Colossus[6]. By December 1943, Colossus, which **incorporated** 1,500 **vacuum tubes**, was operational. It was used by the team headed by Alan Turing[7], in the largely successful attempt to **crack** German radio messages **enciphered** in the Enigma code[8].

**Difference Engine**
差分机

**input stream**
输入（信息）流

**deck** / dek /
n. 卡片叠，卡片组

**analog(ue)**
/ 'ænəlɔg /
a. 模拟的

**shaft** / ʃɑːft /
n. 轴

**approximation**
/ əˌprɔksi'meiʃən /
n. 近似（值）

**torpedo** / tɔː'piːdəu /
n. 鱼雷

**submarine**
/ ˌsʌbmə'riːn /
n. 潜艇

**bombsight**
/ 'bɔmsait /
n. 轰炸瞄准器

**incorporate**
/ in'kɔːpəreit /
v. 包含；把…合并；使并入

**vacuum tube**
真空管

**crack** / kræk /
v. 破译

**encipher** / in'saifə /
v. 把…译成密码

---

1 *Charles Babbage*：查尔斯·巴比奇（1792—1871），英国数学家和发明家。

2 *Augusta Ada Byron*：奥古斯塔·埃达·拜伦（1815—1852），英国数学家，诗人拜伦之女。

3 *It had an input stream in the form of a deck of punched cards, a "store" for saving data, a "mill" for arithmetic operations, and a printer that made a permanent record.*：它有一个以一叠穿孔卡片的形式存在的输入流、一个保存数据的"仓库"、一个进行算术运算的"工厂"和一个产生永久性记录的打印机。

4 *Lord Kelvin*：开尔文勋爵（1824—1907），全名威廉·汤姆森·开尔文（William Thomson Kelvin），英国物理学家，发展了热力学理论，创立了热力学绝对温标（即开尔文温标）。

5 *the Mississippi River*：密西西比河，发源于美国中北部的湖沼区，南注墨西哥湾，系美国主要河流。

6 *Colossus*：该词读作 /kə'lɔsəs/，有"巨像""巨人""巨物"等意。

7 *Alan Turing*：艾伦·图灵（1912—1954），英国数学家和逻辑学家。

8 *Enigma code*：恩尼格码，德军在第二次世界大战期间采用的一种密码。

**prototype** / ˈprəutətaip /
n. 原型；样机
**overshadow** / ˌəuvəˈʃædəu /
v. 使相形见绌
**integrator** / ˈintigreitə /
n. 积分器
**patent** / ˈpeitənt /
n. 专利（权）
**overturn** / ˌəuvəˈtə:n /
v. 推翻；废除

Independently of this, in the United States, a **prototype** electronic machine had been built as early as 1939, by John Atanasoff[1] and Clifford Berry[2] at Iowa State College[3]. This prototype and later research were completed quietly and later **overshadowed** by the development of the Electronic Numerical **Integrator** And Computer (ENIAC[4]) in 1945. ENIAC was granted a **patent**, which was **overturned** decades later, in 1973, when the machine was revealed to have incorporated principles first used in the Atanasoff-Berry Computer.

**Hungarian** / hʌŋˈgɛəriən /
a. 匈牙利的
**memory** / ˈmeməri /
n. 存储器，内存
**paper-tape reader**
纸带阅读器
**execution** / ˌeksiˈkjuʃən /
n. 执行，运行

ENIAC (see Figure 1A-1) contained 18,000 vacuum tubes and had a speed of several hundred multiplications per minute, but originally its program was wired into the processor[5] and had to be manually altered. Later machines were built with program storage, based on the ideas of the **Hungarian**-American mathematician John von Neumann[6]. The instructions, like the data, were stored within a "**memory**", freeing the computer from the speed limitations of the **paper-tape reader** during **execution** and permitting problems to be solved without rewiring the computer.

Figure 1A-1: ENIAC was one of the first fully electronic digital computers

---

1 *John Atanasoff*：约翰·阿塔纳索夫（1903—1995），美国物理学家。
2 *Clifford Berry*：克利福德·贝里（1918—1963），美国物理学家。
3 *Iowa State College*：（美国）艾奥瓦州立学院。
4 *ENIAC*：电子数字积分计算机（*E*lectronic *N*umerical *I*ntegrator *A*nd *C*omputer 的首字母缩略），读作 / ˈi:niæk /。
5 *originally its program was wired into the processor*：其程序最初是通过导线传送到处理器内的。
6 *John von Neumann*：约翰·冯·诺依曼（1903—1957），美籍匈牙利数学家，对量子物理、数学逻辑和高速计算机的发展均有贡献。

**transistor** / træn'zistə, -'sis- / n. 晶体管
**advent** / 'ædvent / n. 出现，到来
**versatile** / 'vəːsətail / a. 多用途的，通用的
**logical element** 逻辑元件
**spacing** / 'speisiŋ / n. 间隔
**integrated** / 'intigreitid / a. 集成的，综合的，一体化的
**integrated circuit** 集成电路
**fabricate** / 'fæbrikeit / v. 制作
**substrate** / 'sʌbstreit / n. 衬底，基底
**microprocessor** / ˌmaikrəu'prəusesə / n. 微处理器
**microchip** / 'maikrəutʃip / n. 微芯片
**etch** / etʃ / v. 蚀刻

The use of the **transistor** in computers in the late 1950s marked the **advent** of smaller, faster, and more **versatile logical elements** than were possible with vacuum-tube machines. Because transistors use much less power and have a much longer life, this development alone was responsible for the improved machines called second-generation computers. Components became smaller, as did inter-component **spacings**, and the system became much less expensive to build.

### 4. Integrated Circuits

Late in the 1960s the integrated circuit, or IC[1] (see Figure 1A-2), was introduced, making it possible for many transistors to be **fabricated** on one silicon **substrate**, with interconnecting wires plated in place[2]. The IC resulted in a further reduction in price, size, and failure rate. The **microprocessor** became a reality in the mid-1970s with the introduction of the large-scale integrated (LSI[3]) circuit and, later, the very large-scale integrated (VLSI[4]) circuit (**microchip**), with many thousands of interconnected transistors **etched** into a single silicon substrate.

Figure 1A-2: An Integrated Circuit

To return, then, to the switching capabilities of a modern computer: computers in the 1970s were generally able to handle eight switches at a

[1] *IC*：集成电路（*i*ntegrated *c*ircuit 的首字母缩略）。
[2] *making it possible for many transistors to be fabricated on one silicon substrate, with interconnecting wires plated in place*：从而有可能将许多晶体管制作在一块硅衬底上，晶体管之间用覆镀在适当位置的导线相连接。
[3] *LSI*：大规模集成的（*l*arge-*s*cale *i*ntegrated 的首字母缩略）。
[4] *VLSI*：超大规模集成的（*v*ery *l*arge-*s*cale *i*ntegrated 的首字母缩略）。

time. That is, they could deal with eight **binary** digits, or **bits**, of data, at every cycle. A group of eight bits is called a **byte**, each byte containing 256 possible patterns of ONs and OFFs (or 1s and 0s). Each pattern is the equivalent of an instruction, a part of an instruction, or a particular type of **datum**, such as a number or a character or a **graphics** symbol. The pattern 11010010, for example, might be binary data—in this case, the **decimal** number 210—or it might be an instruction telling the computer to compare data stored in its **switches** to data stored in a certain memory-**chip** location.

The development of processors that can handle 16, 32, and 64 bits of data at a time has increased the speed of computers. The complete collection of recognizable patterns—the total list of operations—of which a computer is capable is called its **instruction set**. Both factors—the number of bits that can be handled at one time, and the size of instruction sets—continue to increase with the ongoing development of modern digital computers.

## III. Hardware

Modern digital computers are all **conceptually** similar, regardless of size. Nevertheless, they can be divided into several categories on the basis of cost and performance: the personal computer or **microcomputer**, a relatively low-cost machine, usually of **desktop** size (though "**laptops**" are small enough to fit in a briefcase, and "**palmtops**" can fit into a pocket); the **workstation**, a microcomputer with enhanced graphics and communications capabilities that make it especially useful for office work; the **minicomputer**, generally too expensive for personal use, with capabilities suited to a business, school, or laboratory; and the **mainframe** computer, a large, expensive machine with the capability of serving the needs of major business enterprises, government departments, scientific research establishments, or the like (the largest and fastest of these are called **supercomputers**).

A digital computer is not a single machine: rather, it is a system composed of five distinct elements: (1) a **central processing unit**; (2) input devices; (3) memory storage devices; (4) output devices; and (5) a communications network, called a **bus**, which links all the elements of the system and connects the system to the external world.

## IV. Programming

A program is a sequence of instructions that tells the hardware of a

**binary** /ˈbainəri/ a. 二进制的；二元的
**bit** /bit/ n. 位，比特
**byte** /bait/ n. 字节
**datum** /ˈdeitəm/ ([复] **data**) n. 数据
**graphics** /ˈgræfiks/ n. 图形，图形显示
**decimal** /ˈdesiməl/ a. 十进制的
**switch** /ˈswitʃ/ n. 交换（设备）；开关；转换
**chip** /tʃip/ n. 芯片
**instruction set** 指令集
**conceptually** /kənˈseptjuəli/ ad. 概念上
**microcomputer** /ˈmaikrəukəmˌpju:tə/ n. 微型计算机
**desktop** /ˈdesktɔp/ a. 桌面的；台式（计算机）的
**laptop** /ˈlæptɔp/ n. 膝上型计算机
**palmtop** /ˈpɑ:mtɔp/ n. 掌上型计算机
**workstation** /ˈwə:kˌsteiʃən/ n. 工作站
**minicomputer** /ˈminikəmˌpju:tə/ n. 小型计算机
**mainframe** /ˈmeinfreim/ n. 主机，大型机
**supercomputer** /ˈsju:pəkəmˌpju:tə/ n. 超级计算机，巨型计算机
**central processing unit** 中央处理器
**bus** /bʌs/ n. 总线

computer what operations to perform on data. Programs can be built into the hardware itself, or they may exist independently in a form known as software. In some specialized, or "**dedicated**", computers the operating instructions are **embedded** in their **circuitry**; common examples are the microcomputers found in calculators, **wristwatches**, car engines, and **microwave ovens**. A general-purpose computer, on the other hand, although it contains some **built-in** programs (in ROM[1]) or instructions (in the processor chip), depends on external programs to perform useful tasks. Once a computer has been programmed, it can do only as much or as little as the software controlling it at any given moment enables it to do. Software in widespread use includes a wide range of applications programs—instructions to the computer on how to perform various tasks.

**dedicated** /ˈdedikeitid/
a. 专用的
**embed** /imˈbed/
v. 把…嵌入
**circuitry** /ˈsəːkitri/
n. 电路
**wristwatch** /ˈristwɔtʃ/
n. 手表
**microwave oven**
微波炉
**built-in** /ˈbiltˈin/
a. 内置的，内部的

## V. Future Developments

There is active research to make computers out of many promising new types of technology, such as optical computers, DNA[2] computers, **neural** computers, and **quantum** computers. Most computers are universal, and are able to calculate any computable function, and are limited only by their memory capacity and operating speed. However, different designs of computers can give very different performance for particular problems; for example, quantum computers can potentially break some modern **encryption algorithms** (by quantum **factoring**) very quickly.

A computer will solve problems in exactly the way it is programmed to, without regard to efficiency, alternative solutions, possible **shortcuts**, or possible errors in the code. Computer programs that learn and adapt are part of the emerging field of **artificial intelligence** and machine learning. Artificial intelligence-based products generally fall into two major categories: rule-based systems and pattern recognition systems. Rule-based systems attempt to represent the rules used by human experts and tend to be expensive to develop. Pattern-based systems use data about a problem to generate conclusions. Examples of pattern-based systems include voice recognition, **font** recognition, translation and the emerging field of on-line marketing.

**neural** /ˈnjuərəl/
a. 神经的
**quantum** /ˈkwɔntəm/
（[复] **-ta** /-tə/）
n. 量子
**encryption** /inˈkripʃən/
n. 加密
**algorithm** /ˈælgəriðəm/
n. 算法
**factoring** /ˈfæktəriŋ/
n. 因子分解，因式分解
**shortcut** /ˈʃɔːtkʌt/
n. 近路，捷径
**artificial intelligence**
人工智能
**font** /fɔnt/
n. 字体

[1] *ROM*：只读存储器（*r*ead-*o*nly *m*emory 的首字母缩略），读作 /rɔm/。
[2] *DNA*：脱氧核糖核酸（*d*eoxyribo*n*ucleic *a*cid 的缩略）。

## Exercises

**I. Fill in the blanks with the information given in the text:**

1. According to many historians, the true pioneers of the modern digital computer are ________ and ________.

2. A digital computer is generally made up of five distinct elements: a central processing unit, ________ devices, memory storage devices, ________ devices, and a bus.

3. The microprocessor is a central processing unit on a single chip. It was made possible in the mid-1970s with the introduction of the LSI circuit and the ________ circuit.

4. According to the text, modern digital computers can be divided into four major categories on the basis of cost and performance. They are microcomputers, ________, minicomputers, and ________.

5. The first electronic computers, such as Colossus and ENIAC created in Britain and the United States respectively, used ________ tubes, which later gave place to ________.

6. A program is a sequence of ________ that can be executed by a computer. It can either be built into the hardware or exist independently in the form of ________.

7. The smallest unit of information handled by a computer is bit, which is the abbreviation of binary ________. A group of ________ bits is called a(n) ________.

8. Active research is being conducted to use promising new types of technology to make new types of computers, such as ________ computers, DNA computers, neural computers, and ________ computers.

**II. Translate the following terms or phrases from English into Chinese and vice versa:**

1. artificial intelligence
2. paper-tape reader
3. optical computer
4. neural computer
5. instruction set
6. quantum computer
7. difference engine
8. versatile logical element

9. silicon substrate
10. vacuum tube
11. 数据的存储与处理
12. 超大规模集成电路
13. 中央处理器
14. 个人计算机
15. 模拟计算机
16. 数字计算机
17. 通用计算机
18. 处理器芯片
19. 操作指令
20. 输入设备

**III. Fill in each of the blanks with one of the words given in the following list, making changes if necessary:**

| | | | |
|---|---|---|---|
| *microcomputer* | *computing* | *digital* | *base* |
| *advent* | *mode* | *circuit* | *significance* |
| *chip* | *appear* | *speed* | *transistor* |
| *minicomputer* | *combine* | *categorization* | *integration* |

We can define a computer as a device that accepts input, processes data, stores data, and produces output. According to the ________ of processing, computers are either analog or ________. They can also be classified as mainframes, ________, workstations, or microcomputers. All else (for example, the age of the machine) being equal, this ________ provides some indication of the computer's ________, size, cost, and abilities.

Ever since the ________ of computers, there have been constant changes. First-generation computers of historic ________, such as *UNIVAC* (通用自动计算机), introduced in the early 1950s, were ________ on vacuum tubes. Second-generation computers, ________ in the early 1960s, were those in which ________ replaced vacuum tubes. In third-generation computers, dating from the 1960s, integrated ________ replaced transistors. In fourth-generation computers such as ________, which first appeared in the mid-1970s, large-scale ________ enabled thousands of circuits to be incorporated on one ________. Fifth-generation computers are expected to ________ very-large-scale integration with sophisticated approaches to ________, including artificial intelligence and true distributed processing.

**IV. Translate the following passage from English into Chinese:**

Computers will become more advanced and they will also become easier to use. Improved speech recognition will make the operation of a computer easier. *Virtual reality* (虚拟现实), the technology of interacting with a computer using all of the human senses, will also contribute to better human and computer *interfaces* (界面，接口). Other, *exotic* (奇异的) models of computation are being developed, including biological computing that uses living organisms, *molecular* (分子的) computing that uses molecules with particular properties, and computing that uses deoxyribonucleic acid (DNA), the basic unit of *heredity* (遗传), to store data and carry out operations. These are examples of possible future computational platforms that, so far, are limited in abilities or are strictly theoretical. Scientists investigate them because of the physical limitations of *miniaturizing* (使小型化) circuits embedded in silicon. There are also limitations related to heat generated by even the tiniest of transistors.

# Section B

# What Is Computer Science

## I. Introduction

Computer science is the study of the theory, experimentation, and engineering that form the basis for the design and use of computers—devices that automatically process information. Computer science traces its roots to work done by English mathematician Charles Babbage, who first proposed a programmable mechanical calculator in 1837. Until the advent of electronic digital computers in the 1940s, computer science was not generally distinguished as being separate from mathematics and engineering. Since then it has **sprouted** numerous branches of research that are unique to the discipline.

**sprout** / spraut /
v. 发芽；长出

## II. The Development of Computer Science

Early work in the field of computer science during the late 1940s and early 1950s focused on automating the process of making calculations for use in science and engineering. Scientists and engineers developed theoretical models of computation that enabled them to analyze how

**overlap** / ˌəuvəˈlæp /
v. 重叠，层叠
**numerical analysis**
数值分析

efficient different approaches were in performing various calculations. Computer science **overlapped** considerably during this time with the branch of mathematics known as **numerical analysis**, which examines the accuracy and precision of calculations.

**interface** / ˈintəfeis /
n. 界面；接口
**application**
/ ˌæpliˈkeiʃən /
n. 应用程序，应用软件
**mass-produce**
/ ˈmæsprəˌdjuːs /
v.（使用机器）大量生产

As the use of computers expanded between the 1950s and the 1970s, the focus of computer science broadened to include simplifying the use of computers through programming languages—artificial languages used to program computers, and operating systems—computer programs that provide a useful **interface** between a computer and a user. During this time, computer scientists were also experimenting with new **applications** and computer designs, creating the first computer networks, and exploring relationships between computation and thought.

In the 1970s, computer chip manufacturers began to **mass-produce** microprocessors—the electronic circuitry that serves as the main information processing center in a computer. This new technology revolutionized the computer industry by dramatically reducing the cost of building computers and greatly increasing their processing speed. The microprocessor made possible the advent of the personal computer, which resulted in an explosion in the use of computer applications. Between the early 1970s and 1980s, computer science rapidly expanded in an effort to develop new applications for personal computers and to drive the technological advances in the computing industry. Much of the earlier research that had been done began to reach the public through personal computers, which derived most of their early software from existing concepts and systems.

**integral** / ˈintigrəl /
a. 不可缺少的，必要的
**strive** / straiv /
v. 努力；力争
**speculative**
/ ˈspekjulətiv, -lei- /
a. 推测的；投机性的
**viable** / ˈvaiəbəl /
a. 切实可行的；可实施的
**underlie** / ˌʌndəˈlai /
v. 构成…的基础（或起因）；支持

Computer scientists continue to expand the frontiers of computer and information systems by pioneering the designs of more complex, reliable, and powerful computers; enabling networks of computers to efficiently exchange vast amounts of information; and seeking ways to make computers behave intelligently. As computers become an increasingly **integral** part of modern society, computer scientists **strive** to solve new problems and invent better methods of solving current problems.

The goals of computer science range from finding ways to better educate people in the use of existing computers to highly **speculative** research into technologies and approaches that may not be **viable** for decades. **Underlying** all of these specific goals is the desire to better the human condition today and in the future through the improved use of information[1].

## III. Theory and Experiment

Computer science is a combination of theory, engineering, and experimentation. In some cases, a computer scientist develops a theory, then engineers a combination of computer hardware and software based on that theory, and experimentally tests it. An example of such a theory-driven approach is the development of new software engineering tools that are then evaluated in actual use. In other cases, experimentation may result in new theory, such as the discovery that an artificial **neural network** exhibits behavior similar to **neurons** in the brain, leading to a new theory in **neurophysiology**.

**neural network**
神经网络
**neuron** / 'njuərɔn /
n. 神经元，神经细胞
**neurophysiology**
/ 'njuərəu,fizi'ɔlədʒi /
n. 神经生理学
**unforeseen**
/ ,ʌnfɔ:'si:n /
a. 未预见到的；意料之外的

It might seem that the predictable nature of computers makes experimentation unnecessary because the outcome of experiments should be known in advance. But when computer systems and their interactions with the natural world become sufficiently complex, **unforeseen** behavior can result. Experimentation and the traditional scientific method are thus key parts of computer science.

## IV. Major Branches of Computer Science

Computer science can be divided into four main fields: software development, computer **architecture** (hardware), human-computer **interfacing** (the design of the most efficient ways for humans to use computers), and artificial intelligence (the attempt to make computers behave intelligently). Software development is concerned with creating computer programs that perform efficiently. Computer architecture is concerned with developing **optimal** hardware for specific **computational** needs. The areas of artificial intelligence (AI[2]) and human-computer interfacing often involve the development of both software and hardware to solve specific problems.

**architecture**
/ 'ɑ:kitektʃə /
n. 体系结构
**interfacing**
/ 'intə,feisiŋ /
n. 接口技术
**optimal** / 'ɔptiməl /
a. 最优的，最佳的
**computational**
/ ,kɔmpju:'teiʃənəl /
a. 计算（机）的

### 1. Software Development

In developing computer software, computer scientists and engineers study various areas and techniques of software design, such as the best

---

[1] *Underlying all of these specific goals is the desire to better the human condition today and in the future through the improved use of information.*：这是一个倒装句，正常语序应为 The desire to better … information is underlying all of these specific goals.。主语部分的中心词desire后面跟了一个较长的不定式短语作定语，采用倒装语序是为了避免头重脚轻。

[2] *AI*：人工智能（*a*rtificial *i*ntelligence 的首字母缩略）。

**retrieve** / ri'tri:v /
v. 检索

types of programming languages and algorithms to use in specific programs, how to efficiently store and **retrieve** information, and the computational limits of certain software-computer combinations. Software designers must consider many factors when developing a program. Often, program performance in one area must be sacrificed for the sake of the general performance of the software. For instance, since computers have only a limited amount of memory, software designers must limit the number of features they include in a program so that it will not require more memory than the system it is designed for can supply.

**robust**
/ rəu'bʌst, 'rəubʌst /
a. 健壮的，坚固的
**life cycle**
生命周期
**specification**
/ ˌspesifi'keiʃən /
n. 明确说明；[常作~s] 规格，规范
**implementation**
/ ˌimplimen'teiʃən /
n. 实现，执行

Software engineering is an area of software development in which computer scientists and engineers study methods and tools that facilitate the efficient development of correct, reliable, and **robust** computer programs. Research in this branch of computer science considers all the phases of the software **life cycle**, which begins with a formal problem **specification**, and progresses to the design of a solution, its **implementation** as a program, testing of the program, and program maintenance. Software engineers develop software tools and collections of tools called programming environments to improve the development process. For example, tools can help to manage the many components of a large program that is being written by a team of programmers.

**2. Computer Architecture**

Computer architecture is the design and analysis of new computer systems. Computer architects study ways of improving computers by increasing their speed, storage capacity, and reliability, and by reducing their cost and power consumption. Computer architects develop both software and hardware models to analyze the performance of existing and proposed computer designs, and then use this analysis to guide the development of new computers. They are often involved with the engineering of a new computer because the accuracy of their models depends on the design of the computer's circuitry. Many computer architects are interested in developing computers that are specialized for particular applications such as image processing, signal processing, or the control of mechanical systems. The **optimization** of computer architecture to specific tasks often yields higher performance, lower cost, or both[1].

**optimization**
/ ˌɔptimai'zeiʃən /
n.（最）优化，最佳化

---

[1] *The optimization of computer architecture to specific tasks often yields higher performance, lower cost, or both.*：按照特定任务对计算机体系结构进行的优化，常常带来性能的提高、成本的降低或两者。本句中的介词 to 表示"按""按照"。

**mimic** / 'mimik /
v. 模仿
**sensory** / 'sensəri /
a. 感觉的；感官的
**cognition** / kɔg'niʃən /
n. 认识，认知
**computer vision**
计算机视觉
**robotics** / rəu'bɔtiks /
n. 机器人学；机器人技术
**complexity** / kəm'pleksiti /
n. 复杂性，错综性
**automated** / 'ɔ:təmeitid /
a. 自动化的
**repetitive** / ri'petitiv /
a. 重复的
**roboticist** / rəu'bɔtisist /
n. 机器人专家
**sensor** / 'sensə /
n. 传感器；敏感元件
**dexterity** / dek'steriti /
n. 灵巧，敏捷
**diversity** / dai'və:siti /
n. 多样性；差异（性）
**unify** / 'ju:nifai /
v. 使联合，统一；使一致
**disability** / ,disə'biliti /
n. 无能力；残疾
**three-dimensional** / 'θri:di'menʃənəl /
a. 三维的，立体的
**virtual** / 'və:tʃuəl /
a. 虚拟的
**virtual reality**
虚拟现实
**head-up display**
前导显示器，平视显示器

### 3. Artificial Intelligence

Artificial intelligence (AI) research seeks to enable computers and machines to **mimic** human intelligence and **sensory** processing ability, and models human behavior with computers to improve our understanding of intelligence. The many branches of AI research include machine learning, inference, **cognition**, knowledge representation, problem solving, case-based reasoning, natural language understanding, speech recognition, **computer vision**, and artificial neural networks.

### 4. Robotics

Another area of computer science that has found wide practical use is robotics—the design and development of computer controlled mechanical devices. Robots range in **complexity** from toys to **automated** factory assembly lines, and relieve humans from tedious, **repetitive**, or dangerous tasks. Robots are also employed where requirements of speed, precision, consistency, or cleanliness exceed what humans can accomplish. **Roboticists**—scientists involved in the field of robotics—study the many aspects of controlling robots. These aspects include modeling the robot's physical properties, modeling its environment, planning its actions, directing its mechanisms efficiently, using **sensors** to provide feedback to the controlling program, and ensuring the safety of its behavior. They also study ways of simplifying the creation of control programs. One area of research seeks to provide robots with more of the **dexterity** and adaptability of humans, and is closely associated with AI.

### 5. Human-Computer Interfacing

Human-computer interfaces provide the means for people to use computers. An example of a human-computer interface is the keyboard, which lets humans enter commands into a computer and enter text into a specific application. The **diversity** of research into human-computer interfacing corresponds to the diversity of computer users and applications. However, a **unifying** theme is the development of better interfaces and experimental evaluation of their effectiveness. Examples include improving computer access for people with **disabilities**, simplifying program use, developing **three-dimensional** input and output devices for **virtual reality**, improving handwriting and speech recognition, and developing **head-up displays** for aircraft instruments in which critical information such as speed, altitude, and heading are displayed on a screen in front of the pilot's

**visualization**
/ ˌvizjuəlaiˈzeiʃ ən /
n. 可视化，直观化
**graphic(al)**
/ ˈgræfik(əl) /
a. 图形的，图示的
**comprehend**
/ ˌkɔmpriˈhend /
v. 理解，领会

window. One area of research, called **visualization**, is concerned with **graphically** presenting large amounts of data so that people can **comprehend** its key properties.

## V. Connection of Computer Science to Other Disciplines

Theoretical computer science draws many of its approaches from mathematics and logic. Research in numerical computation overlaps with mathematics research in numerical analysis. Computer architects work closely with the electrical engineers who design the circuits of a computer.

**linguistics**
/ liŋˈgwistiks /
n. 语言学
**physiologist**
/ ˌfiziˈɔlədʒist /
n. 生理学家

Beyond these historical connections, there are strong ties between AI research and psychology, neurophysiology, and **linguistics**. Human-computer interface research also has connections with psychology. Roboticists work with both mechanical engineers and **physiologists** in designing new robots.

**collaboration**
/ kəˌlæbəˈreiʃ ən /
n. 合作；协作
**interdisciplinary**
/ ˌintəˈdisiplinəri /
a. 学科间的，跨学科的

Computer science also has indirect relationships with virtually all disciplines that use computers. Applications developed in other fields often involve **collaboration** with computer scientists, who contribute their knowledge of algorithms, data structures, software engineering, and existing technology. In return, the computer scientists have the opportunity to observe novel applications of computers, from which they gain a deeper insight into their use. These relationships make computer science a highly **interdisciplinary** field of study.

---

# Exercises

**I. Fill in the blanks with the information given in the text:**

1. Computer science, which traces its roots to work done by English mathematician Charles Babbage, is a combination of theory, engineering, and ________.

2. Computer science can be divided into four main fields: software development, computer architecture, human-computer ________, and artificial intelligence.

3. Computer science is said to be a highly ________ field of study because it has strong ties with many disciplines and indirect relationships with virtually all disciplines that use computers.

4. According to the text, it was the ________ that made possible the advent of the personal computer.

**II. Translate the following terms or phrases from English into Chinese and vice versa:**

1. artificial neural network
2. computer architecture
3. robust computer program
4. human-computer interface
5. knowledge representation
6. 数值分析
7. 程序设计环境
8. 数据结构
9. 存储和检索信息
10. 虚拟现实

# Section C

## Digital Devices

At one time, it was possible to define three distinct categories of computers. Mainframes were housed in large, closet-sized metal frames. Minicomputers were smaller, less expensive, and less powerful, but they could support multiple users and run business software. Microcomputers were clearly **differentiated** from computers in other categories because they were **dedicated** to a single user and their CPUs[1] consisted of a single microprocessor chip.

**differentiate** /ˌdifəˈrenʃieit/ v. 区分，区别

**dedicate** /ˈdedikeit/ v. 把…献给；把…用于（*to*）

Today, microprocessors are no longer a distinction between computer categories because just about every computer uses one or more microprocessors as its CPU. The term *minicomputer* has fallen into disuse, and the terms *microcomputer* and *mainframe* are used with less and less

[1] *CPU*：中央处理器（*c*entral *p*rocessing *u*nit 的首字母缩略）。

frequency.

## I. Enterprise Computers

Today's most powerful computers include supercomputers, mainframes, and **servers**. These devices are generally used in businesses and government agencies. They have the ability to service many **simultaneous** users and process data at very fast speeds.

**server** / 'sə:və /
n. 服务器；服务程序
**simultaneous** / ˌsiməl'teiniəs; ˌsai- /
a. 同时（发生）的

### 1. Supercomputers

A computer falls into the supercomputer category if it is, at the time of construction, one of the fastest computers in the world. Because of their speed, supercomputers can tackle complex tasks that just would not be practical for other computers. Typical uses for supercomputers include breaking codes, modeling worldwide weather systems, and **simulating** nuclear explosions.

Computer manufacturers such as IBM[1], Cray[2], and China's NRCPC[3] have in recent years held top honors for the world's fastest computer. Supercomputer speeds are measured in **petaflops** (PFLOPS[4]). One petaflop is an **astounding** 1,000,000,000,000,000 (**quadrillion**) calculations per second. That's about 20,000 times faster than your **laptop** computer.

**simulate** / 'simjuleit /
v. 模拟，仿真
**petaflops** / 'petəflɔps /
n. 每秒千万亿次浮点运算，每秒 $10^{15}$ 次浮点运算
**astounding** / ə'staundiŋ /
a. 令人震惊的；使人惊骇的
**quadrillion** / kwɔ'driljən /
n.〈英〉$10^{24}$（100万的4次幂）；〈美〉$10^{15}$（1000的5次幂）
**laptop** / 'læptɔp /
a. 膝上型的

### 2. Mainframes

A mainframe computer (or simply a mainframe) is a large and expensive computer capable of simultaneously processing data for hundreds or thousands of users. Its main processing circuitry is housed in a closet-sized cabinet; but after large components are added for storage and output, a mainframe installation can fill a **good-sized** room.

**good-sized** / 'gud'saizd /
a.（相当）大的；宽大的

Mainframes are generally used by businesses and government agencies to provide centralized storage, processing, and management for large amounts of data. For example, banks depend on mainframes as their

---

[1] *IBM*: 国际商用机器公司（*I*nternational *B*usiness *M*achines 的首字母缩略），一家集信息技术、咨询服务和业务解决方案于一体的公司，创立于 1911 年，总部位于美国纽约州阿蒙克市（Armonk）。

[2] *Cray*: 克雷公司，一家超级计算机制造商，创立于 1972 年，总部位于美国华盛顿州西雅图市，2019 年被慧与公司（Hewlett-Packard Enterprise）收购。

[3] *NRCPC*:（中国）国家并行计算机工程技术研究中心（*N*ational *R*esearch *C*enter of *P*arallel *C*omputer Engineering and Technology 的缩略）。

[4] *PFLOPS*: 每秒千万亿次浮点运算（*petaflops* 的缩略）。flops 系 *f*loating-point *o*perations *p*er *s*econd 的首字母缩略，表示“每秒浮点运算次数”。

computer of choice[1] to ensure reliability, data security, and centralized control. The price of a mainframe computer **typically** starts at several hundred thousand dollars and can easily exceed $1 million.

**typically** / 'tipikəli /
ad. 一般地，通常

### 3. Servers

The purpose of a server is to "serve" data to computers connected to a network. When you search Google[2] or access a **Web**[3] **site**, the information you obtain is provided by servers. At **ecommerce** sites, the store's **merchandise** information is housed in database servers. Email, chat, Skype[4], and online multiplayer games are all operated by servers.

**Web site**
网站，站点
**ecommerce**
/ i:'kɔmə:s /
n. 电子商务
**merchandise**
/ 'mə:tʃəndaiz, -dais /
n. [总称] 商品
**configure**
/ kən'figə /
v. 配置

Technically, just about any computer can be **configured** to perform the work of a server. However, computer manufacturers such as IBM and Dell[5] offer devices classified as servers that are especially suited for storing and distributing data on networks. These devices are about the size of a desk drawer and are often mounted in racks of multiple servers.

## II. Personal Computers

A personal computer is designed to meet the computing needs of an individual. These computers were originally referred to as microcomputers. Personal computers provide access to a wide variety of computing applications, such as word processing, photo editing, email, and Internet access.

The term *personal computer* is sometimes **abbreviated** as PC[6]. However, *PC* can also refer to a specific type of personal computer that descended from the original IBM PC and runs Windows software.

Personal computers can be classified as desktop, **portable**, or mobile devices. The lines that **delineate** these categories are sometimes a bit **blurry**, but the general characteristics of each category are described below.

**abbreviate**
/ ə'bri:vieit /
v. 缩写
**portable** / 'pɔ:təbəl /
a. 便携式的；可移植的
**delineate** / di'linieit /
v. 勾画出…的轮廓；画出；描绘
**blurry** / 'blə:ri /
a. 模糊的，难辨认的

---

1 *of choice*：应先选择的；首选的；广受欢迎的。

2 *Google*：谷歌公司，全球最大的搜索引擎公司，业务范围包括网络信息服务、搜索引擎、移动操作系统等，创立于1998年，总部位于美国加利福尼亚州芒廷维尤（Mountain View，亦译作"山景城"）。

3 *Web*：万维网，WWW网（全称为World Wide Web或World-Wide Web，常缩略为WWW、$w^3$、W3和Web）。

4 *Skype*：讯佳普，一款即时通信软件，由Skype Limited公司开发、运行。Skype Limited创立于1998年，总部位于卢森堡，2011年被微软公司收购，成为微软的一个独立部门。Skype读作 / skaip /。

5 *Dell*：戴尔公司，一家开发、制造和销售计算机及相关产品的公司，创立于1984年，总部位于美国得克萨斯州朗德罗克（Round Rock）。

6 *PC*：个人计算机，个人电脑（personal computer的首字母缩略）。

**desktop** / 'desktɔp /
n. 台式（计算）机；桌面

### 1. Desktops

A desktop computer fits on a desk and runs on power from an electrical wall outlet. The keyboard is typically a separate component connected to the main unit by a cable. A desktop computer can be housed in a vertical case or in a horizontal case. In some modern desktops, called all-in-one units, the system board is incorporated into the display device.

**portability**
/ ˌpɔːtə'biliti /
n. 便于携带；可移植性

**portable** / 'pɔːtəbəl /
n. 便携式计算机

Desktop computers are popular for offices and schools where **portability** is not important. Their operating systems include Microsoft[1] Windows, macOS[2], and Linux[3]. The price of an entry-level desktop computer starts at $500 or a bit less.

### 2. Portables

A portable computer runs on battery power. Its screen, keyboard, camera, storage devices, and speakers are fully contained in a single case so that the device can be easily transported from one place to another. Portable computers include laptops, **tablets**, and smartphones.

**tablet** / 'tæblit /
n. 平板电脑

#### (1) Laptops

A laptop computer (also referred to as a notebook computer) is a small, lightweight personal computer designed like a **clamshell** with a keyboard as the base and a screen on the **hinged** cover. Most laptops use the same operating systems as desktop computers, with the exception of Chromebooks[4], which use Google's Chrome OS[5] as their operating system.

**clamshell**
/ 'klæmʃel /
n. 蛤壳

**hinged** / hindʒd /
a. 有铰链的

**hard disk**
硬（磁）盘

**form factor**
（电子产品等的）物理尺寸和形状；规格

**app** / æp /
n. 应用程序（*app*lication 的缩略）

A Chromebook is a special category of laptop, designed to be connected to the Internet for most routine computing activities. Chromebook owners use Web-based software and store all their data in the cloud rather than on a local **hard disk**. Chromebooks use a standard clamshell **form factor**, so they look very much like a laptop. Their sub-$300 price tags are attractive to consumers who primarily browse the Web and use Web-based **apps**.

---

[1] *Microsoft*：微软公司，全球最大的电脑软件提供商，创立于 1975 年，总部位于美国华盛顿州雷德蒙德市（Redmond）。

[2] *macOS*：Mac 操作系统，苹果公司为 Mac 系列产品开发的专属操作系统。*OS*：操作系统（*o*perating *s*ystem 的首字母缩略）。

[3] *Linux*：Linux 操作系统，一种自由和开放源码的类 UNIX 操作系统，其内核由芬兰人林纳斯 • 托瓦兹（Linus Torvalds）于 1991 年 10 月首次发布。

[4] *Chromebook*：谷歌公司推出的网络笔记本（上网本）。

[5] *Chrome OS*：Chrome 操作系统，谷歌公司为上网本设计的轻量级开源操作系统。

**touch-sensitive**
/ 'tʌtʃ,sensitiv /
a. 触敏的
**touch-sensitive screen**
触摸屏
**cell phone**
蜂窝电话，
移动电话，手机
**slate** / sleit /
n. & a. 石板（的）
**slate tablet**
平板电脑
**configuration**
/ kən,figju'reiʃən /
n. 配置
**convertible**
/ kən'və:tibəl /
a. 可转变的；可转换的；可折叠的
**smartphone**
/ 'sma:tfəun /
n. 智能手机
**touchscreen**
/ 'tʌtʃskri:n /
n. 触摸屏，触屏

**(2) Tablets**

A tablet computer is a portable computing device featuring a **touch-sensitive screen** that can be used for input as well as for output. Tablet computers use specialized operating systems, such as iOS[1] and Android[2], or special operating system modes, such as Windows 10 Tablet mode. Some models support **cell phone** network data plans but require apps such as Google Voice[3] or Skype for voice calls.

A **slate tablet configuration** is basically a screen in a narrow frame that lacks a physical keyboard (although one can be attached). The Apple[4] iPad[5] and Samsung[6] Galaxy Tab[7] are popular slate tablets. A 2-in-1 (or **convertible** tablet)[8] can be operated using its touch-sensitive screen or with a physical keyboard that can be folded out of the way or removed.

**(3) Smartphones**

Smartphones are mobile devices that have features similar to tablet computers, but also provide telecommunications capabilities over cell phone networks. They can make voice calls, send text messages, and access the Internet. Unlike a basic mobile phone, smartphones are programmable, so they can download, store, and run software.

Smartphones are the most commonly used digital device in the world. A smartphone features a small keyboard or **touchscreen** and is designed to fit into a pocket, run on batteries, and be used while you are holding it in your hands.

Smartphones are equipped with built-in speech recognition that allows you to ask questions and control the device using spoken commands. Smartphones also include GPS[9] capability so that apps are able to provide

---

1 *iOS*：iOS 操作系统，原名 iPhone OS，系苹果公司为其移动设备开发的操作系统。

2 *Android*：安卓操作系统，一种基于 Linux 的自由且开放源代码的操作系统，主要用于移动设备，最初由安迪 • 鲁宾（Andy Rubin）等开发，后由谷歌公司和开放手机联盟改进。

3 *Google Voice*：谷歌之音，简称 GV，谷歌公司推出的一种电话服务，通过所提供的一个独立的美国电话号码来管理用户所有的手机号码、电话号码、语音邮件、短信等。

4 *Apple*：苹果公司，主要业务包括消费电子产品、计算机软件、在线服务和个人计算机，创立于1976年，总部位于美国加利福尼亚州丘珀蒂诺（Cupertino）。

5 *iPad*：苹果公司推出的平板电脑产品系列。

6 *Samsung*：三星集团，创立于 1938 年，总部位于韩国京畿道城南市，三星电子是其核心子公司。

7 *Galaxy Tab*：三星公司推出的平板电脑产品系列。

8 *2-in-1 (or convertible tablet)*：二合一平板电脑（或变形本）。convertible tablet 可译为“可变形平板电脑”“变形平板”“变形本”等。

9 *GPS*：全球（卫星）定位系统（*G*lobal *P*ositioning *S*ystem 的首字母缩略）。

**navigation** / ˌnævi'geiʃ ən /
n. 导航；航行

location-based services such as a route **navigation** map or a list of nearby restaurants.

Smartphones evolved from basic cell phones and PDAs[1]. A PDA (personal digital assistant) was a handheld device used as an electronic appointment book, calculator, and **notepad**. Modern smartphones include a similar **suite** of applications, but they also have access to a huge variety of mobile apps that help you calculate tips, play your favorite music, and entertain you with games.

**notepad** / 'nəutpæd /
n. 记事本；便笺簿；拍纸簿

**suite** / swi:t /
n.（同类物的）系列，组，套

The operating systems for smartphones are similar to those used for tablet computers. iOS is used on the iPad and iPhone[2]. Microsoft Windows 10 Mobile[3] is used on smartphones that offer a similar user experience as Windows laptops. The Android operating system used on Samsung tablets is also used for Samsung Galaxy[4] and Motorola[5] Droid[6] smartphones.

**niche** / nitʃ, ni:ʃ /
n. 缝隙市场，利基市场，小众市场

**fitness tracker**
健身跟踪器，健身追踪器

**smartwatch** / 'smɑ:twɔtʃ /
n. 智能手表

**wearable computer**
可穿戴计算机，穿戴式计算机

## III. Niche Devices

The list of digital devices is long. Many devices, such as **fitness trackers**, cameras, and handheld GPSs, are dedicated to specific tasks. Other devices perform a wider variety of tasks.

Niche devices all have one thing in common: They contain a microprocessor. Some of these devices, such as **smartwatches** and fitness trackers, can be classified as **wearable computers**.

### 1. Raspberry Pi[7]

**tad** / tæd /
n.〈口〉微量

A full computer system unit that is just a **tad**[8] larger than a deck of cards, the Raspberry Pi can be connected to a keyboard and screen for a full computer experience. These little devices cost less than $50 and provide an inexpensive platform for experimenting with programming, robotics, and just about any creative computer application you can imagine.

---

1 *PDA*：个人数字助理（*p*ersonal *d*igital *a*ssistant 的首字母缩略）。

2 *iPhone*：苹果手机，苹果公司研发及销售的智能手机系列。

3 *Windows 10 Mobile*：Windows 10 移动版，微软开发的手机操作系统。

4 *Galaxy*：三星公司推出的智能手机系列。

5 *Motorola*：摩托罗拉公司，一家涉及消费电子、移动通信、互联网领域的公司，创立于 1928 年，总部位于美国伊利诺伊州肖姆堡（Schaumburg），2011 年拆分为摩托罗拉系统公司和摩托罗拉移动公司，后者现为联想集团旗下的全资子公司。

6 *Droid*：摩托罗拉与美国电信公司 Verizon（威瑞森）及谷歌公司合作的智能手机品牌。

7 *Raspberry Pi*：树莓派，由英国慈善组织 Raspberry Pi Foundation 推出的一款针对电脑业余爱好者、教师、小学生以及小型企业等用户的迷你电脑。

8 *a tad*：稍微，略微。

**console** / ˈkɔnsəul /
n. 控制台，操纵台

**stream** / striːm /
v. 流播

### 2. Game Consoles

Devices for playing computer games include Sony's[1] PlayStation[2], Nintendo's[3] Wii[4], and Microsoft's Xbox[5]. They feature powerful processing capability and excellent graphics, but they are generally used for dedicated game playing and **streaming** videos rather than running application software.

### 3. Portable Media Players

Media players, such as the iPod Touch[6], revolutionized the music industry by providing consumers with a handheld device that can store and play thousands of songs. These devices are controlled by touchscreens or simple click-wheel mechanisms[7].

### 4. Smartwatches

**dub** / dʌb /
v. 给…起绰号；把…称为

Watches and clocks were some of the first devices to go digital. Mass produced in the 1970s with a price as low as $10, these watches were limited to time and date functions. In 2013, Samsung, Google, and Qualcomm[8] introduced a new breed of digital watch. **Dubbed** smartwatches, these multifunction devices can include a camera, thermometer, compass, calculator, cell phone, GPS, media player, and fitness tracker. Some smartwatch functions are onboard the device, whereas other functions require access to the Internet or to the wearer's smartphone.

### 5. Activity Trackers

**calorie** / ˈkæləri /
n. 卡（路里）；热量

**graph** / græf, grɑːf /
v. 用图（或图表、曲线图等）表示

To monitor activity throughout the day, you can wear a fitness tracker. These devices, worn on the wrist or clipped to a pocket, monitor your steps and heart rate. They can calculate **calories**, **graph** your fitness achievements, and share information with your Facebook[9] friends.

---

[1] *Sony*：索尼公司，经营范围包括电子、娱乐、金融、信息技术等，创立于1946年，总部位于日本东京。
[2] *PlayStation*：简称PS，索尼推出的家用游戏机品牌。
[3] *Nintendo*：任天堂株式会社，一家游戏娱乐企业，现主要生产和销售家用游戏机和游戏软件，创立于1889年，总部位于日本京都市。
[4] *Wii*：任天堂推出的家用游戏机。
[5] *Xbox*：微软开发的家用游戏机系列。
[6] *iPod Touch*：iPod系苹果公司推出的系列便携式数字多媒体播放器。iPod Touch属于iPod系列的分支。
[7] *click-wheel mechanisms*：点按式选盘机制，由一个转轮和五个物理按钮组成，菜单按钮位于转轮上方，前进/后退按钮位于左右两侧，播放/暂停按钮位于下方，确定按钮位于中间。
[8] *Qualcomm*：高通公司，一家无线电通信技术和芯片研发公司，创立于1985年，总部位于美国加利福尼亚州圣迭戈市。
[9] *Facebook*：脸书，脸谱，一家社交网络服务网站，创立于2004年，总部位于美国加利福尼亚州门洛帕克（Menlo Park）。

**6. Smart Appliances**

**microcontroller**
/ 'maikrəukən'trəulə /
n. 微控制器

Modern refrigerators, washing machines, and other appliances are controlled by integrated circuits called **microcontrollers** that combine sensors with processing circuitry. Microcontrollers can monitor energy efficiency, offer programmed start times, and may be controlled remotely from a smartphone or laptop.

---

## Exercises

**I. Fill in the blanks with the information given in the text:**

1. A computer is classified as a(n) ________ if it is one of the fastest computers in the world when constructed. The speed of such computers is measured in ________.

2. A(n) ________ computer is the only type of personal computer that must remain plugged into an electrical source during operation.

3. Portable computers include laptops, ________, and smartphones, all of which run on battery power.

4. Some niche devices, such as smartwatches and fitness trackers, can be classified as ________ computers.

**II. Translate the following terms or phrases from English into Chinese and vice versa:**

1. niche device
2. fitness tracker
3. touch-sensitive screen
4. mainframe computer
5. 膝上型计算机
6. 平板电脑
7. 便携式媒体播放器
8. 硬盘

# Unit 2 Computer Architecture
# （计算机体系结构）

## Section A

## Computer Hardware

### I. Introduction

Computer hardware is the equipment involved in the function of a computer and consists of the components that can be physically handled. The function of these components is typically divided into three main categories: input, output, and storage. Components in these categories connect to microprocessors, specifically, the computer's central processing unit (CPU), the electronic circuitry that provides the computational ability and control of the computer, via wires or circuitry called a bus.

Software, on the other hand, is the set of instructions a computer uses to manipulate data, such as a word-processing program or a **video game**. These programs are usually stored and transferred via the computer's hardware to and from the CPU. Software also governs how the hardware is utilized; for example, how information is retrieved from a storage device. The interaction between the input and output hardware is controlled by software called the Basic Input/Output System (BIOS[1]) software.

Although microprocessors are still technically considered to be hardware, portions of their function are also associated with computer software. Since microprocessors have both hardware and software aspects, they are therefore often referred to as **firmware**.

## II. Input Hardware

Input hardware consists of external devices—that is, components outside of the computer's CPU—that provide information and instructions to the computer. A **light pen** is a **stylus** with a light sensitive tip that is used to draw directly on a computer's video screen or to select information on the screen by pressing a clip in the light pen or by pressing the light pen against the surface of the screen. The pen contains light sensors that identify which portion of the screen it is passed over. A mouse is a pointing device designed to be gripped by one hand. It has a detection device on the bottom that enables the user to control the motion of an on-screen **pointer**, or **cursor**, by moving the mouse on a flat surface. As the device moves across the surface, the cursor moves across the screen. To select items or choose commands on the screen, the user presses a button on the mouse. A **joystick** is a pointing device composed of a **lever** that moves in multiple directions to **navigate** a cursor or other graphical object on a computer screen. A keyboard is a typewriter-like device that allows the user to type in text and commands to the computer. Some keyboards have special **function keys** or integrated pointing devices, such as a **trackball** or touch-sensitive regions that let the user's finger motions move an on-screen cursor.

An optical **scanner** uses light-sensing equipment to convert images such as a picture or text into electronic signals that can be manipulated by a computer. For example, a photograph can be scanned into a computer and then included in a text document created on that computer. The two most common scanner types are the **flatbed scanner**, which is similar to an office **photocopier**, and the **hand-held scanner**, which is passed manually

**video game**
电子游戏

**firmware**
/ ˈfəːmwεə /
n. ［总称］固件

**light pen**
光笔

**stylus** / ˈstailəs /
（［复］-**luses** 或-**li** / -lai /）
n. 输入笔，光笔

**pointer** / ˈpɔintə /
n. 指针（光标）；指示字，指示符

**cursor** / ˈkəːsə /
n. 光标

**joystick** / ˈdʒɔistik /
n. 控制杆，操纵杆，游戏杆

**lever** / ˈliːvə; ˈle- /
n.（杠）杆；控制杆

**navigate** / ˈneivigeit /
v. 航行（于）；（为…）领航；指引

**function key**
功能键，操作键，函数键

**trackball** / ˈtrækbɔːl /
n. 跟踪球，轨迹球

**scanner** / ˈskænə /
n. 扫描仪；扫描程序

**flatbed scanner**
平板扫描仪

**photocopier**
/ ˈfəutəuˌkɔpiə /
n. 复印机

**hand-held scanner**
手持式扫描仪

[1] *BIOS*：基本输入 / 输出系统（*B*asic *I*nput/*O*utput *S*ystem 的首字母缩略），读作 / ˈbaiəus /。

across the image to be processed. A microphone is a device for converting sound into signals that can then be stored, manipulated, and played back by the computer. A voice recognition **module** is a device that converts spoken words into information that the computer can recognize and process.

A modem, which stands for **modulator-demodulator**, is a device that connects a computer to a telephone line and allows information to be transmitted to or received from another computer. Each computer that sends or receives information must be connected to a modem. The information sent from one computer is converted by the modem into an audio signal, which is then transmitted by telephone lines to the receiving modem, which converts the signal into information that the receiving computer can understand.

## III. Output Hardware

A computer display device that simply displays text and images is classified as an output device. Touchscreens, however, can be classified as both input and output devices because they accept input and also display output. **Standalone** display devices, sometimes called monitors, are popular for desktop computers. Display devices for laptops, tablets, and handheld devices are built into the system units, but these devices may also accept an external monitor. Two technologies are commonly used for display devices: LCD[1] and LED[2]. LCD (**liquid crystal display**) technology produces an image by filtering light through a layer of liquid crystal cells. The advantages of LCD screens include display **clarity**, low radiation **emission**, portability, and compactness. The source of the light that filters through the LCD is referred to as **backlighting**. Most modern screens are **backlit** with LED (**light-emitting diode**) technology and are marketed as LED displays.

Today's best-selling multifunction printers use ink jet or laser technology and can also serve as scanners, **copiers**, and fax machines. An **ink jet printer** has a **nozzle**-like print head that sprays ink onto paper to form characters and graphics. The print head in a color ink jet printer consists of a series of nozzles, each with its own **ink cartridge**. A **laser printer** uses the same technology as a photocopier to paint dots of light on a light-sensitive **drum**. **Electrostatically** charged ink[3] is applied to the drum and then transferred to paper. A basic laser printer produces only

**module** / 'mɔdju:l /
n. 模块
**modulator**
/ 'mɔdjuleitə /
n. 调制器
**demodulator**
/ di:'mɔdjuleitə /
n. 解调器
**standalone**
/ 'stændə,ləun /
a. 独立的
**liquid crystal display**
液晶显示(器)
**clarity** / 'klæriti /
n. 清晰,明晰
**emission** / i'miʃən /
n. 发出;射出;发射
**backlighting**
/ 'bæklaitiŋ /
n. 背后照明;背光
**backlight** / 'bæklait /
( -lighted 或 -lit / -lit / )
v. 从背后照亮
**diode** / 'daiəud /
n. 二极管
**light-emitting diode**
发光二极管
**copier** / 'kɔpiə /
n. 复印机
**ink jet printer**
喷墨打印机
**nozzle** / 'nɔzəl /
n. 喷嘴;管嘴
**cartridge** / 'kɑ:tridʒ /
n. 盒,匣
**ink cartridge**
墨盒
**laser printer**
激光打印机
**drum** / drʌm /
n. 磁鼓
**electrostatic**
/ i,lektrə'stætik /
a. 静电的;静电学的

[1] *LCD*:液晶显示(器)(*l*iquid *c*rystal *d*isplay 的首字母缩略)。
[2] *LED*:发光二极管(*l*ight-*e*mitting *d*iode 的缩略)。
[3] *electrostatically charged ink*:带静电的墨水。

black-and-white printouts. Color laser printers are available, but they are somewhat more costly than basic black-and-white models. The technology that deposits ink on paper is the foundation for 3D[1] printers that deposit layers of plastic, **resin**, or metal that build into a three-dimensional object. 3D printing is technically called **additive manufacturing**. There are several additive manufacturing technologies, but most consumer-grade 3D printers use a technology called **fused** deposition modeling (FDM[2]) that melts a coiled **filament** and deposits it in layers that harden and form an object.

## IV. Storage Hardware

A computer's **main memory** is organized in manageable units called cells. To reflect the ability to access cells in any order, a computer's main memory is often called **random access memory** (RAM[3]). RAM is the "waiting room" for the microprocessor. It holds raw data waiting to be processed, the program instructions for processing that data, and operating system instructions that control the basic functions of a computer system. It also holds the results of processing until they can be moved to a more permanent location.

ROM (**read-only memory**) contains a small set of instructions and data called the **boot loader**. The boot loader instructions tell a digital device how to start. Typically, the boot loader performs self-tests to find out if the hardware is operating properly and may also verify that essential programs have not been **corrupted**. It then loads the operating system into RAM. Whereas RAM is temporary and **volatile**, ROM is more permanent and non-volatile. The contents of ROM remain in place even when the device is turned off.

Most computers have additional memory devices called **mass storage** (or secondary storage) systems. Three types are commonly used for personal computers: magnetic, optical, and solid state.

Magnetic storage represents data by **magnetizing microscopic** particles on a disk or tape surface. The first personal computers used **cassette tapes** for storage, though **floppy disk** storage was soon available. Today, the most common example of magnetic storage technology is the

**resin** / 'rezin /
n. 树脂；合成树脂
**additive** / 'æditiv /
a. 添加的
n. 添加物；添加剂
**additive manufacturing**
增材制造
**fuse** / fju:z /
v. 熔合；熔凝
**filament** / 'filəmənt /
n. 细丝；灯丝
**main memory**
主存（储器）
**random** / 'rændəm /
a. 随机的；任意的
**random access memory** 随机（存取）存储器
**read-only memory**
只读存储器
**boot** / bu:t /
v. 启动，引导（*up*）
**boot loader**
引导装入程序，引导加载程序
**corrupt** / kə'rʌpt /
v. 破坏，损坏；腐蚀
**volatile** / 'vɔlətail /
a. 易失（性）的
**mass storage**
海量存储器，大容量存储器
**magnetize** / 'mægnitaiz /
v.（使）磁化
**microscopic** / ˌmaikrə'skɔpik /
a. 非用显微镜不可见的；微小的
**cassette tape**
盒式磁带
**floppy** / 'flɔpi /
a.（松）软的
**floppy disk**
软（磁）盘

1 *3D*：三维的，立体的（*three-d*imensional 的缩略），读作 / 'θri:'di: /。
2 *FDM*：熔融沉积成型（*f*used *d*eposition *m*odeling 的首字母缩略）。
3 *RAM*：随机（存取）存储器（*r*andom *a*ccess *m*emory 的首字母缩略），读作 / ræm /。

**magnetic disk**
磁盘
**drive** / draiv /
n. 驱动器
**hard disk drive**
硬(磁)盘驱动器
**flash memory**
闪存,快闪存储器
**erasable** / iˈreizəbəl /
a. 可擦(除)的;可消除的
**memory card**
存储卡
**solid state drive**
固态驱动器
**flash drive**
闪存驱动器

**magnetic disk** or **hard disk drive** (HDD[1]).

CD[2], DVD[3], and Blu-ray (BD[4]) technologies are classified as optical storage, which represents data as microscopic light and dark spots on the disc[5] surface. Optical technologies are grouped into three categories: read-only (ROM), recordable (R), and rewritable (RW)[6].

Solid state storage (sometimes called **flash memory**) stores data in **erasable**, rewritable circuitry. A **memory card** is a flat, solid state storage medium commonly used to transfer files from digital cameras and media players to computers. A **solid state drive** (SSD[7]) is a package of flash memory that can be used as a substitute for a hard disk drive. A USB[8] **flash drive** is a portable storage device that plugs directly into a computer's system unit using a built-in USB connector.

## V. Hardware Connections

**address bus**
地址总线
**data bus**
数据总线
**serial** / ˈsiəriəl /
a. 串行的;连续的

To function, hardware requires physical connections that allow components to communicate and interact. A bus provides a common interconnected system composed of a group of wires or circuitry that coordinates and moves information between the internal parts of a computer. A computer bus consists of two channels: one that the CPU uses to locate data, called the **address bus**, and another to send the data to that address, called the **data bus**. A bus is characterized by two features: how much information it can manipulate at one time, called the bus width, and how quickly it can transfer these data.

A **serial** connection is a wire or set of wires used to transfer information from the CPU to an external device such as a mouse, keyboard, modem, scanner, and some types of printers. This type of connection

---

1 *HDD*:硬(磁)盘驱动器(*h*ard *d*isk *d*rive 的首字母缩略)。hard disk 和 hard disk drive 经常互换使用,但严格说来,hard disk 指密封在 hard disk drive 内的盘片。

2 *CD*:光盘(*c*ompact *d*isc 的首字母缩略)。

3 *DVD*:数字多功能光盘(*d*igital *v*ersatile *d*isc 的首字母缩略)。

4 *BD*:蓝光光盘(*B*lu-ray *d*isc 的首字母缩略)。

5 *disc*:disc 本来是英国英语的拼法,其对应的美国英语的拼法为 disk。但是,在计算机英语中,现在公认的标准做法是将 disc 用于表示光盘的场合,而把 disk 用于所有其他的场合。因此,本文出现了 disc 和 disk 两种拼写形式。

6 *read-only (ROM), recordable (R), and rewritable (RW)*:read-only (ROM) 表示"只读",即此类光盘批量生产,内容只能读,不能修改;recordable (R) 表示"可写",即此类光盘允许用户自己刻录,但内容只能写一次,此后不能修改;rewritable (RW) 表示"可重写",即此类光盘允许用户重写无数次,其内容可修改。

7 *SSD*:固态驱动器(*s*olid *s*tate *d*rive 的首字母缩略)。

8 *USB*:通用串行总线(*u*niversal *s*erial *b*us 的首字母缩略)。

transfers only one piece of data at a time, and is therefore slow. The advantage of using a serial connection is that it provides effective connections over long distances.

A parallel connection uses multiple sets of wires to transfer blocks of information simultaneously. Most scanners and printers use this type of connection. A parallel connection is much faster than a serial connection, but it is limited to distances of less than 3 m (10 ft) between the CPU and the external device.

---

## Exercises

**I. Fill in the blanks with the information given in the text:**

1. The function of computer hardware is typically divided into three main categories. They are ________, ________, and ________.

2. The software that controls the interaction between the input and output hardware is called BIOS, which stands for ________.

3. The two most common types of scanners are ________ scanners and ________ scanners.

4. The two technologies commonly used today for display devices are ________ and ________.

5. 3D printing is technically called ________ manufacturing, and the technology used for most consumer-grade 3D printers is called ________.

6. A computer's main memory is often called ________.

7. Three types of mass storage systems are commonly used for personal computers: ________, optical, and ________ state.

8. A(n) ________ connection transfers only one piece of data at a time while a(n) ________ connection transfers blocks of information at the same time.

**II. Translate the following terms or phrases from English into Chinese and vice versa:**

1. function key　　　　2. voice recognition module

| | |
|---|---|
| 3. multifunction printer | 12. 液晶显示(器) |
| 4. address bus | 13. 喷墨打印机 |
| 5. additive manufacturing | 14. 数据总线 |
| 6. memory card | 15. 串行连接 |
| 7. parallel connection | 16. 闪存 |
| 8. solid state drive | 17. 激光打印机 |
| 9. boot loader | 18. 蓝光光盘 |
| 10. ink cartridge | 19. 发光二极管 |
| 11. 只读存储器 | 20. 随机存取存储器 |

**III. Fill in each of the blanks with one of the words given in the following list, making changes if necessary:**

| | | | |
|---|---|---|---|
| *cloud* | *school* | *local* | *memory* |
| *file* | *computer* | *store* | *external* |
| *flash* | *access* | *device* | *offer* |
| *network* | *drive* | *Internet* | *storage* |

Today's digital devices may use local storage and remote storage. Local storage refers to storage ________ and media that can be directly attached to a(n) ________, smartphone, or appliance. Local storage options include hard ________, CDs, DVDs, flash drives, solid state drives, and ________ cards. Most digital devices have some type of ________ storage that is permanently available as you use the device. Built-in ________ can be supplemented by removable storage, such as ________ drives and memory cards.

In contrast, remote storage is housed on a(n) ________ device that can be accessed from a(n) ________. Remote storage may be available on a home, ________, or work network. It can also be available as a(n) ________ service, in which case it is called ________ storage. The basic concept is that files can be ________ in a *subscriber's* (用户) cloud-based storage area and ________ by logging in from any device. Some cloud implementations ________ a *synchronization* (同步化) feature that automatically *duplicates* (复制) ________ stored on a local device by also saving them in the cloud.

**IV. Translate the following passage from English into Chinese:**

A modem is a device that converts between analog and digital signals. Digital signals, which are used by computers, are made up of separate units, usually represented by a series of 1's and 0's. Analog signals vary continuously; an example of an analog signal is a sound wave. Modems are often used to enable computers to communicate with each other across telephone lines. A modem converts the digital signals of the sending computer to analog signals that can be transmitted through telephone lines. When the signal reaches its destination, another modem reconstructs the original digital signal, which is processed by the receiving computer. If both modems can transmit data to each other simultaneously, the modems are operating in full *duplex* (双工的) mode; if only one modem can transmit at a time, the modems are operating in half duplex mode.

# Section B

# Components of an Operating System

## I. The User Interface of an Operating System

**user interface**
用户界面
**textual**
/ 'tekstjuəl, -tʃuəl /
a. 文本的，正文的
**graphical user interface**
图形用户界面

In order to perform the actions requested by the computer's users, an operating system must be able to communicate with those users. The portion of an operating system that handles this communication is often called the **user interface**. Older user interfaces, called shells, communicated with users through **textual** messages using a keyboard and monitor screen. More modern systems perform this task by means of a **graphical user interface** (GUI[1]—pronounced "GOO–ee") in which objects

[1] *GUI*：图形用户界面（graphical *u*ser *i*nterface 的首字母缩略），读作 / 'gu:i /。

to be manipulated, such as files and programs, are represented **pictorially** on the display as **icons**. These systems allow users to issue commands by using one of several common input devices. For example, a computer mouse can be used to **click** or drag icons on the screen. In place of a mouse, special-purpose pointing devices or styluses are often used by graphic artists or on several types of handheld devices. More recently, advances in **fine-grained touch screens** allow users to manipulate icons directly with their fingers. Whereas today's GUIs use two-dimensional image projection systems, three-dimensional interfaces that allow human users to communicate with computers by means of 3D projection systems, **tactile** sensory devices, and **surround sound** audio reproduction systems are subjects of current research.

**pictorial** / pik'tɔ:riəl /
a. 用图表示的，图示的；形象化的
**icon** / 'aikɔn /
n. 图标，图符
**click** / klik /
v.（鼠标）单击
**fine-grained** / 'fain'greind /
a. 细粒度的
**touch screen**
触摸屏，触屏
**tactile** / 'tæktail; -til /
a.（有）触觉的
**surround sound**
环绕立体声
**functionality** / ˌfʌŋkʃə'næliti /
n. 功能性
**intermediary** / ˌintə'mi:diəri /
n. 媒介（物）；调解人；中间人

Although an operating system's user interface plays an important role in establishing a machine's **functionality**, this framework merely acts as an **intermediary** between the computer's user and the real heart of the operating system (Figure 2B-1). This distinction between the user interface and the internal parts of the operating system is emphasized by the fact that some operating systems allow a user to select among different interfaces to obtain the most comfortable interaction for that particular user.

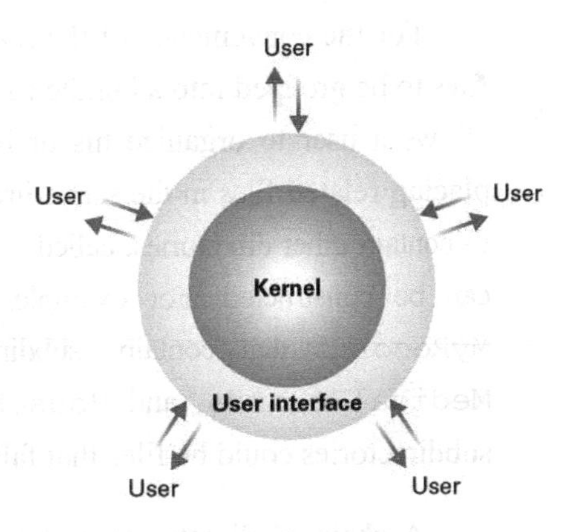


**Figure 2B-1: The user interface acts as an intermediary between users and the operating system's kernel**

An important component within today's GUI shells is the window **manager**, which **allocates** blocks of space on the screen, called windows, and keeps track of which application is associated with each window. When an application wants to display something on the screen, it notifies the

**manager** / 'mænidʒə /
n. 管理程序，管理器
**allocate** / 'æləkeit /
v. 分配；分派

window manager, and the window manager places the desired image in the window assigned to the application. In turn, when a mouse button is clicked, it is the window manager that computes the mouse's location on the screen and notifies the appropriate application of the mouse action.

**kernel** / 'kə:nəl /
n. 内核，内核程序，核心程序

## II. The Kernel of an Operating System

In contrast to an operating system's user interface, the internal part of an operating system is called the kernel. An operating system's kernel contains those software components that perform the very basic functions required by the computer installation.

### 1. File Manager

**extension** / ik'stenʃən /
n. 扩展；扩展名
**directory** / di'rektəri /
n. 目录
**folder** / 'fəuldə /
n. 文件夹
**subdirectory** / ˌsʌbdi'rektəri /
n. 子目录
**hierarchical** / ˌhaiə'rɑ:kikəl /
a. 分级的，分层的，层次的
**directory path**
目录路径
**slash** / slæʃ /
n. 斜线，斜杠
**prehistoric** / ˌpri:hi'stɔrik /
a. 史前的
**dinosaur** / 'dainəsɔ: /
n. 恐龙；(尤指废弃过时的）庞然大物
**terminate** / 'tə:mineit /
v. (使）终止
**path expression**
路径表达式

One such unit is the file manager, whose job is to coordinate the use of the machine's mass storage facilities. More precisely, the file manager maintains records of all the files stored in mass storage, including where each file is located, which users are allowed to access the various files, and which portions of mass storage are available for new files or **extensions** to existing files. These records are kept on the individual storage medium containing the related files so that each time the medium is placed online, the file manager can retrieve them and thus know what is stored on that particular medium.

For the convenience of the machine's users, most file managers allow files to be grouped into a bundle called a **directory** or **folder**. This approach allows a user to organize his or her files according to their purposes by placing related files in the same directory. Moreover, by allowing directories to contain other directories, called **subdirectories**, a **hierarchical** organization can be constructed. For example, a user may create a directory called `MyRecords` that contains subdirectories called `FinancialRecords`, `MedicalRecords`, and `HouseHoldRecords`. Within each of these subdirectories could be files that fall within that particular category.

A chain of directories within directories is called a **directory path**. Paths are often expressed by listing the directories along the path separated by **slashes**. For instance, `animals/`**`prehistoric/dinosaurs`** would represent the path starting at the directory named `animals`, passing through its subdirectory named `prehistoric`, and **terminating** in the sub-subdirectory `dinosaurs`. (For Windows users the slashes in such a **path expression** are reversed as in `animals\prehistoric\dinosaurs`.)

Any access to a file by other software units is obtained at the **discretion** of[1] the file manager. The procedure begins by requesting that the file manager grant access to the file through a procedure known as opening the file. If the file manager approves the requested access, it provides the information needed to find and to manipulate the file.

**discretion** / disˈkreʃən /
n. 斟酌决定（或处理）的自由
**driver** / ˈdraivə /
n. 驱动程序，驱动器
**device driver**
设备驱动程序
**controller**
/ kənˈtrəulə /
n. 控制器
**peripheral**
/ pəˈrifərəl /
a. 外围的，外部的
**peripheral device**
外围设备，外部设备
**generic** / dʒiˈnerik /
a. 类属的；一般的；通用的
**decode** / ˌdiːˈkəud /
v. 译（码），解（码）
**status word**
状态字
**handshaking**
/ ˈhændˌʃeikiŋ /
n. 握手，信号交换
**customize**
/ ˈkʌstəmaiz /
v. 定制，使用户化
**minimal** / ˈminiməl /
a. 最小的；最低限度的
**multitasking**
/ ˈmʌltiˌtɑːskiŋ /
n. 多任务（处理）
**reside** / riˈzaid /
v. 驻留，常住
**concurrent**
/ kənˈkʌrənt /
a. 同时发生的，并发的

## 2. Device Driver

Another component of the kernel consists of a collection of **device drivers**, which are the software units that communicate with the **controllers** (or at times, directly with **peripheral devices**) to carry out operations on the peripheral devices attached to the machine. Each device driver is uniquely designed for its particular type of device (such as a printer, disk drive, or monitor) and translates **generic** requests into the more technical steps required by the device assigned to that driver. For example, a device driver for a printer contains the software for reading and **decoding** that particular printer's **status word** as well as all the other **handshaking** details. Thus, other software components do not have to deal with those technicalities in order to print a file. Instead, the other components can merely rely on the device driver software to print the file, and let the device driver take care of the details. In this manner, the design of the other software units can be independent of the unique characteristics of particular devices. The result is a generic operating system that can be **customized** for particular peripheral devices by merely installing the appropriate device drivers.

## 3. Memory Manager

Still another component of an operating system's kernel is the memory manager, which is charged with the task of coordinating the machine's use of main memory. Such duties are **minimal** in an environment in which a computer is asked to perform only one task at a time. In these cases, the program for performing the current task is placed at a predetermined location in main memory, executed, and then replaced by the program for performing the next task. However, in multiuser or **multitasking** environments in which the computer is asked to address many needs at the same time, the duties of the memory manager are extensive. In these cases, many programs and blocks of data must **reside** in main memory **concurrently**. Thus, the memory manager must find and assign memory

[1] *at the discretion of*: 由…斟酌决定；随…的意见。

**allot** / ə'lɔt /
v. 分配；分派

space for these needs and ensure that the actions of each program are restricted to the program's **allotted** space. Moreover, as the needs of different activities come and go, the memory manager must keep track of those memory areas no longer occupied.

**paging** / 'peidʒiŋ /
n. 页面调度（技术或方法），分页

**bit pattern**
位模式

**shuffle** / 'ʃʌfl /
v. 洗（牌）；混洗

**fictional** / 'fikʃənəl /
a. 小说的；虚构的

**virtual memory**
虚拟内存，虚拟存储器

The task of the memory manager is complicated further when the total main memory space required exceeds the space actually available in the computer. In this case the memory manager may create the illusion of additional memory space by rotating programs and data back and forth between main memory and mass storage (a technique called **paging**). Suppose, for example, that a main memory of 8GB[1] is required but the computer only has 4GB. To create the illusion of the larger memory space, the memory manager reserves 4GB of storage space on a magnetic disk. There it records the **bit patterns** that would be stored in main memory if main memory had an actual capacity of 8GB. This data is divided into uniform sized units called pages, which are typically a few KB[2] in size. Then the memory manager **shuffles** these pages back and forth between main memory and mass storage so that the pages that are needed at any given time are actually present in the 4GB of main memory. The result is that the computer is able to function as though it actually had 8GB of main memory. This large "**fictional**" memory space created by paging is called **virtual memory**.

**scheduler**
/ 'ʃedju:ələ; 'ske-dʒu:ələr /
n. 调度程序

**dispatcher**
/ dis'pætʃə /
n. 分派程序

**time-sharing**
/ 'taim,ʃεəriŋ /
n. & a. 分时（的）

### 4. Scheduler and Dispatcher

One of the most fundamental concepts of modern operating systems is the distinction between a program and the activity of executing a program. The former is a static set of directions, whereas the latter is a dynamic activity whose properties change as time progresses. The activity of executing a program under the control of the operating system is known as a process. Typical **time-sharing**/multitasking computers are running many processes, all competing for the computer's resources. The tasks associated with coordinating the execution of processes are handled by the scheduler and dispatcher within the operating system's kernel.

The scheduler maintains a record of the processes present in the computer system, introduces new processes to this pool, and removes completed processes from the pool. Thus when a user requests the

1 *GB*：吉字节，千兆字节（*giga*byte 的缩略）。

2 *KB*：千字节（*kilo*byte 的缩略）。

execution of an application, it is the scheduler that adds the execution of that application to the pool of current processes.

To keep track of all the processes, the scheduler maintains a block of information in main memory called the **process table**. Each time the execution of a program is requested, the scheduler creates a new entry for that process in the process table. This entry contains such information as the memory area assigned to the process (obtained from the memory manager), the priority of the process, and whether the process is ready or waiting. A process is ready if it is in a state in which its progress can continue; it is waiting if its progress is currently delayed until some external event occurs, such as the completion of a mass storage operation, the pressing of a key at the keyboard, or the arrival of a message from another process.

The dispatcher is the component of the kernel that **oversees** the execution of the scheduled processes. In a time-sharing/multitasking system, this task is accomplished by **multiprogramming**; that is, dividing time into short segments, each called a **time slice** (typically measured in **milliseconds** or **microseconds**), and then switching the CPU's attention among the processes as each is allowed to execute for one time slice. The procedure of changing from one process to another is called a **process switch** (or a **context switch**).

Each time the dispatcher awards a time slice to a process, it **initiates** a **timer** circuit that will indicate the end of the slice by generating a signal called an **interrupt**. When the CPU receives an **interrupt signal**, it completes its current machine cycle, saves its position in the current process, and begins executing a program, called an **interrupt handler**, which is stored at a predetermined location in main memory. This interrupt handler is a part of the dispatcher, and it describes how the dispatcher should respond to the interrupt signal.

Thus, the effect of the interrupt signal is to **preempt** the current process and transfer control back to the dispatcher. At this point, the dispatcher selects the process from the process table that has the highest priority among the ready processes (as determined by the scheduler), restarts the timer circuit, and allows the selected process to begin its time slice.

**process table**
进程表
**oversee** / ˌəuvəˈsiː /
v. 监视；监督
**multiprogramming**
/ ˈmʌltiˌprəugræmiŋ /
n. 多（道）程序设计
**time slice**
时间片
**millisecond**
/ ˈmiliˌsekənd /
n. 毫秒
**microsecond**
/ ˈmaikrəuˌsekənd /
n. 微秒
**process switch**
进程转换
**context switch**
上下文转换，语境转换
**initiate** / iˈniʃieit /
v. 开始；发起
**timer** / ˈtaimə /
n. 计时器，定时器；定时程序
**interrupt**
/ ˌintəˈrʌpt /
n. 中断
**interrupt signal**
中断信号
**handler** / ˈhændlə /
n. 处理程序，处理器
**interrupt handler**
中断处理程序
**preempt**
/ ˌpriːˈempt, pri- /
v. 预先制止；抢先，先占

## Exercises

**I. Fill in the blanks with the information given in the text:**

1. A modern operating system performs the task of communicating with computer users by means of a(n) ________ user interface.

2. The kernel of a typical operating system contains such software components as the ________ manager, various device drivers, the memory manager, the ________, and the dispatcher.

3. When the total main memory space required exceeds the space actually available in the computer, the memory manager may resort to ________ memory.

4. In a time-sharing/multitasking system, time is divided into short segments, each of which is called a time ________.

**II. Translate the following terms or phrases from English into Chinese and vice versa:**

| | |
|---|---|
| 1. interrupt handler | 6. 外围设备 |
| 2. virtual memory | 7. 进程表 |
| 3. context switch | 8. 时间片 |
| 4. main memory | 9. 图形用户界面 |
| 5. bit pattern | 10. 海量存储器 |

# Section C

# System Organization

## I. Introduction

The organization of a system reflects the basic strategy that is used to structure a system. You have to make decisions on the overall organizational model of a system early in the architectural design process. The system organization may be directly reflected in the sub-system structure. However, it is often the case that the sub-system model includes more detail than the organizational model, and there is not always a simple **mapping** from sub-systems to organizational structure[1].

This section discusses three organizational styles that are very widely used. These are a shared data **repository** style, a shared services and servers style and an **abstract machine** or **layered** style where the system is

**map** / mæp /
v. 映射；变换，变址
**repository**
/ ri'pɔzitəri /
n. 存储库，仓库
**abstract machine**
抽象机
**layered** / 'leiəd /
a. 分层的

[1] *there is not always a simple mapping from sub-systems to organizational structure*：从子系统到组织结构并非总是一种简单的映射。这是一个部分否定的结构。

**tier** / tiə /
n.（一）层

organized as a **tier** of functional layers. These styles can be used separately or together. For example, a system may be organized around a shared data repository but may construct layers around this to present a more abstract view of the data.

## II. The Repository Model

Sub-systems making up a system must exchange information so that they can work together effectively. There are two fundamental ways in which this can be done.

**interchange**
/ ˌintəˈtʃeindʒ /
v. 交换，互换

- All shared data is held in a central database that can be accessed by all sub-systems. A system model based on a shared database is sometimes called a *repository model*.
- Each sub-system maintains its own database. Data is **interchanged** with other sub-systems by passing messages to them.

The majority of systems that use large amounts of data are organized around a shared database or repository. This model is therefore suited to applications where data is generated by one sub-system and used by another. Examples of this type of system include command and control systems, management information systems, CAD[1] systems and CASE[2] **toolsets**.

**toolset** / ˈtu:lset /
n. 成套工具，工具箱
**translator**
/ trænsˈleitə /
n. 翻译程序，翻译器
**editor** / ˈeditə /
n. 编辑程序，编辑器
**generator**
/ ˈdʒenəreitə /
n. 生成程序，生成器
**analyzer** / ˈænəlaizə /
n. 分析程序，分析器

Figure 2C-1 is an example of a CASE toolset architecture based on a shared repository. The first shared repository for CASE tools was probably developed in the early 1970s by a UK company called ICL[3] to support their

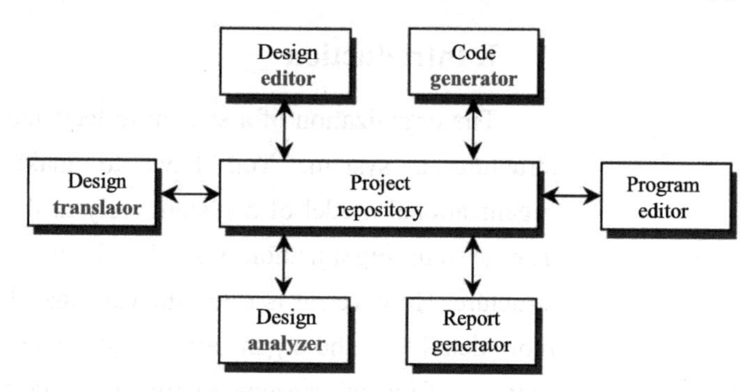


Figure 2C-1: The Architecture of an Integrated CASE Toolset

[1] *CAD*：计算机辅助设计（computer-*a*ided *d*esign 的首字母缩略）。

[2] *CASE*：计算机辅助软件工程（computer-*a*ided *s*oftware *e*ngineering 的首字母缩略）。

[3] *ICL*：国际计算机有限公司（*I*nternational *C*omputers *L*imited 的首字母缩略）。英国的一家大型计算机硬件、软件及服务公司，创立于 1968 年，1981 年被日本富士通株式会社兼并，2002 年更名为 Fujitsu Services Limited（富士通服务有限公司）。

operating system development. This model became more widely known when Buxton[1] made proposals to support the development of systems written in Ada[2]. Since then, many CASE toolsets have been developed around a shared repository.

The advantages and disadvantages of a shared repository are as follows:

- It is an efficient way to share large amounts of data. There is no need to transmit data explicitly from one sub-system to another.
- However, sub-systems must agree on the repository data model. Inevitably, this is a compromise between the specific needs of each tool. Performance may be **adversely** affected by this compromise. It may be difficult or impossible to integrate new sub-systems if their data models do not fit the agreed **schema**.
- Sub-systems that produce data need not be concerned with how that data is used by other sub-systems.
- However, evolution may be difficult as a large volume of information is generated according to an agreed data model. Translating this to a new model will certainly be expensive; it may be difficult or even impossible.
- Activities such as **backup**, security, access control and recovery from error are centralized. They are the responsibility of the repository manager. Tools can focus on their principal function rather than be concerned with these issues.
- However, different sub-systems may have different requirements for security, recovery and backup policies. The repository model forces the same policy on all sub-systems.
- The model of sharing is visible through the repository schema. It is **straightforward** to integrate new tools given that they are **compatible** with the agreed data model[3].
- However, it may be difficult to distribute the repository over a number of machines. Although it is possible to distribute a logically centralized repository, there may be problems with data **redundancy** and **inconsistency**.

**adverse** / 'ædvə:s, æd'və:s /
a. 不利的，有害的

**schema** / 'ski:mə /
n. 模式；纲要

**backup** / 'bækʌp /
n. 备份，后备

**straightforward** / ˌstreit'fɔ:wəd /
a. 径直的；简单的

**compatible** / kəm'pætəbl /
a. 兼容的

**redundancy** / ri'dʌndənsi /
n. 冗余

**inconsistency** / ˌinkən'sistənsi /
n. 不一致

---

[1] *Buxton*：即 John N. Buxton，约翰 N. 巴克斯顿，英国计算机科学家。

[2] *Ada*：Ada 语言。

[3] *It is straightforward to integrate new tools given that they are compatible with the agreed data model.*：只要新工具与商定的数据模型兼容，将新工具结合进去就会很容易。句中的 given 为介词，given 作介词时后面可跟 that 从句，表示“考虑到”“假设”“倘若”等意思。

In the above model, the repository is passive and control is the responsibility of the sub-systems using the repository. An alternative approach has been derived for AI systems that use a "**blackboard**" **model**, which **triggers** sub-systems when particular data becomes available. This is appropriate when the form of the repository data is less well structured. Decisions about which tool to **activate** can only be made when the data has been analyzed.

**blackboard model**
黑板法模型
**trigger** / ˈtrigə /
v. 触发
**activate** / ˈæktiveit /
v. 激活，启动

## III. The Client-Server Model

**client** / ˈklaiənt /
n. 客户机；
客户程序

The client-server architectural model is a system model where the system is organized as a set of services and associated servers and clients that access and use the services. The major components of this model are:

- A set of servers that offer services to other sub-systems. Examples of servers are print servers that offer printing services, **file servers** that offer file management services and a compile server, which offers programming language **compilation** services.
- A set of clients that call on the services offered by servers. These are normally sub-systems in their own right[1]. There may be several instances of a client program executing concurrently.
- A network that allows the clients to access these services. This is not strictly necessary as both the clients and the servers could run on a single machine. In practice, however, most client-server systems are implemented as **distributed** systems.

**file server**
文件服务器
**compilation**
/ ˌkɔmpiˈleiʃən /
n. 编译；汇编

**distributed**
/ diˈstributid /
a. 分布（式）的
**procedure call**
过程调用
**protocol**
/ ˈprəutəkɔl /
n. 协议
**hypertext**
/ ˈhaipətekst /
n. 超（级）文本
**Hypertext Transfer Protocol**
超文本传送协议，
超文本传输协议
**library** / ˈlaibrəri /
n. 库，程序（或文件、对象）库

Clients may have to know the names of the available servers and the services that they provide. However, servers need not know either the identity of clients or how many clients there are. Clients access the services provided by a server through remote **procedure calls** using a request-reply **protocol** such as the http[2] (**Hypertext Transfer Protocol**) protocol used in the WWW[3]. Essentially, a client makes a request to a server and waits until it receives a reply.

Figure 2C-2 shows an example of a system that is based on the client-server model. This is a multi-user, Web-based system to provide a film and photograph **library**[4]. In this system, several servers manage and

---

1 *These are normally sub-systems in their own right.*：这些（客户机）通常本身就是子系统。短语 in one's own right 表示"凭本身的权利（或能力、实力、资格等）"。

2 *http*：亦作 HTTP，超文本传送协议，超文本传输协议（*H*ypertext *T*ransfer *P*rotocol 的缩略）。

3 *WWW*：万维网（*W*orld *W*ide *W*eb 的首字母缩略）。

4 *This is a multi-user, Web-based system to provide a film and photograph library.*：这是一个多用户的网基系统，用于提供一个电影与照片库。

**frame** / freim /
n. 帧，画面
**synchrony** / 'siŋkrəni /
n. 同时（性）；同步（性）
**store** / stɔː /
n. 存储（器）
**compression** / kəm'preʃ ən /
n. 压缩
**decompression** / ˌdiːkəm'preʃ ən /
n. 解压缩
**clip** / klip /
n. 剪下来的东西；电影（或电视）片段
**film clip**
剪片
**digitize** / 'didʒitaiz /
v. 使数字化
**info** / 'infəu /
n.〈口〉信息（= information）
**query** / 'kwiəri /
n. 查询

display the different types of media. Video **frames** need to be transmitted quickly and in **synchrony** but at relatively low resolution. They may be compressed in a **store**, so the video server may handle video **compression** and **decompression** into different formats. Still pictures, however, must be maintained at a high resolution, so it is appropriate to maintain them on a separate server.

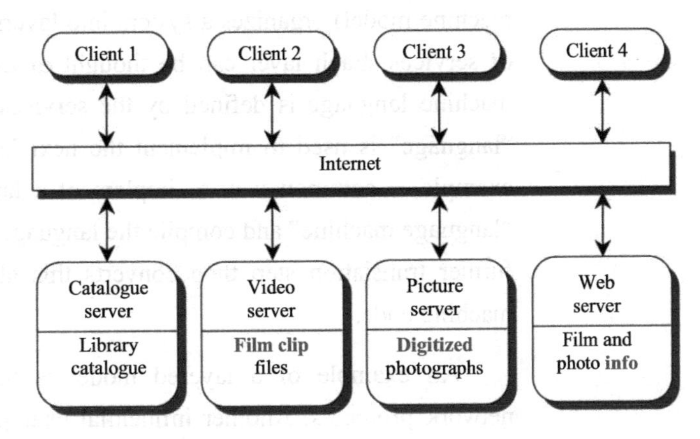


**Figure 2C-2: The Architecture of a Film and Picture Library System**

The catalogue must be able to deal with a variety of **queries** and provide links into the Web information system that includes data about the film and video clip, and an e-commerce system that supports the sale of film and video clips. The client program is simply an integrated user interface, constructed using a **Web browser**, to these services[1].

**Web browser**
网络浏览器
**network** / 'netwəːk /
v. 连网，联网，建网
**upgrade** / ʌp'greid, 'ʌpg- /
v. 使升级；改善

The most important advantage of the client-server model is that it is a distributed architecture. Effective use can be made of **networked** systems with many distributed processors. It is easy to add a new server and integrate it with the rest of the system or to **upgrade** servers transparently without affecting other parts of the system.

**optimize** / 'ɔptimaiz /
v. 使优化，使最佳化

However, changes to existing clients and servers may be required to gain the full benefits of integrating a new server. There may be no shared data model across servers and sub-systems may organize their data in different ways. This means that specific data models may be established on each server to allow its performance to be **optimized**. Of course, if an

[1] *The client program is simply an integrated user interface, constructed using a Web browser, to these services.*：客户程序只是一个用网络浏览器构建的连接这些服务的综合用户接口。interface 后面跟有两个定语：constructed using a Web browser 和 to these services。

XML[1]-based representation of data is used, it may be relatively simple to convert from one schema to another. However, XML is an inefficient way to represent data, so performance problems can arise if this is used.

## IV. The Layered Model

The layered model of an architecture (sometimes called an abstract machine model) organizes a system into layers, each of which provide a set of services. Each layer can be thought of as an abstract machine whose machine language is defined by the services provided by the layer. This "language" is used to implement the next level of abstract machine. For example, a common way to implement a language is to define an ideal "language machine" and compile the language into code for this machine. A further translation step then converts this abstract machine code to real machine code.

An example of a layered model is the OSI[2] reference model of network protocols. Another influential example was proposed by Buxton, who suggested a three-layer model for an Ada Programming Support Environment (APSE[3]). Figure 2C-3 reflects the APSE structure and shows how a configuration management system might be integrated using this abstract machine approach.

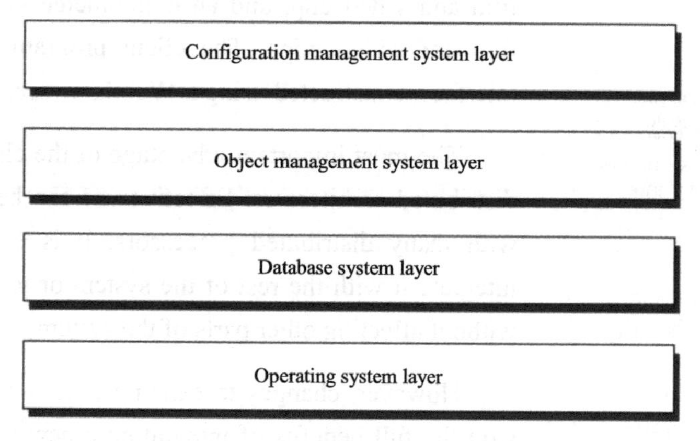


**Figure 2C-3: The Layered Model of a Version Management System**

---

[1] *XML*：可扩展标记语言（e*X*tensible *M*arkup *L*anguage 的缩略）。

[2] *OSI*：开放式系统互联（*O*pen *S*ystems *I*nterconnection 的首字母缩略）。OSI 模型是由国际标准化组织（ISO）设计的。该模型理论上允许任意两个不同系统（如计算机）互相通信而无须考虑它们的基本体系结构。

[3] *APSE*：Ada 程序设计支持环境（*A*da *P*rogramming *S*upport *E*nvironment 的首字母缩略）。

The configuration management system manages versions of objects and provides general configuration management facilities. To support these configuration management facilities, it uses an object management system that provides information storage and management services for **configuration items** or objects. This system is built on top of a database system to provide basic data storage and services such as transaction management, **rollback** and recovery, and access control. The database management uses the **underlying** operating system facilities and **filestore** in its implementation.

**configuration item** 配置项
**rollback** / 'rəulbæk / n. 回滚，回退
**underlying** / ˌʌndə'laiiŋ / a. 基本的，根本的
**filestore** / 'failstɔː / n. 文件存储（器）
**incremental** / ˌinkri'mentəl / a. 增量的；递增的

The layered approach supports the **incremental** development of systems. As a layer is developed, some of the services provided by that layer may be made available to users. This architecture is also changeable and portable. So long as its interface is unchanged, a layer can be replaced by another, equivalent layer. Furthermore, when layer interfaces change or new facilities are added to a layer, only the **adjacent** layer is affected. As layered systems **localize** machine **dependencies** in inner layers, this makes it easier to provide multi-platform implementations of an application system. Only the inner, machine-dependent layers need be reimplemented to take account of the facilities of a different operating system or database.

**adjacent** / ə'dʒeisənt / a. 相邻的，毗连的
**localize** / 'ləukəlaiz / v. 使局部化，使本地化
**dependency** / di'pendənsi / n. 依靠，依赖

A disadvantage of the layered approach is that structuring systems in this way can be difficult. Inner layers may provide basic facilities, such as file management, that are required at all levels. Services required by a user of the top level may therefore have to "punch through" adjacent layers to get access to services that are provided several levels beneath it. This **subverts** the model, as the outer layer in the system does not just depend on its immediate **predecessor**.

**subvert** / səb'vəːt, sʌb- / v. 颠覆；（暗中）破坏
**predecessor** / 'priːdisesə; 'pre- / n. 前任；前身；先驱

Performance can also be a problem because of the multiple levels of command interpretation that are sometimes required. If there are many layers, a service request from a top layer may have to be interpreted several times in different layers before it is processed. To avoid these problems, applications may have to communicate directly with inner layers rather than use the services provided by the adjacent layer.

## Exercises

**I. Fill in the blanks with the information given in the text:**

1. The three widely used organizational models for systems discussed in the text are the ________ model, client-server model, and layered model.

2. There are two fundamental ways in which sub-systems making up a system exchange information: all shared data is held in a(n) ________ database accessible to all of them; and each maintains its own database and interchanges data with other ________.

3. The major components of the client-server model usually include a set of servers, a set of clients, and a(n) ________.

4. The OSI reference model of network protocols is an example of the ________ model.

**II. Translate the following terms or phrases from English into Chinese and vice versa:**

| | |
|---|---|
| 1. code generator | 5. 计算机辅助设计 |
| 2. abstract machine | 6. 数据冗余 |
| 3. program editor | 7. 指挥与控制系统 |
| 4. configuration item | 8. 视频压缩与解压缩 |

# Unit 3 Computer Language and Programming

（计算机语言与编程）

## Section A

## Programming Language

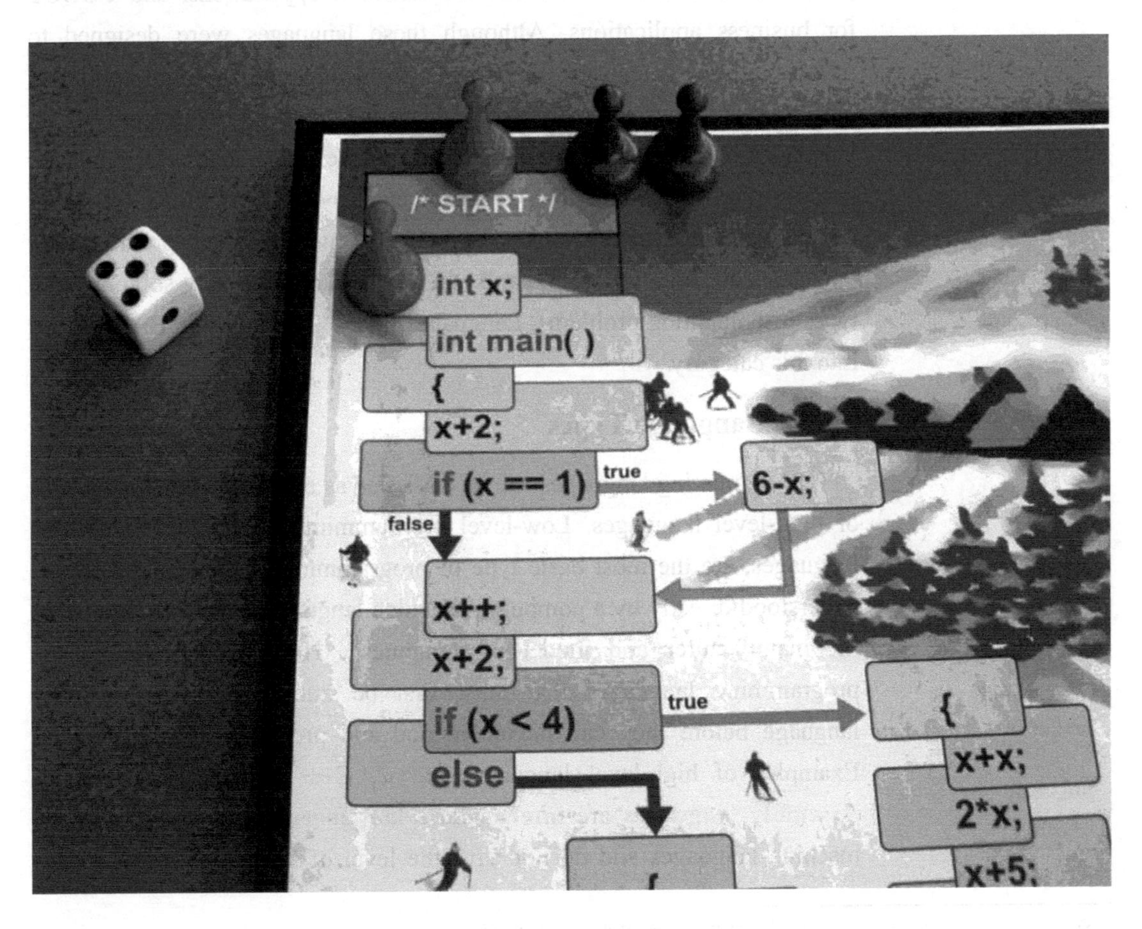


### I. Introduction

Programming languages, in computer science, are the artificial languages used to write a sequence of instructions (a computer program) that can be run by a computer. Similar to natural languages, such as English, programming languages have a vocabulary, grammar, and **syntax**. However, natural languages are not suited for programming computers because they are **ambiguous**, meaning that their vocabulary and **grammatical** structure

**syntax** / 'sintæks /
n. 句法

**ambiguous**
/ æm'bigjuəs /
a. 含糊不清的，模棱两可的

**grammatical**
/ grə'mætikəl /
a.（符合）语法的

may be interpreted in multiple ways. The languages used to program computers must have simple logical structures, and the rules for their grammar, spelling, and **punctuation** must be precise.

**punctuation**
/ ˌpʌŋktju'eiʃən, -tʃu- /
n. 标点符号
**sophistication**
/ səˌfisti'keiʃən /
n. 复杂性；尖端性
**versatility**
/ ˌvə:sə'tiliti /
n. 多用途；通用性

Programming languages vary greatly in their **sophistication** and in their degree of **versatility**. Some programming languages are written to address a particular kind of computing problem or for use on a particular model of computer system. For instance, programming languages such as Fortran[1] and COBOL[2] were written to solve certain general types of programming problems—Fortran for scientific applications, and COBOL for business applications. Although these languages were designed to address specific categories of computer problems, they are highly portable, meaning that they may be used to program many types of computers. Other languages, such as machine languages, are designed to be used by one specific model of computer system, or even by one specific computer in certain research applications. The most commonly used programming languages are highly portable and can be used to effectively solve diverse types of computing problems. Languages like C[3], Pascal[4], and BASIC[5] fall into this category.

## II. Language Types

Programming languages can be classified as either low-level languages or high-level languages. Low-level programming languages, or machine languages, are the most basic type of programming languages and can be understood directly by a computer. Machine languages differ depending on the manufacturer and model of computer[6]. High-level languages are programming languages that must first be translated into a machine language before they can be understood and processed by a computer. Examples of high-level languages are C, C++[7], Pascal, and Fortran. **Assembly languages** are **intermediate languages** that are very close to machine languages and do not have the level of **linguistic** sophistication

**assembly language**
汇编语言
**intermediate language**
中间语言，中级语言
**linguistic**
/ liŋ'gwistik /
a. 语言（学）的

[1] *Fortran*：Fortran 语言，公式翻译程序语言，亦译为“福传”（*For*mula *Tran*slation 的缩合），读作 /'fɔ:træn/。
[2] *COBOL*：COBOL 语言，面向商业的通用语言（*Co*mmon *B*usiness-*O*riented *L*anguage 的缩略），读作 /'kəubɔl/。
[3] *C*：C 语言，一种面向过程的程序设计语言。
[4] *Pascal*：Pascal 语言，为纪念法国数学家和哲学家布莱斯·帕斯卡（Blaise Pascal）而命名，读作 /'pæskəl/ 或 /pɑ:s'kɑ:l/。
[5] *BASIC*：BASIC 语言，初学者通用符号指令码（*B*eginner's *A*ll-purpose *S*ymbolic *I*nstruction *C*ode 的首字母缩略），读作 /'beisik/。
[6] *Machine languages differ depending on the manufacturer and model of computer.*：机器语言视计算机制造商与型号不同而有所区别。
[7] *C++*：C++语言，一种面向对象的程序设计语言。

exhibited by other high-level languages, but must still be translated into machine language.

**1. Machine Languages**

In machine languages, instructions are written as sequences of 1s and 0s, called bits, that a computer can understand directly. An instruction in machine language generally tells the computer four things: (1) where to find one or two numbers or simple pieces of data in the main computer memory (Random Access Memory, or RAM), (2) a simple operation to perform, such as adding the two numbers together, (3) where in the main memory to put the result of this simple operation, and (4) where to find the next instruction to perform. While all executable programs are eventually read by the computer in machine language, they are not all programmed in machine language. It is extremely difficult to program directly in machine language because the instructions are sequences of 1s and 0s. A typical instruction in a machine language might read *10010 1100 1011* and mean add the contents of **storage register** A to the contents of storage register B.

**register** / 'redʒistə /
n. 寄存器
**storage register**
存储寄存器

**2. High-Level Languages**

**statement**
/ 'steitmənt /
n. 语句

High-level languages are relatively sophisticated sets of **statements** utilizing words and syntax from human language. They are more similar to normal human languages than assembly or machine languages and are therefore easier to use for writing complicated programs. These programming languages allow larger and more complicated programs to be developed faster. However, high-level languages must be translated into machine language by another program called a **compiler** before a computer can understand them. For this reason, programs written in a high-level language may take longer to execute and use up more memory than programs written in an assembly language.

**compiler**
/ kəm'pailə /
n. 编译程序，
编译器

**3. Assembly Languages**

Computer programmers use assembly languages to make machine-language programs easier to write. In an assembly language, each statement corresponds roughly to one machine language instruction. An assembly language statement is composed with the aid of easy to remember commands. The command to add the contents of storage register A to the contents of storage register B might be written *ADD B, A* in a typical assembly language statement. Assembly languages share certain features with machine languages. For instance, it is possible to manipulate specific bits in

**minimize** / 'minimaiz /
v. 使减少到最低限度；使最小化

both assembly and machine languages. Programmers use assembly languages when it is important to **minimize** the time it takes to run a program, because the translation from assembly language to machine language is relatively simple. Assembly languages are also used when some part of the computer has to be controlled directly, such as individual dots on a monitor or the flow of individual characters to a printer.

## III. Classification of High-Level Languages

**-oriented** / 'ɔːrientid /
comb. form 表示“面向…的”
**logic language** 逻辑语言
**program module** 程序模块

High-level languages are commonly classified as procedure-**oriented**, functional, object-oriented, or **logic languages**. The most common high-level languages today are procedure-oriented languages. In these languages, one or more related blocks of statements that perform some complete function are grouped together into a **program module**, or procedure, and given a name such as “procedure A.” If the same sequence of operations is needed elsewhere in the program, a simple statement can be used to refer back to the procedure. In **essence**[1], a procedure is just a mini-program. A large program can be constructed by grouping together procedures that perform different tasks. **Procedural languages** allow programs to be shorter and easier for the computer to read, but they require the programmer to design each procedure to be general enough to be used in different situations.

**essence** / 'esəns /
n. 本质，实质
**procedural language**
过程语言

**functional language**
函数式语言
**rigorous** / 'rigərəs /
a. 严密的；严格的

**Functional languages** treat procedures like mathematical functions and allow them to be processed like any other data in a program. This allows a much higher and more **rigorous** level of program construction. Functional languages also allow variables—symbols for data that can be specified and changed by the user as the program is running—to be given values only once. This simplifies programming by reducing the need to be concerned with the exact order of statement execution, since a variable does not have to be redeclared, or restated, each time it is used in a program statement. Many of the ideas from functional languages have become key parts of many modern procedural languages.

**object-oriented language**
面向对象语言
**outgrowth** / 'autgrəuθ /
n. 发展结果；产物

**Object-oriented languages** are **outgrowths** of functional languages. In object-oriented languages, the code used to write the program and the data processed by the program are grouped together into units called objects. Objects are further grouped into classes, which define the attributes objects must have. A simple example of a class is the class Book. Objects within

1 *in essence*：本质上，实质上。

this class might be Novel and Short Story. Objects also have certain functions associated with them, called methods. The computer accesses an object through the use of one of the object's methods. The method performs some action to the data in the object and returns this value to the computer. Classes of objects can also be further grouped into **hierarchies**, in which objects of one class can inherit methods from another class. The structure provided in object-oriented languages makes them very useful for complicated programming tasks.

**hierarchy** / 'haiəˌrɑ:ki /
n. 层次，分层（结构），分级（结构）

Logic languages use logic as their mathematical base. A logic program consists of sets of facts and if-then rules, which specify how one set of facts may be **deduced** from others, for example:

**deduce** / di'dju:s /
v. 推论，推断

If the statement X is true, then the statement Y is false.

In the execution of such a program, an input statement can be logically deduced from other statements in the program. Many artificial intelligence programs are written in such languages.

## IV. Language Structure and Components

Programming languages use specific types of statements, or instructions, to provide functional structure to the program. A statement in a program is a basic sentence that expresses a simple idea—its purpose is to give the computer a basic instruction. Statements define the types of data allowed, how data are to be manipulated, and the ways that procedures and functions work. Programmers use statements to manipulate common components of programming languages, such as variables and **macros** (mini-programs within a program).

**macro** / 'mækrəu /
n. 宏，宏指令

Statements known as **data declarations** give names and properties to elements of a program called variables. Variables can be assigned different values within the program. The properties variables can have are called types, and they include such things as what possible values might be saved in the variables, how much numerical accuracy is to be used in the values, and how one variable may represent a collection of simpler values in an organized fashion, such as a table or **array**. In many programming languages, a key data type is a pointer. Variables that are pointers do not themselves have values; instead, they have information that the computer can use to locate some other variable—that is, they point to another variable.

**data declaration**
数据声明

**array** / ə'rei /
n. 数组；阵列；一系列

**expression**
/ ik'spreʃən /
n. 表达式
**assignment statement**
赋值语句
**conditional statement**
条件语句
**procedure statement**
过程语句
**function statement**
函数语句
**parameter**
/ pə'ræmitə /
n. 参数

An **expression** is a piece of a statement that describes a series of computations to be performed on some of the program's variables, such as *X*+*Y*/*Z*, in which the variables are *X*, *Y*, and *Z* and the computations are addition and division. An **assignment statement** assigns a variable a value derived from some expression, while **conditional statements** specify expressions to be tested and then used to select which other statements should be executed next.

**Procedure** and **function statements** define certain blocks of code as procedures or functions that can then be returned to later in the program. These statements also define the kinds of variables and **parameters** the programmer can choose and the type of value that the code will return when an expression accesses the procedure or function[1]. Many programming languages also permit minitranslation programs called macros. Macros translate segments of code that have been written in a language structure defined by the programmer into statements that the programming language understands.

## V. History

Programming languages date back almost to the invention of the digital computer in the 1940s. Initially, programmers used machine languages for programming computers, and these languages are sometimes referred to as first-generation languages. Second-generation languages added a level of **abstraction** to machine languages by substituting abbreviated command words for the strings of 1s and 0s used in machine languages. Languages in this new generation were called assembly languages.

**abstraction**
/ æb'strækʃən /
n. 抽象；提取

When high-level languages were originally conceived in the 1950s, they were dubbed third-generation languages because they seemed to be a major improvement over machine and assembly languages. Third-generation languages used easy-to-remember command words, such as PRINT and INPUT. Third-generation languages, such as COBOL and Fortran, were used extensively for business and scientific applications. Pascal and BASIC were popular teaching languages. C and C++ remain popular today for system and application software development—for example, to develop Microsoft Windows and Linux. Recently developed

1 *These statements also define the kinds of variables and parameters the programmer can choose and the type of value that the code will return when an expression accesses the procedure or function.*：这些语句也定义程序员可选的变量和参数种类，以及表达式访问过程或函数时，代码所返回的值的类型。

third-generation languages are important for modern apps. Objective-C[1] and Swift[2] are programming languages used to develop iPhone and iPad apps. Java[3] is used for Android apps.

In 1969, computer scientists began to develop high-level languages, called fourth-generation languages, which more closely resembled human languages than did third-generation languages. Fourth-generation languages, such as SQL[4] and RPG[5], eliminate many of the strict punctuation and grammar rules that complicate third-generation languages. Today, fourth-generation languages are primarily used for database applications.

In 1982, a group of Japanese researchers began work on a fifth-generation computer project that used Prolog[6]—a computer programming language based on a declarative programming **paradigm**. Prolog and other **declarative languages** became closely identified with the fifth-generation project and were classified as fifth-generation languages. Some experts disagree with this classification, however, and instead define fifth-generation languages as those that allow programmers to use graphical or visual tools to construct programs rather than typing lines of statements.

**paradigm**
/ ˈpærədaim /
n. 范例；范式
**declarative language**
声明式语言

## Exercises

**I. Fill in the blanks with the information given in the text:**

1. A programming language is any ________ language that can be used to write a sequence of ________ that can ultimately be processed and executed by a computer.

2. We can classify programming languages into two types: ________ languages and ________ languages.

3. A(n) ________ language is a low-level language in binary code that a computer can understand and execute directly.

4. High-level languages must first be translated into a(n) ________ language before they can be understood and processed by a computer.

---

1 *Objective-C*：扩充 C 的面向对象编程语言，一种通用、高级、面向对象的编程语言。
2 *Swift*：Swift 语言（中文名称为“雨燕”），一种支持多编程范式和编译式的开源编程语言。
3 *Java*：Java 语言，一种面向对象的程序设计语言，可以跨平台运行。
4 *SQL*：结构化查询语言（*S*tructured *Q*uery *L*anguage 的首字母缩略）。
5 *RPG*：RPG 语言，报表程序生成器（*R*eport *P*rogram *G*enerator 的首字母缩略）。
6 *Prolog*：Prolog 语言，逻辑程序设计语言（*Pro*gramming in *Log*ic 的缩合）。

5. High-level languages are commonly classified as procedure-oriented, ________, object-oriented, or ________ languages.

6. In an assembly language, each ________ corresponds roughly to one machine language instruction.

7. In procedure-oriented languages, one or more related blocks of statements that perform some complete function are grouped together into a program ________, or procedure.

8. The history of programming languages can be traced back almost to the invention of the ________ computer in the 1940s.

**II. Translate the following terms or phrases from English into Chinese and vice versa:**

| | |
|---|---|
| 1. storage register | 11. 可执行程序 |
| 2. function statement | 12. 程序模块 |
| 3. program statement | 13. 条件语句 |
| 4. object-oriented language | 14. 赋值语句 |
| 5. assembly language | 15. 逻辑语言 |
| 6. intermediate language | 16. 机器语言 |
| 7. declarative language | 17. 结构化查询语言 |
| 8. artificial language | 18. 程序设计语言 |
| 9. data declaration | 19. 运行计算机程序 |
| 10. functional language | 20. 计算机程序员 |

**III. Fill in each of the blanks with one of the words given in the following list, making changes if necessary:**

| | | | |
|---|---|---|---|
| *reuse* | *translate* | *step* | *memory* |
| *high-level* | *computer* | *machine* | *execution* |
| *program* | *processing* | *inconvenient* | *separate* |
| *programming* | *combine* | *programmer* | *powerful* |

A programming language is a language used to write instructions for the computer. It lets the programmer express data ________ in a symbolic manner without regard to machine-specific details.

The difficulty of writing programs in the ________ language of 0s and 1s led first to the development of assembly language, which allows

________ to use *mnemonics* (助记符) for instructions and symbols for variables. Such programs are then ________ by a program known as an *assembler* (汇编程序) into the binary encoding used by the ________. Other pieces of system software known as *linking loaders* (链接装入程序) ________ pieces of assembled code and load them into the machine's main ________ unit, where they are then ready for execution. The concept of linking ________ pieces of code was important, since it allowed "libraries" of ________ to be built up to carry out common tasks—a first ________ toward the increasingly emphasized notion of software ________. Assembly language was found to be sufficiently ________ that higher-level languages (closer to natural languages) were invented in the 1950s for easier, faster ________; along with them came the need for compilers, programs that translate ________ language programs into machine code. As programming languages became more ________ and abstract, building efficient compilers that create high-quality code in terms of ________ speed and storage consumption became an interesting computer science problem in itself.

**IV. Translate the following passage from English into Chinese:**

Object-oriented programming (OOP) languages, such as C++ and Java, are based on traditional high-level languages, but they enable a programmer to think in terms of collections of cooperating objects instead of lists of commands. Objects, such as a circle, have properties such as the *radius* (半径) of the circle and the command that draws it on the computer screen. Classes of objects can inherit features from other classes of objects. For example, a class defining squares can inherit features such as right angles from a class defining *rectangles* (长方形). This set of programming classes simplifies the programmer's task, resulting in more "reusable" computer code. Reusable code allows a programmer to use code that has already been designed, written, and tested. This makes the programmer's task easier, and it results in more reliable and efficient programs.

# Section B

# The Java Language

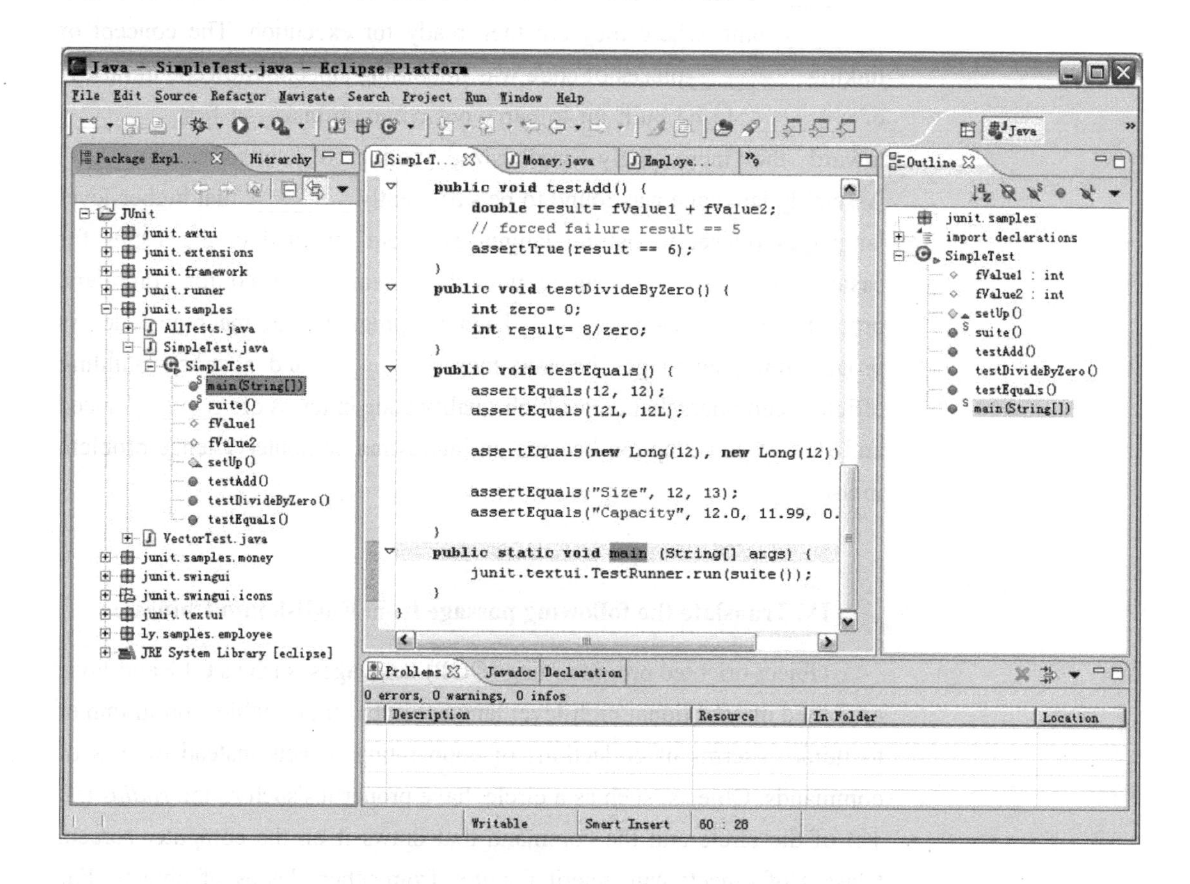


The Java language is a remarkable example of programming language evolution. Java builds on the familiar and useful features of C++ while removing its complex, dangerous, and **superfluous** elements. The result is a language that is safer, simpler, and easier to use. The following sections describe Java in contrast to C++.

## 1. Java Is Familiar and Simple

If you have ever programmed in C++, you will find Java's appeal to be **instantaneous**. Since Java's syntax mirrors that of C++, you will be able to write Java programs within minutes. Your first programs will come quickly and easily, with very little programming **overhead**.

You will have the feeling that you have eliminated a lot of **clutter** from

**superfluous** /sjuːˈpəːfluəs/
a. 多余的；过剩的

**instantaneous** /ˌinstənˈteiniəs/
a. 瞬间的，即刻的

**overhead** /ˈəuvəhed/
n. 经常（或管理、间接）费用

**clutter** /ˈklʌtə/
n. 凌乱，杂乱；杂乱的东西

your programs—and you will have. All the **cryptic header files** and **preprocessor** statements of C and C++ are gone. All the **arcane** #define[1] statements and typedefs[2] have been taken away. You will no longer have to **delve** through several levels of header files to correctly **reference** API[3] calls. And no one will have to suffer to figure out how to use your software.

Java programs simply import the **software packages** they need. These packages may be in another directory, on another drive, or on a machine on the other side of the Internet. The Java compiler and **interpreter** figure out what objects are referenced and supply the necessary linkage.

**2. Java Is Object-Oriented**

If you think C++ is an object-oriented programming language, you are in for a big surprise[4]. After using Java to write a few programs, you'll get a better feeling for what object-oriented software is all about.

Java deals with classes and objects, pure and simple[5]. They aren't just more data structures that are available to the programmer—they are the basis for the entire programming language.

In C++, you can declare a class, but you don't have to. You can declare a structure or a **union** instead. You can declare a whole bunch of loosely associated variables and use them with C-style functions. In Java, classes and objects are at the center of the language. Everything else revolves around them. You can't declare functions and procedures. They don't exist. You can't use structures, unions, or typedefs. They're gone, too. You either use classes and objects or you don't use Java. It's that simple.

Java provides all the luxuries of object-oriented programming: **class hierarchy**, **inheritance**, **encapsulation**, and **polymorphism**—in a context that is truly useful and efficient.

The main reason for developing object-oriented software, besides clarity and simplicity, is the desperate hope that somehow the objects you develop will be reused. Java not only encourages software reuse, it demands it. To write any sort of Java program, no matter how simple, you must build

**cryptic** / 'kriptik /
a. 隐秘的；令人困惑的
**header** / 'hedə /
n. 标题；头标；页眉
**header file**
头文件
**preprocessor**
/ pri:'prəusesə /
n. 预处理程序，预处理器
**arcane** / ɑ:'kein /
a. 神秘的，晦涩难解的
**delve** / delv /
v. 搜索，翻查
**reference** / 'refərəns /
v. 引用，参考
**package** / 'pækidʒ /
n. 程序包，软件包
**software package**
软件包
**interpreter**
/ in'tə:pritə /
n. 解释程序，解释器
**union** / 'ju:njən /
n. 共用体（定义关键字），共用（数据类型）
**class hierarchy**
类层次
**inheritance**
/ in'heritəns /
n. 继承（性）
**encapsulation**
/ in,kæpsju'leiʃən; -sə- /
n. 封装
**polymorphism**
/ ,pɔli'mɔ:fizəm /
n. 多态性，多形性

[1] *#define*：C 语言中的特定字符，只有编程上的含义。
[2] *typedef*：C 语言中的特定字符，只有编程上的含义。
[3] *API*：应用程序接口（*A*pplication *P*rogram *I*nterface 的首字母缩略）。
[4] *you are in for a big surprise*：你一定会大吃一惊。句中的 be in for 为固定搭配，意为“肯定会经历”“注定要遭受”。
[5] *pure and simple*：纯粹的，不折不扣的。

on the classes and methods of the Java API.

Once you have begun developing software in Java, you have two choices:

- Build on the classes you have developed, thereby reusing them.
- Rewrite your software from scratch, copying and tailoring useful parts of existing software.

**appealing** / ə'pi:liŋ /
a. 吸引人的；有感染力的

With Java, the temptation to start from scratch is no longer **appealing**. Java's object-oriented structure forces you to develop more useful, more tailorable, and much simpler software the first time around[1].

### 3. Java Is Safer and More Reliable

Java is safer to use than C++ because it keeps you from doing the things that you do badly, while making it easier to do the things that you do well.

**assembly code**
汇编代码

Java won't automatically convert data types. You have to explicitly convert from one class to another. C++, under the most undesirable conditions, will automatically convert one type to another. It has all the flexibility of **assembly code**. Java doesn't assume that you know what you are doing. It makes sure that you do.

C++ pointers don't exist in Java. You can no longer access objects indirectly or by chance. You don't need to. You declare objects and reference those objects directly. Complex pointer arithmetic is avoided. If you need an indexed set of objects, you can use an array of objects. The concept of "the address of an object" is eliminated from the programming model, and another assembly language dinosaur is laid to rest. As a result, it becomes much easier to do things correctly in Java.

**runtime** / 'rʌntaim /
n. 运行时刻
**syntactic** / sin'tæktik /
a.（按照）句法的
**semantically** / si'mæntikəli /
ad. 在语义上
**mom** / mɔm /
n.〈主美口〉妈妈

Java's reliability extends beyond the language level to the compiler and the **runtime** system. Compile-time checks identify many programming errors that go undetected in other programming languages. These checks go beyond **syntactic** checking to ensure that statements are **semantically** correct.

Runtime checks are also more extensive and effective. Remember your teacher or **mom** telling you to "Check your work twice to make sure it's

---

1 *the first time around*：这里的 around 为副词，用于描述 something that has happened before or things that happen regularly，只是表示所描述的事情以前发生过或经常发生，省略了对意思影响不大。

right"? The Java **linker** understands class types and performs compiler-level type checking, adding redundancy to reliability. It also performs **bounds checking** and eliminates indirect object access, even under error conditions.

**linker** / 'liŋkə /
n. 连接程序，链接程序
**bounds checking**
边界检查

**4. Java Is Secure**

If you gave a skilled hacker a program written in C or C++ and told him to find any security **flaws**, there are half a dozen things that he would immediately look for: gaining access to the operating system, causing an unexpected return of control, overwriting critical memory areas, acquiring the ability to **spoof** or modify other programs, **browsing** for security information, and gaining **unauthorized** access to the file system.

**flaw** / flɔː /
n. 缺点，瑕疵
**spoof** / spuːf /
v. 哄骗，欺骗
**browse** / brauz /
v. 浏览
**unauthorized** / ˌʌn'ɔːθəraizd /
a. 未经授权的

Why is C or C++ more **vulnerable** than Java? When a programmer develops software, he or she usually focuses on how to get the software to work correctly and efficiently. C and C++ do not **constrain** the programmer from meeting these goals and provide a number of flexible features that enable the programmer to meet his end. The hacker is also able to take advantage of these features and use them in ways that weren't originally intended, causing the undesirable consequences identified in the previous paragraph. In short, C and C++ provide a great offense, but no defense. Java, on the other hand, is defensive by nature. Every time a Java-enabled browser downloads a compiled Java class, such as an **applet**, it runs the risk of running **Trojan horse** code. Because of this ever-present threat, it subjects the code to a series of checks that ensure that it is correct and secure.

**vulnerable** / 'vʌlnərəbl /
a. 脆弱的；易受攻击的
**constrain** / kən'strein /
v. 约束，限制
**applet** / 'æplət /
n. 小应用程序
**Trojan (horse)** / 'trəudʒən /
n. 特洛伊木马（程序或病毒），木马

The Java runtime system is designed to enforce a security policy that prevents execution of **malicious** code. It does this by remembering how objects are stored in memory and enforcing correct and secure access to those objects according to its security rules. It performs **bytecode** verification by passing compiled classes through a simple **theorem prover** that either proves that the code is secure or prevents the code from being loaded and executed. The class is Java's basic execution unit and security is implemented at the class level.

**malicious** / mə'liʃəs /
a. 恶意的
**bytecode** / 'baitkəud /
n. 字节码
**theorem** / 'θiərəm /
n. 定理
**prover** / 'pruːvə /
n. 证明程序
**theorem prover**
定理证明程序，定理证明器

The Java runtime system also **segregates** software according to its origin. Classes from the local system are processed separately from those of other systems. This prevents remote systems from replacing local system software with code that is less **trustworthy**.

**segregate** / 'segrigeit /
v. 隔离；分开
**trustworthy** / 'trʌstˌwəːði /
a. 值得信任的；可信的；可靠的

Java-enabled browsers, such as HotJava[1], allow the user to control the accesses that Java software may make of the local system. When a Java applet needs permission to access local resources, such as files, a security **dialog box** is presented to the user, requesting explicit user permission. This "Mother may I?" approach ensures that the user always has the final say in the security of his system.

**dialog box**
对话框
**multithreaded**
/ 'mʌlti,θredid /
a. 多线程的
**multithreading**
/ 'mʌlti,θrediŋ /
n. 多线程操作
**thread** / θred /
n. 线程，线索
**duke** / dju:k /
n. 公爵
**convoluted**
/ 'kɔnvəlu:tid /
a. 盘绕的；错综复杂的
**gaggle** / 'gægl /
n.（紊乱而有联系的）一堆
**intertwine**
/ ,intə'twain /
v.（使）缠结，（使）缠绕在一起

### 5. Java Is Multithreaded

Java, like Ada, and unlike other languages, provides built-in language support for **multithreading**. Multithreading allows more than one **thread** of execution to take place within a single program. This allows your program to do many things at once: make the **Duke** dance, play his favorite tune, and interact with the user, seemingly all at the same time. Multithreading is an important asset because it allows the programmer to write programs as independent threads, rather than as a **convoluted gaggle** of **intertwined** activities. Multithreading also allows Java to use idle CPU time to perform necessary garbage collection and general system maintenance, enabling these functions to be performed with less impact on program performance.

Writing multithreaded programs is like dating several people concurrently. Everything works fine until the threads start to interact with each other in unexpected ways. Java provides the support necessary to make multithreading work safely and correctly. Java supports multithreading by providing **synchronization** capabilities that ensure that threads share information and execution time in a way that is thread safe.

**synchronization**
/ ,siŋkrənai'zeiʃən /
n. 同步（化）

### 6. Java Is Interpreted and Portable

While it is true that **compiled code** will almost always run more quickly than **interpreted code**, it is also true that interpreted code can usually be developed and **fielded** more inexpensively, more quickly, and in a more flexible manner. It is also usually much more portable.

Java, in order to be a truly platform-independent programming language, must be interpreted. It does not run as fast as compiled **native code**, but it doesn't run much slower, either. For the cases where execution in native machine code is absolutely essential, work is **underway** to translate Java bytecode into machine code as it is loaded.

**compiled code**
编译执行的代码
**interpreted code**
解释执行的代码
**field** / fi:ld /
v. 派…上场；实施
**native code**
本机（代）码
**underway**
/ 'ʌndə'wei /
a. 在进行中的

[1] *HotJava*：由美国Sun公司（太阳微系统公司）开发的支持Java开发环境的Internet浏览器。

**outweigh** /ˌaut'wei/
v. 在价值（或重要性、影响等）方面超过

The advantages of being interpreted **outweigh** any performance impacts. Because Java is interpreted, it is much more portable. If an operating system can run the Java interpreter and support the Java API, then it can faithfully run all Java programs.

**source code**
源（代）码

Interpreted programs are much more easily kept up-to-date. You don't have to recompile them for every change. In Java, recompilation is automatic. The interpreter detects the fact that a program's bytecode file is out-of-date with respect to its **source code** file and recompiles it as it is loaded.

Because of Java's interpreted nature, linking is also more powerful and flexible. Java's runtime system supports dynamic linking between local class files and those that are downloaded from across the Internet. This feature provides the basis for Web programming.

**de facto** /diːˈfæktəu/
a.〈拉〉实际的，事实上的

**offshoot** /ˈɔfʃuːt/
n. 支族，旁系；衍生事物

**release** /riˈliːs/
n.（程序或软件的）版本；发布

**vendor** /ˈvendə/
n. 卖主；厂家，厂商

**broker** /ˈbrəukə/
n. 代理者；代理程序

**7. Java Is the Programming Language of the Web**

Java has become the **de facto** programming language of the Web. It is being licensed by nearly every major software company. It has some **offshoots** and potential competition, such as JavaScript[1] and VBScript[2], but it remains the first Web programming language and the most powerful language for developing platform-independent software.

Java is also evolving beyond the Web and becoming a key component in distributed application development. Some **releases** of Sun's products emphasize Java's importance to distributed object-based software development. Several other **vendors** have introduced products that enable Java to be integrated into the Common Object Request **Broker** Architecture (CORBA[3]), which is the framework for distributed object communication.

[1] *JavaScript*：Java 脚本语言。

[2] *VBScript*：Visual Basic 脚本语言。

[3] *CORBA*：公用对象请求代理(程序)体系结构(*C*ommon *O*bject *R*equest *B*roker *A*rchitecture 的首字母缩略)。

## Exercises

**I. Fill in the blanks with the information given in the text:**

1. Java is an object-oriented programming language, for which classes and ________ are the basis.

2. Java is designed to be ________, which makes it a useful language for programming Web applications, since users access the Web from many types of computers.

3. Java provides built-in language support for ________. That is, it allows more than one thread of execution to take place within a single program.

4. The Java ________ system is designed to enforce a security policy that prevents execution of malicious code.

**II. Translate the following terms or phrases from English into Chinese and vice versa:**

| | |
|---|---|
| 1. native code | 6. 机器码 |
| 2. header file | 7. 汇编码 |
| 3. multithreaded program | 8. 特洛伊木马程序 |
| 4. Java-enabled browser | 9. 软件包 |
| 5. malicious code | 10. 类层次 |

# Section C

# Arrays

Imagine you have a problem that requires 20 numbers to be processed. You need to read them, process them, and print them. You must also keep these 20 numbers in memory for the duration of the program. You can define 20 variables, each with a different name, as shown in Figure 3C-1.

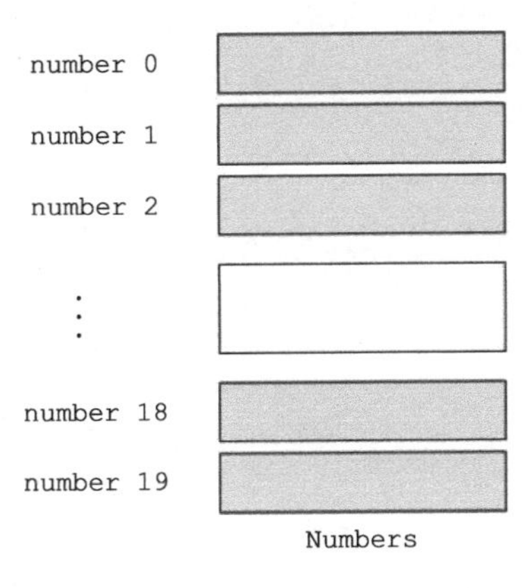


**Figure 3C-1: Twenty Individual Variables**

But having 20 different names creates another problem. How can you read 20 numbers from the keyboard and store them? To read 20 numbers from the keyboard, you need 20 **references**, one to each variable[1]. Furthermore, once you have them in memory, how can you print them? To print them, you need another 20 references. In other words, you need the **flowchart** in Figure 3C-2 to read, process, and print these 20 numbers.

**reference**
/ ˈrefərəns /
n. 引用，参考；基准

**flowchart**
/ ˈfləutʃɑːt /
n. 流程图

Although this may be acceptable for 20 numbers, it is definitely not acceptable for 200 or 2000 or 20,000 numbers. To process large amounts of data, you need a powerful data structure such as an array. An array is a fixed-size, sequenced collection of elements of the same data type. Since an array is a sequenced collection, you can refer to the elements in the array as the first element, the second element, and so forth until you get to the last element. If you were to put your 20 numbers into an array, you could

[1] *To read 20 numbers from the keyboard, you need 20 references, one to each variable.*：为了从键盘上读入20个数，你需要20个引用，每个引用对应着一个变量。

**designate** /ˈdezigneit/
v. 指定；命名；指派
**subscript** /ˈsʌbskript/
n. 下标，脚注
**ordinal** /ˈɔːdinəl/
a. 顺序的
**ordinal number**
序数

**designate** the first element $\texttt{number}_0$ as shown in Figure 3C-1. In a similar fashion, you could refer to the second number as $\texttt{number}_1$ and the third number as $\texttt{number}_2$. Continuing the series, the last number would be $\texttt{number}_{19}$. The **subscripts** indicate the **ordinal number** of the element counting from the beginning of the array[1].

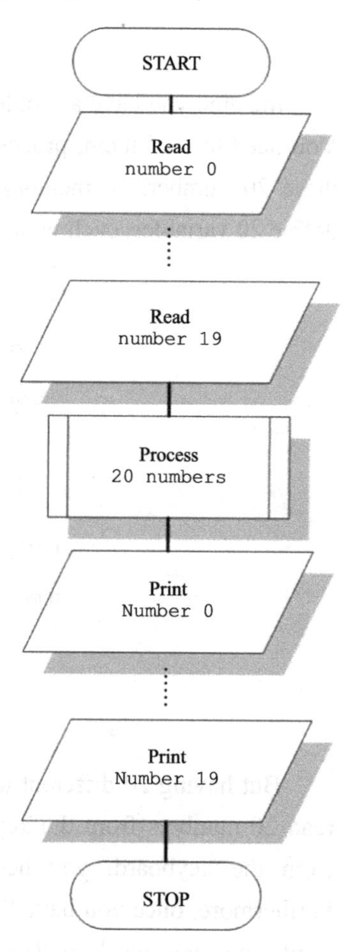


Figure 3C-2: Processing Individual Variables

**address** /əˈdres/
v. 编址；寻址

What you have seen is that the elements of the array are individually **addressed** through their subscripts (Figure 3C-3)[2]. The array as a whole has a name, `number`, but each member can be accessed individually using its subscript.

The advantages of the array would be limited if you didn't also have

1 *The subscripts indicate the ordinal number of the element counting from the beginning of the array.*：下标表示从数组开头数的元素序数。

2 *the elements of the array are individually addressed through their subscripts*：数组的元素通过其下标给出各自的地址。

**programming construct**
编程结构

**programming constructs** that allow you to process the data more conveniently. Fortunately, there is a powerful set of programming constructs, loops, that makes array processing easy.

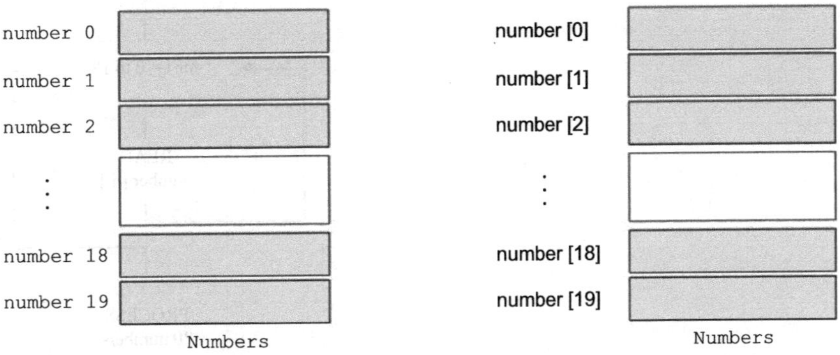

number 0
number 1
number 2
⋮
number 18
number 19
Numbers

a) Subscript Form

number [0]
number [1]
number [2]
⋮
number [18]
number [19]
Numbers

b) Index Form

**Figure 3C-3: Arrays with Subscripts and Indexes**

You can use loops to read and write the elements in an array. You can use loops to add, subtract, multiply, and divide the elements. You can also use loops for more complex processing such as calculating averages. Now it does not matter if there are 2, 20, 200, 2000, or 20,000 elements to be processed. Loops make it easy to handle them all.

**bracket** / ˈbrækit /
n. 括号
**square bracket**
方括号
**notation**
/ nəuˈteiʃən /
n. 标记法；记号
**indexing** / ˈindeksiŋ /
n. 编索引；标引；加下标；变址
**looping** / ˈlu:piŋ /
n. 循环；构成环形
**frequency array**
频率数组

But one question still remains: How can you write an instruction so that one time it refers to the first element of an array and the next time it refers to another element? It is really quite simple: You simply borrow from the subscript concept you have been using. Rather than using subscripts, however, you place the subscript value in **square brackets**. Using this **notation**, you refer to $number_0$ as number[0].

Following the convention, $number_1$ becomes number[1] and $number_{19}$ becomes number[19]. This is known as **indexing**. Using a typical reference, you now refer to your array using the variable[1]. The flowchart to process your 20 numbers using an array and **looping** is in Figure 3C-4.

In the following section, we study one array application: the **frequency array** and its graphical representation.

A frequency array shows the number of elements with the same value found in a series of numbers. For example, suppose you have taken a sample

---

[1] *Using a typical reference, you now refer to your array using the variable.*：使用一个典型的引用，你就可以使用变量来访问数组。

of 100 values between 0 and 19. You want to know how many of the values are 0, how many are 1, how many are 2, and so forth up through 19.

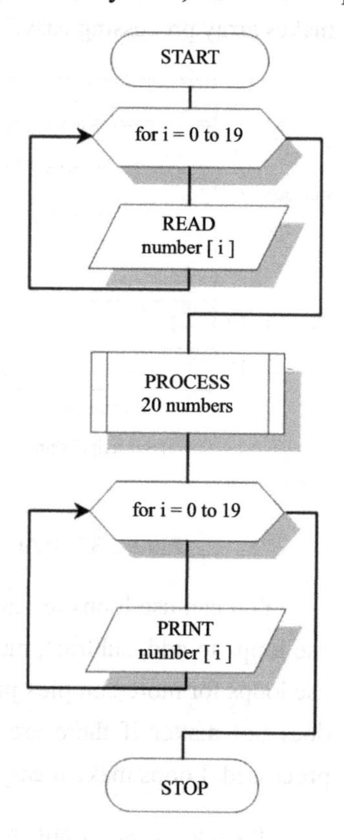


**Figure 3C-4: Processing an Array**

You can read these values into an array called `numbers`. Then you create an array of 20 elements that will show the frequency of each value in the series (Figure 3C-5).

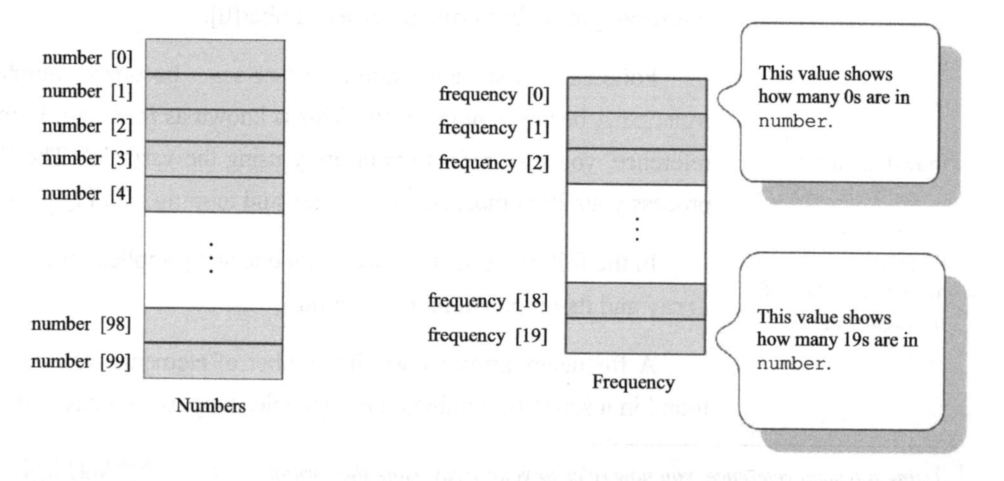


**Figure 3C-5: Frequency Array**

**histogram**
/ 'histəugræm /
n. 直方图，矩形图；频率分布图
**bar** / bɑ: /
n. 条；条形图
**bar chart**
条形图
**asterisk** / 'æstərisk /
n. 星号

A **histogram** is a pictorial representation of a frequency array. Instead of printing the values of the elements to show the frequency of each number, you print a histogram in the form of a **bar chart**. For example, Figure 3C-6 is a histogram for a set of numbers in the range 0 to 19. In this example, **asterisks** (*) are used to build the bar. Each asterisk represents one occurrence of the data value.

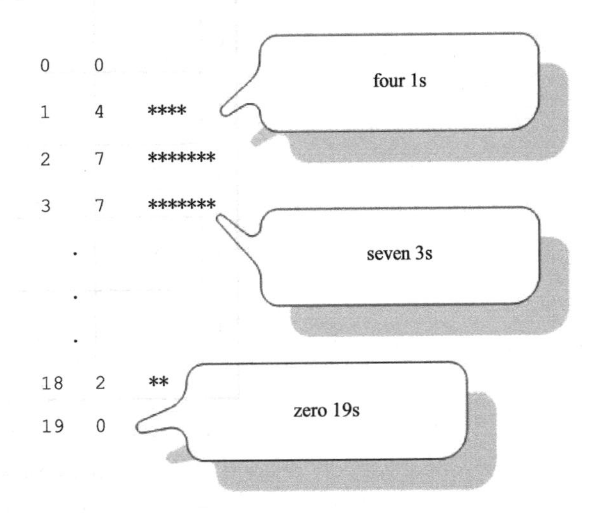


Figure 3C-6: Histogram

**one-dimensional array** 一维数组
**linear** / 'liniə /
a. 线（性）的；直线的

**two-dimensional array** 二维数组
**multidimensional**
/ ˌmʌltidi'menʃ ənəl, -dai'm- /
a. 多维的
**multidimensional array** 多维数组

The arrays discussed so far are known as **one-dimensional arrays** because the data are organized **linearly** in only one direction. Many applications require that data be stored in more than one dimension. One common example is a table, which is an array that consists of rows and columns. Figure 3C-7 shows a table, which is commonly called a **two-dimensional array**. Note that arrays can have three, four, or more dimensions. However, the discussion of **multidimensional arrays** is beyond the scope of this section.

The indexes in the definition of a two-dimensional array represent rows and columns[1]. This format maps the way the data are laid out in memory[2]. If you were to consider memory as a row of bytes with the lowest address on the left and the highest address on the right, then an array would be placed in memory with the first element to the left and the last element to the right. Similarly, if the array is two-dimensional, then the first dimension

[1] *The indexes in the definition of a two-dimensional array represent rows and columns.*：定义二维数组的索引分别表示行和列。
[2] *This format maps the way the data are laid out in memory.*：这种格式映射了数据在存储器中的存储方式。

is a row of elements that is stored to the left. This is known as "row-major" storage (Figure 3C-8)[1].

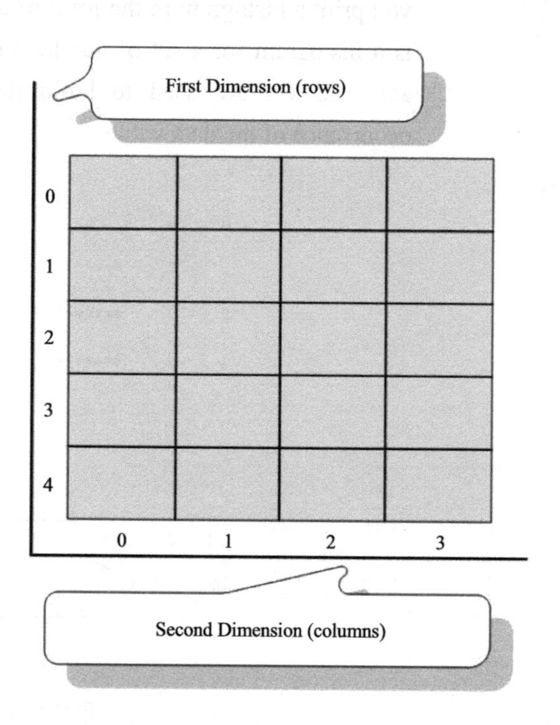


**Figure 3C-7: Two-Dimensional Array**

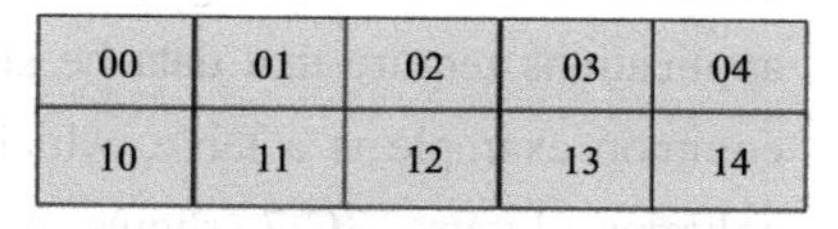


User's View

**view** / vju: /
n. 视图

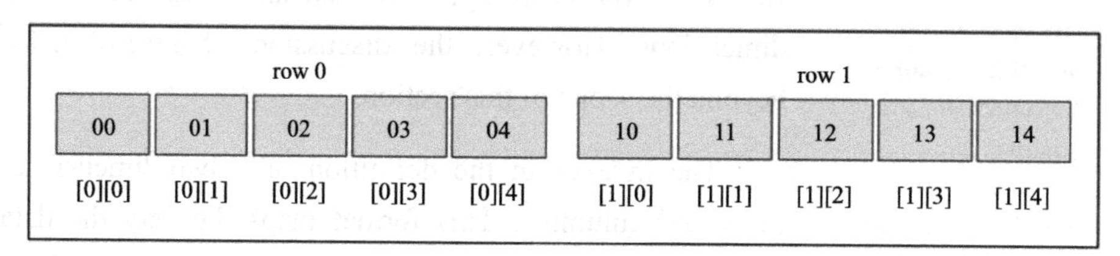


Memory View

**Figure 3C-8: Memory Layout**

---

1 *This is known as "row-major" storage.*：这通常称为“以行为主的”存储。row-major 表示“以行为主的”“行优先的”。

## Exercises

**I. Fill in the blanks with the information given in the text:**

1. Given the array called `object` with 20 elements, if you see the term $\texttt{object}_{10}$, you know the array is in ________ form; if you see the term object[10], you know the array is in ________ form.

2. In most programming languages, an array is a static data structure. When you define an array, the size is ________.

3. A(n) ________ is a pictorial representation of a frequency array.

4. An array that consists of just rows and columns is probably a(n) ________ array.

**II. Translate the following terms or phrases from English into Chinese and vice versa:**

| | |
|---|---|
| 1. bar chart | 5. 用户视图 |
| 2. frequency array | 6. 下标形式 |
| 3. graphical representation | 7. 一维数组 |
| 4. multidimensional array | 8. 编程结构 |

# Unit 4　Software Development

# （软件开发）

## Section A

## Computer Program

### I. Introduction

A computer program is a set of instructions that directs a computer to perform some processing function or combination of functions. For the instructions to be carried out, a computer must execute a program, that is, the computer reads the program, and then follows the steps **encoded** in the program in a precise order until completion. A program can be executed many different times, with each execution yielding a potentially different result depending upon the options and data that the user gives the computer.

**encode** / enˈkəud /
v. 把…编码；把…译成电码（或密码）

Programs fall into two major classes: application programs and operating systems. An application program is one that carries out some function directly for a user, such as word processing or game playing. An operating system is a program that manages the computer and the various resources and devices connected to it, such as RAM (random access memory), **hard drives**, monitors, keyboards, printers, and modems, so that they may be used by other programs.

**hard drive**
硬盘驱动器

## II. Program Development

Software designers create new programs by using special applications programs, often called **utility programs** or development programs. A programmer uses another type of program called a **text editor** to write the new program in a special notation called a programming language. With the text editor, the programmer creates a text file, which is an ordered list of instructions, also called the program **source file**. The individual instructions that make up the program source file are called source code. At this point, a special applications program translates the source code into machine language, or **object code**—a format that the operating system will recognize as a proper program and be able to execute.

**utility** / juːˈtiliti /
n. 实用程序，公用程序
**utility program**
实用程序，公用程序
**text editor**
文本编辑程序，文本编辑器
**source file**
源文件
**object code**
目标（代）码

Three types of applications programs translate from source code to object code: compilers, interpreters, and **assemblers**. The three operate differently and on different types of programming languages, but they serve the same purpose of translating from a programming language into machine language.

**assembler**
/ əˈsemblə /
n. 汇编程序，汇编器

A compiler translates text files written in a high-level programming language—such as Fortran, C, or Pascal—from the source code to the object code all at once. This differs from the approach taken by **interpreted languages** such as BASIC, in which a program is translated into object code statement by statement as each instruction is executed. The advantage of interpreted languages is that they can begin executing the program immediately instead of having to wait for all of the source code to be compiled. Changes can also be made to the program fairly quickly without having to wait for it to be compiled again. The disadvantage of interpreted languages is that they are slow to execute, since the entire program must be translated one instruction at a time, each time the program is run. On the other hand, **compiled languages** are compiled only once and thus can be executed by the computer much more quickly than interpreted languages. For this reason, compiled languages are more common and are almost always used in professional and scientific applications.

**interpreted language**
解释执行的语言，解释型语言

**compiled language**
编译执行的语言，编译型语言

Another type of translator is the assembler, which is used for programs or parts of programs written in assembly language. Assembly language is another programming language, but it is much more similar to machine language than other types of high-level languages. In assembly language, a single statement can usually be translated into a single instruction of machine language. Today, assembly language is rarely used to write an entire program, but is instead most often used when the programmer needs to directly control some aspect of the computer's function.

Programs are often written as a set of smaller pieces, with each piece representing some aspect of the overall application program. After each piece has been compiled separately, a program called a linker combines all of the translated pieces into a single executable program.

**debugger** / diːˈbʌɡə /
n. 调试程序，排错程序
**bug** / bʌɡ /
n.（程序）错误，故障
**debug** / diːˈbʌɡ /
v. 调试，排除（程序）中的错误
**source program**
源程序
**interpreter program**
解释程序

Programs seldom work correctly the first time, so a program called a **debugger** is often used to help find problems called **bugs**. **Debugging** programs usually detect an event in the executing program and point the programmer back to the origin of the event in the program code.

Recent programming systems, such as Java, use a combination of approaches to create and execute programs. A compiler takes a Java **source program** and translates it into an intermediate form. Such intermediate programs are then transferred over the Internet into computers where an **interpreter program** then executes the intermediate form as an application program.

## III. Program Elements

Most programs are built from just a few kinds of steps that are repeated many times in different contexts and in different combinations throughout the program. The most common step performs some computation, and then proceeds to the next step in the program, in the order specified by the programmer.

Programs often need to repeat a short series of steps many times, for instance in looking through a list of game scores and finding the highest score. Such repetitive sequences of code are called loops.

One of the capabilities that make computers so useful is their ability to make conditional decisions and perform different instructions based on the values of data being processed. *If-then-else* statements implement this function by testing some piece of data and then selecting one of two sequences of instructions on the basis of the result. One of the instructions in these alternatives may be a *goto* statement that directs the computer to select its next instruction from a different part of the program. For example, a program might compare two numbers and branch to a different part of the program depending on the result of the comparison:

*If* x is greater than y
*then*
*goto* instruction #10
*else* continue

Programs often use a specific sequence of steps more than once. Such a sequence of steps can be grouped together into a **subroutine**, which can then be called, or accessed, as needed in different parts of the main program. Each time a subroutine is called, the computer remembers where it was in the program when the call was made, so that it can return there upon completion of the subroutine. **Preceding** each call, a program can specify that different data be used by the subroutine, allowing a very general piece of code to be written once and used in multiple ways.

**subroutine** /ˈsʌbruːˌtiːn/
n. 子例程，子（例行）程序

**precede** /ˌpriːˈsiːd, pri-/
v. 处在…之前，先于

Most programs use several varieties of subroutines. The most common of these are functions, procedures, **library routines**, **system routines**, and device drivers. Functions are short subroutines that compute some value, such as computations of angles, which the computer cannot compute with a single basic instruction. Procedures perform a more complex function, such as sorting a set of names. Library routines are subroutines that are written for use by many different programs. System routines are similar to library routines but are actually found in the operating system. They provide some service for the application programs, such as printing a line of text. Device drivers are system routines that are added to an operating system to allow the computer to communicate with a new device, such as a scanner, modem, or printer. Device drivers often have features that can be executed directly as applications programs. This allows the user to directly control the device, which is useful if, for instance, a color printer needs to be **realigned** to attain the best printing quality after changing an ink cartridge.

**routine** /ruːˈtiːn/
n. 例程，例行程序
**library routine**
库例程，程序库例行程序
**system routine**
系统例程

**realign** /ˌriːəˈlain/
v. 重新调整

## IV. Program Function

Modern computers usually store programs on some form of magnetic storage media that can be accessed randomly by the computer, such as the hard drive disk permanently located in the computer. Additional information on such disks, called directories, indicates the names of the various programs on the disk, when they were written to the disk, and where the program begins on the disk media. When a user directs the computer to execute a particular application program, the operating system looks through these directories, locates the program, and reads a copy into RAM. The operating system then directs the CPU (central processing unit) to start executing the instructions at the beginning of the program. Instructions at the beginning of the program prepare the computer to process information by locating free **memory locations** in RAM to hold working data, retrieving copies of the standard options and **defaults** the user has indicated from a disk, and drawing initial displays on the monitor.

**memory location**
存储单元

**default** / di'fɔ:lt /
n. 默认，缺省，系统设定值

The application program requests a copy of any information the user enters by making a call to a system routine. The operating system converts any data so entered into a standard internal form. The application then uses this information to decide what to do next—for example, perform some desired processing function such as **reformatting** a page of text, or obtain some additional information from another file on a disk. In either case, calls to other system routines are used to actually carry out the display of the results or the accessing of the file from the disk.

**reformat**
/ ˌri:'fɔ:mæt /
v. 重新格式化

When the application reaches completion or is **prompted** to quit, it makes further **system calls** to make sure that all data that needs to be saved has been written back to disk. It then makes a final system call to the operating system indicating that it is finished. The operating system then frees up the RAM and any devices that the application was using and awaits a command from the user to start another program.

**prompt** / prɔmpt /
v. 提示，提醒

**system call**
系统调用

## V. History

People have been storing sequences of instructions in the form of a program for several centuries. Music boxes of the 18th century and **player pianos** of the late 19th and early 20th centuries played musical programs stored as series of metal pins, or holes in paper, with each line (of pins or holes) representing when a note was to be played, and the pin or hole indicating what note was to be played at that time. More elaborate control of physical devices became common in the early 1800s with French inventor Joseph-Marie Jacquard's invention of the **punch-card** controlled weaving loom. In the process of weaving a particular pattern, various parts of the loom had to be mechanically positioned. To automate this process, Jacquard used a single paper card to represent each positioning of the loom,

**player piano**
自动钢琴

**punch card**
穿孔卡片

with holes in the card to indicate which loom actions should be done. An entire **tapestry** could be encoded onto a deck of such cards, with the same deck yielding the same tapestry design each time it was used. Programs of over 24,000 cards were developed and used.

**tapestry** / 'tæpistri /
n. 花毯；挂毯

The world's first programmable machine was designed—although never fully built—by the English mathematician and inventor, Charles Babbage. This machine, called the Analytical Engine, used punch cards similar to those used in the Jacquard loom to select the specific arithmetic operation to apply at each step. Inserting a different set of cards changed the computations the machine performed. This machine had **counterparts** for almost everything found in modern computers, although it was mechanical rather than electrical. Construction of the Analytical Engine was never completed because the technology required to build it did not exist at the time.

**counterpart**
/ 'kauntəpɑ:t /
n. 对应的物（或人）

The first card deck programs for the Analytical Engine were developed by British mathematician Augusta Ada Byron, daughter of the poet Lord Byron[1]. For this reason, she is recognized as the world's first programmer.

The modern concept of an internally stored computer program was first proposed by Hungarian-American mathematician John von Neumann in 1945. Von Neumann's idea was to use the computer's memory to store the program as well as the data. In this way, programs can be viewed as data and can be processed like data by other programs. This idea greatly simplifies the role of program storage and execution in computers.

## VI. The Future

The field of computer science has grown rapidly since the 1950s due to the increase in the use of computers. Computer programs have undergone many changes in response to user need and advances in technology. Newer ideas in computing, such as parallel computing, distributed computing, and artificial intelligence, have radically altered the traditional concepts that once determined program form and function.

Computer scientists working in the field of parallel computing, in which multiple CPUs cooperate on the same problem at the same time, have introduced a number of new program models. In parallel computing, parts of a problem are worked on simultaneously by different processors, and this speeds up the solution of the problem. Many challenges face scientists and engineers who design programs for parallel processing computers, because

[1] *Lord Byron*：拜伦勋爵（1788—1824），英国诗人，全名 George Gordon Byron。

of the extreme complexity of the systems and the difficulty involved in making them operate as effectively as possible.

Another type of parallel computing called distributed computing uses CPUs from many interconnected computers to solve problems. Often the computers used to process information in a distributed computing application are connected over the Internet. Internet applications are becoming a particularly useful form of distributed computing, especially with programming languages such as Java. In such applications, a user logs on[1] to a Web site and downloads a Java program onto their computer. When the Java program is run, it communicates with other programs at its home Web site, and may also communicate with other programs running on different computers or Web sites.

Research into artificial intelligence (AI) has led to several other new styles of programming. Logic programs, for example, do not consist of individual instructions for the computer to follow blindly, but instead consist of sets of rules: if x happens then do y. A special program called an **inference engine** uses these rules to "reason" its way to a conclusion when presented with a new problem. Applications of logic programs include automatic monitoring of complex systems, and proving mathematical theorems.

**inference engine**
推理机

A radically different approach to computing in which there is no program in the conventional sense is called a neural network. A neural network is a group of highly interconnected simple processing elements, designed to mimic the brain. Instead of having a program direct the information processing in the way that a traditional computer does, a neural network processes information depending upon the way that its processing elements are connected. Programming a neural network is accomplished by presenting it with known patterns of input and output data and adjusting the relative importance of the interconnections between the processing elements until the desired pattern matching is accomplished. Neural networks are usually simulated on traditional computers, but unlike traditional computer programs, neural networks are able to learn from their experience.

[1] *log on*：登录，注册，进入系统。

## Exercises

**I. Fill in the blanks with the information given in the text:**

1. Computer programs fall into two major classes: ________ programs and ________ systems.

2. There are three types of application programs to translate source code into object code. They are compilers, interpreters, and ________.

3. A(n) ________ translates all the source code of a program written in a high-level language into object code prior to the execution of the program.

4. In the case of a(n) ________, a program is translated into executable form and executed one statement at a time rather than being translated completely before execution.

5. A(n) ________ is a program that is often used to help find problems in other programs.

6. A(n) ________ is a sequence of code in a program executed repeatedly, either for a fixed number of times or until a certain condition is met.

7. When you install a new device in a computer, you have to add the correct device ________ to the operating system to allow the computer to communicate with the device.

8. The modern concept of an internally stored computer program was first proposed by ________ in 1945.

**II. Translate the following terms or phrases from English into Chinese and vice versa:**

| | |
|---|---|
| 1. inference engine | 11. 解释执行的语言 |
| 2. mathematical theorem | 12. 设备驱动程序 |
| 3. compiled language | 13. 分布式计算 |
| 4. parallel computing | 14. 调试程序 |
| 5. pattern matching | 15. 目标代码 |
| 6. memory location | 16. 应用程序 |
| 7. interpreter program | 17. 实用程序 |
| 8. library routine | 18. 逻辑程序 |
| 9. intermediate program | 19. 系统调用 |
| 10. source file | 20. 程序的存储与执行 |

**III. Fill in each of the blanks with one of the words given in the following list, making changes if necessary:**

| | | | |
|---|---|---|---|
| *similar* | *computer* | *directly* | *compiler* |
| *produce* | *engineer* | *assembler* | *low-level* |
| *program* | *instruction* | *machine* | *different* |
| *translation* | *manufacturer* | *language* | *version* |

A compiler, in computer science, is a computer program that translates source code into object code. Software ________ write source code using high-level programming ________ that people can understand. Computers cannot ________ execute source code, but need a(n) ________ to translate these instructions into a(n) ________ language called machine code.

Compilers collect and reorganize (compile) all the ________ in a given set of source code to ________ object code. Object code is often the same as or ________ to a computer's machine code. If the object code is the same as the ________ language, the computer can run the ________ immediately after the compiler produces its ________. If the object code is not in machine language, other programs—such as ________, *binders* (联编程序), linkers, and *loaders* (装入程序)—finish the translation.

Most computer languages use different ________ of compilers for different types of ________ or operating systems, so one language may have ________ compilers for personal computers (PC) and Apple Macintosh computers. Many different ________ often produce versions of the same programming language, so compilers for a language may vary between manufacturers.

**IV. Translate the following passage from English into Chinese:**

In software, a bug is an error in coding or logic that causes a program to malfunction or to produce incorrect results. Minor bugs—for example, a cursor that does not behave as expected—can be inconvenient or frustrating, but not damaging to information. More severe bugs can cause a program to "hang" (stop responding to commands) and might leave the user with no alternative but to restart the program, losing any previous work that has not been saved. In either case, the programmer must find and correct the error by the process known as debugging. Because of the potential risk to important data, commercial application programs are tested and debugged as completely as possible before release. Minor bugs found after the program becomes available are corrected in the next update; more severe bugs can sometimes be fixed with special software, called patches, that *circumvents* (规避) the problem or otherwise *alleviates* (减轻) its effects.

# Section B

# Model Driven Development

## I. Introduction

One of the oldest and most commonly used approaches to analyzing and designing information systems is based on **modeling**.

Modeling is the act of drawing one or more graphical representations (or pictures) of a system. Modeling is a communication technique based on the old saying, “a picture is worth a thousand words.”

In the FAST[1] **methodology**, system models are used to illustrate and communicate the DATA, PROCESS, or INTERFACE **building blocks** of information systems[2]. This approach is called model-driven development (MDD[3]).

Model-driven development techniques emphasize the drawing of models to help **visualize** and analyze problems, define business requirements, and design information systems.

Model-driven development is illustrated in Figure 4B-1. The model-driven approach usually does not **skip** or **consolidate** the basic phases. The model-driven route takes on the appearance of a **waterfall**, suggesting that phases must generally be completed in sequence[4]. As shown in the figure, it is possible to back up to correct mistakes or **omissions**; however, such rework is often difficult, time consuming, and costly[5].

**modeling** / ˈmɔdəliŋ /
n. 建模，模型化

**methodology** / ˌmeθəˈdɔlədʒi /
n.（学科的）一套方法；方法论

**building block**
积木块，构建模块，构件

**visualize** / ˈvizjuəlaiz /
v. 使可视化；使直观化

**skip** / skip /
v. 略过，跳过

**consolidate** / kənˈsɔlideit /
v.（把…）联为一体，合并

**waterfall** / ˈwɔːtəfɔːl /
n. 瀑布

**omission** / əuˈmiʃən /
n. 省略；遗漏；疏忽

---

[1] *FAST*：系统技术应用框架（*F*ramework for the *A*pplication of *S*ystems *T*echniques 的首字母缩略）。一种软件开发方法。

[2] *In the FAST methodology, system models are used to illustrate and communicate the DATA, PROCESS, or INTERFACE building blocks of information systems.*：在 FAST 方法中，系统模型用于就信息系统的 DATA、PROCESS 或 INTERFACE 构建模块进行解释说明和交流沟通。DATA、PROCESS 和 INTERFACE 指信息系统逻辑上的结构，即（存储）数据层、（运算）处理层和（用户）界面层，是信息系统的构建模块。

[3] *MDD*：模型驱动开发（*m*odel-*d*riven *d*evelopment 的首字母缩略）。

[4] *The model-driven route takes on the appearance of a waterfall, suggesting that phases must generally be completed in sequence.*：模型驱动方法呈现出瀑布的样子，也就是说各阶段一般必须按顺序完成。

[5] *it is possible to back up to correct mistakes or omissions; however, such rework is often difficult, time consuming, and costly*：退回去更改错误和疏忽是能够做到的；然而，这种返工常常困难、耗时和费钱。

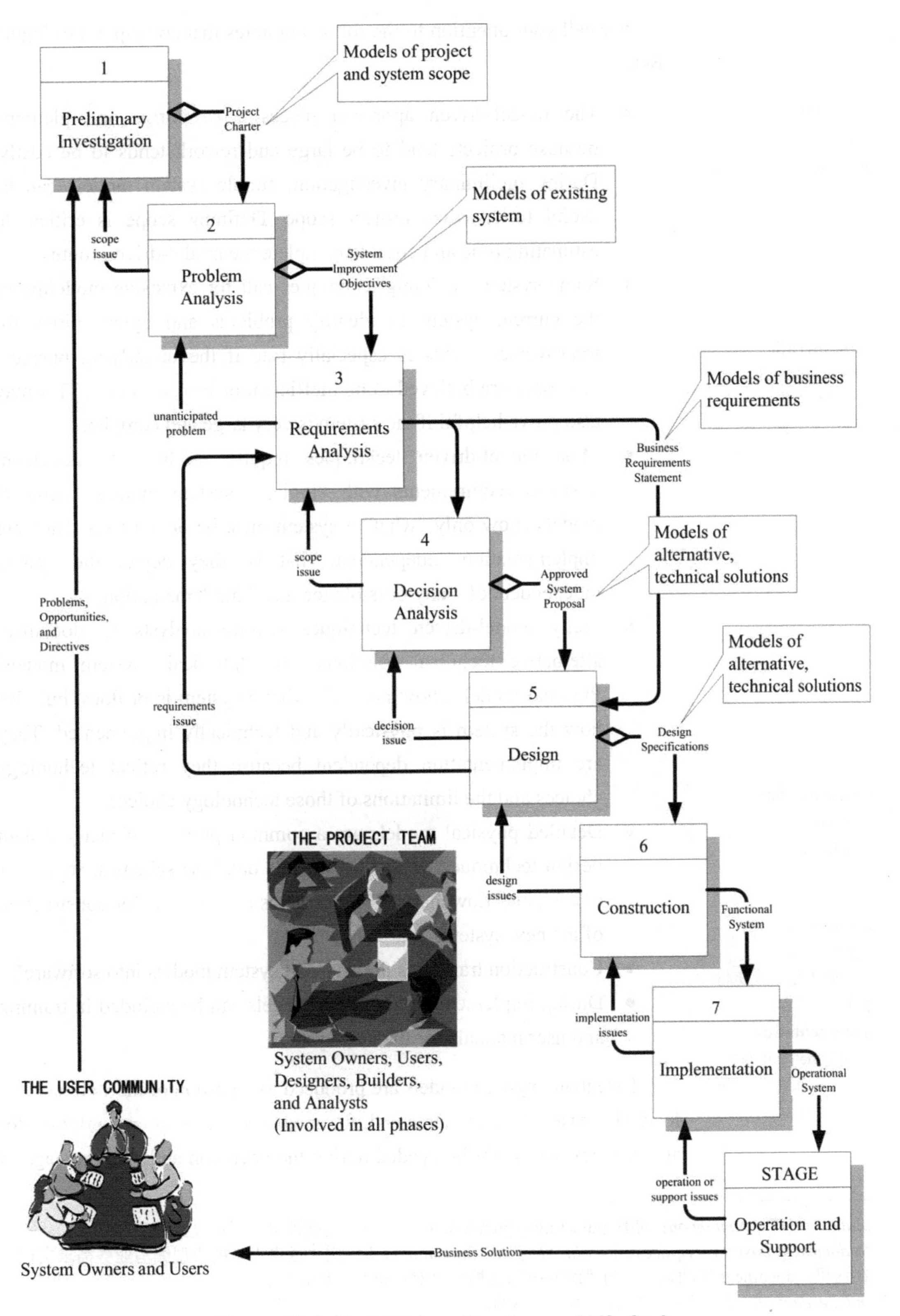


**Figure 4B-1: Model-Driven Development Life Cycle**

We call your attention to the following notes that correspond to Figure 4B-1.

**premium** / 'pri:miəm / n. 奖品；额外费用

- The model-driven approach places a **premium** on[1] planning because projects tend to be large and rework tends to be costly. During preliminary investigation, simple system models can be useful to visualize project scope. Defining scope is critical to estimating time and costs to complete the model-driven route.
- Some system modeling techniques call for extensive modeling of the current system to identify problems and opportunities for improvement. This is especially true if the underlying business processes are believed to be inefficient or **bureaucratic**. This may also prove helpful if the system is very large and complex.

**bureaucratic** / ˌbjuərəu'krætik / a. 官僚（政治）的；官僚主义的

**analyst** / 'ænəlist / n. 分析员，分析师

- Most model-driven techniques require **analysts** to document business requirements with "logical" system models[2]. Logical models show only "what" a system must be or must do. They are implementation independent; that is, they depict the system independent of[3] any possible technical implementation.
- Many model-driven techniques require analysts to document alternative technical solutions with "physical" system models. Physical models show not only what a system is or does, but also how the system is physically and technically implemented. They are implementation dependent because they reflect technology choices and the limitations of those technology choices.

**structure chart** 结构图

**blueprint** / 'blu:print / n. 蓝图

**deliverable** / di'livərəbəl / n. [常作复] 可交付使用的产品

**documentation** / ˌdɔkjumen'teiʃən / n. 文档编制；[总称] 文件证据，文献资料

- Detailed physical models are a common product of many system design techniques. Examples include database schemas, **structure charts**, and flowcharts. They serve as a **blueprint** for construction of the new system.
- Construction translates the physical system models into software[4].
- During implementation, system models can be included in training and user manuals for the new system.

Essentially system models are produced as a portion of the **deliverables** for each phase. Once implemented, the models serve as **documentation** for any changes that might be needed during the operation and support stage of

[1] *place a (high) premium on*：亦作 put a (high) premium on，意为“高度评价”“高度重视”。

[2] *to document business requirements with "logical" system models*：用“逻辑”系统模型编写业务需求文档。在这里，document 作动词，意为“编制…的文档”“对…进行文档化”。

[3] *independent of*：独立于…以外的；不管，不顾。

[4] *Construction translates the physical system models into software.*：系统构造将系统物理模型转变成软件。

the life cycle.

The model-driven approach offers several advantages:

- It minimizes planning overhead because all the phases are planned up front[1]. (This does not mean the project cannot become infeasible and get canceled.)
- Requirements analysis tends to be more thorough and better documented in the model-driven approach.
- Alternative technical solutions tend to be more thoroughly analyzed in the model-driven approach.
- System designs tend to be more sound, stable, adaptable, and flexible because they are model-based and more thoroughly analyzed before they are built.
- The approach is effective for systems that are well understood but so complex that they require large project teams to complete.
- The approach works well when fulfilling user expectations and quality are more important than cost and schedule.

There are also several disadvantages of model-driven development. Most often cited is the long duration of projects; it takes time to collect the facts, draw the models, and **validate** those models. This is especially true if users are uncertain or imprecise about their system requirements. The models can only be as good as the users' understanding of those requirements. Second, pictures are not software; some argue that this reduces the users' role in a project to passive participation. Most users don't get excited about pictures. Instead, they want to see working software and they **gauge** project progress by the existence of software (or its absence). Finally, the model-driven approach is considered by some to be inflexible; users must fully specify requirements before design; design must fully document technical specifications before construction; and so forth. Some view such **rigidity** as impractical. Regardless, model-driven approaches remain popular.

**validate** /ˈvælideit/
v. 确认（…有效），证实，验证

**gauge** /geidʒ/
v. 估计，判断；计量

**rigidity** /riˈdʒiditi/
n. 严格；刻板

There are several different model-driven techniques. They differ primarily in terms of the types of models that they require the analyst to draw and validate.

Let's briefly examine three common model-driven development techniques. Please note that we are only introducing the techniques here,

---

[1] *It minimizes planning overhead because all the phases are planned up front.*：它将计划费用降到最低，因为所有阶段都是预先计划好的。在这句话中，up front 意为“预先”。

not the models.

## II. Structured Analysis and Design

Structured analysis and design were two of the first formal system modeling techniques[1].

Structured analysis is a PROCESS-centered technique that is used to model business requirements for a system. Structured analysis introduced a modeling tool called the **data flow diagram**, used to illustrate business process requirements.

**data flow diagram**
数据流程图

Structured design is a PROCESS-centered technique that transforms the structured analysis models into good software design models. Structured design introduced a modeling tool called structure charts, used to illustrate software structure to fulfill business requirements.

The information system building blocks include several possible focuses: data, processes, and interfaces. Structured analysis and design tend to focus primarily on the process building blocks. Thus, data flow diagrams and structure charts focus on business and software processes in the system. (The techniques have evolved to also consider the data building blocks as a secondary emphasis.)

Data flow diagrams and structure charts contributed significantly to reducing the communication gap that often exists between the nontechnical system owners and users and the technical system designers and builders[2]. For this reason, structured analysis and design are still among the most widely practiced modeling techniques.

## III. Information Engineering (IE[3])

Many organizations have chosen information engineering modeling techniques.

Information engineering is a DATA-centered but PROCESS-sensitive technique used to model business requirements and design systems that fulfill those requirements. Information engineering emphasizes a modeling

---

[1] *Structured analysis and design were two of the first formal system modeling techniques.*：结构化分析与设计是最早的正规的系统建模技术中的两种。

[2] *the communication gap that often exists between the nontechnical system owners and users and the technical system designers and builders*：不懂技术的系统拥有者和用户与懂技术的系统设计者和构建者之间，在交流上常常存在的隔阂。

[3] *IE*：信息工程（*i*nformation *e*ngineering 的首字母缩略）。

tool called **entity relationship diagrams** to model business requirements.

With respect to the information system building blocks, information engineering tends to focus primarily on the data building blocks. Thus, entity relationship diagrams focus on data required in the system. Information engineering subsequently uses data flow diagrams and structure charts to model the system's processes that capture, store, and use that data.

Technically, information engineering is more ambitious than we have described here. At the risk of oversimplification, information engineering attempts to use its techniques on the organization as a whole—modeling the organization's entire data requirements, as opposed to a single information system in the organization.

Regardless, the DATA-centered technique in information engineering remains a practical option for developing individual information systems.

**entity** /ˈentiti/
n. 实体
**entity relationship diagram**
实体关系图，实体联系图，E-R 图

## IV. Object-Oriented Analysis and Design (OOAD[1])

Techniques such as structured analysis and design and information engineering have deliberately separated concerns of data from those of processes. In other words, data and process models are separate and distinct. But these concerns must be **synchronized**. Object techniques are an attempt to eliminate the separation of concerns about data and process.

Object-oriented analysis and design attempt to **merge** the data and process concerns into singular constructs called objects[2]. Object-oriented analysis and design introduced **object diagrams** that document a system in terms of its objects and their interactions.

**synchronize** /ˈsiŋkrənaiz/
v.（使）同步；（使）协调
**merge** /məːdʒ/
v. 合并；结合
**object diagram**
对象图

Business objects might correspond to real things of importance in the business such as customers and the orders they place for products. Each object consists of both the data that describes the object and the processes that can create, read, **update**, and delete that object.

**update** /ʌpˈdeit/
v. 更新；修改

With respect to the information system building blocks, object-oriented analysis and design significantly change the paradigm. The data and process columns are essentially merged into a single object column. The models then focus on identifying objects, building objects, and assembling appropriate objects into useful information systems.

---

1 *OOAD*：面向对象分析与设计（*o*bject-*o*riented *a*nalysis and *d*esign 的首字母缩略）。

2 *Object-oriented analysis and design attempt to merge the data and process concerns into singular constructs called objects.*：面向对象分析与设计试图将与数据和处理过程有关的事务融合到称为对象的单一结构中。

**frame** / freim /
n. 图文框
**drop-down menu**
下拉（式）菜单
**radio button**
单选按钮
**check box**
复选框
**scroll** / skrəul /
v. & n. 滚动
**scroll bar**
滚动条
**maximize**
/ ˈmæksimaiz /
v. 使增加到最大限度
**resize** / riːˈsaiz /
v. 调整大小

Object models are also extendable to the interface building blocks of the information system framework. Most contemporary computer user interfaces are already based on object technology. For example, both the Microsoft Windows and Netscape Navigator[1] interfaces use standard objects such as windows, **frames**, **drop-down menus**, **radio buttons**, **check boxes**, **scroll bars**, and the like. Consider a window object. It can be characterized in terms of data such as height, width, color, and so on. And it has methods such as minimizing, **maximizing**, **resizing**, and so forth. Object programming technologies such as C++, Java, Smalltalk, and Visual Basic[2] are used to construct and assemble such interface objects.

## Exercises

**I. Fill in the blanks with the information given in the text:**

1. Information engineering emphasizes a modeling tool called ________ relationship diagrams.

2. One of the disadvantages of model-driven development is the long ________ of projects.

3. Unlike structured analysis and design and information engineering, object-oriented analysis and design attempt to merge the ________ and ________ concerns into singular constructs called objects.

4. Unlike logical models, physical models show not only what a system is or does, but also how the system is physically and technically ________.

**II. Translate the following terms or phrases from English into Chinese and vice versa:**

| | |
|---|---|
| 1. check box | 6. 系统建模技术 |
| 2. structured design | 7. 模型驱动开发 |
| 3. building block | 8. 数据流程图 |
| 4. database schema | 9. 下拉菜单 |
| 5. radio button | 10. 滚动条 |

[1] *Netscape Navigator*：网景浏览器，美国网景公司开发的网页浏览器。网景通信公司（Netscape Communications Corporation），简称网景公司，创立于1994年，以其同名的网页浏览器Netscape Navigator而闻名。

[2] *Visual Basic*：可视化Basic语言，简称VB，系微软公司开发的一种事件驱动编程语言和集成开发环境（IDE），最初于1991年发行。

# Section C
# Software Process Models

## I. Introduction

A software process is a set of activities that leads to the production of a software product. These activities may involve the development of software from scratch in a standard programming language like Java or C. Increasingly, however, new software is developed by extending and modifying existing systems and by configuring and integrating **off-the-shelf** software or system components.

**off-the-shelf**
/ ˈɔfðəˈʃelf /
a. 现成的，非专门设计（或定制）的

A software process model is an abstract representation of a software process. Each process model represents a process from a particular perspective, and thus provides only partial information about that process. This section introduces a number of very general process models (sometimes called *process paradigms*) and presents them from an architectural perspective. That is, we see the framework of the process but not the details of specific activities.

**definitive** / di'finitiv / a. 决定性的；确定的；规定的

These generic models are not **definitive** descriptions of software processes. Rather, they are abstractions of the process that can be used to explain different approaches to software development. You can think of them as process frameworks that may be extended and adapted to create more specific software engineering processes.

**waterfall model** 瀑布模型

The process models covered here are the **waterfall model**, evolutionary development and component-based software engineering. These three generic process models are widely used in current software engineering practice. They are not mutually exclusive and are often used together, especially for large systems development. Sub-systems within a larger system may be developed using different approaches. Therefore, although it is convenient to discuss these models separately, you should understand that, in practice, they are often combined.

## II. The Waterfall Model

**cascade** / kæ'skeid / n. 小瀑布；级联；层叠

The first published model of the software development process was derived from more general system engineering processes. This is illustrated in Figure 4C-1. Because of the **cascade** from one phase to another, this model is known as the waterfall model or software life cycle. The principal stages of the model map onto fundamental development activities:

**constraint** / kən'streint / n. 约束，限制

**partition** / pɑ:'tiʃən / v. 划分，分区，分割

**unit testing** 单元测试

**program unit** 程序单元

- *Requirements analysis and definition.* The system's services, **constraints** and goals are established by consultation with system users. They are then defined in detail and serve as a system specification.
- *System and software design.* The systems design process **partitions** the requirements to either hardware or software systems. It establishes an overall system architecture. Software design involves identifying and describing the fundamental software system abstractions and their relationships.
- *Implementation and unit testing.* During this stage, the software design is realized as a set of programs or **program units**. Unit testing involves verifying that each unit meets its specification.
- *Integration and system testing.* The individual program units or programs are integrated and tested as a complete system to ensure that the software requirements have been met. After testing, the software system is delivered to the customer.

- *Operation and maintenance.* Normally (although not necessarily) this is the longest life-cycle phase. The system is installed and put into practical use. Maintenance involves correcting errors which were not discovered in earlier stages of the life cycle, improving the implementation of system units and enhancing the system's services as new requirements are discovered.

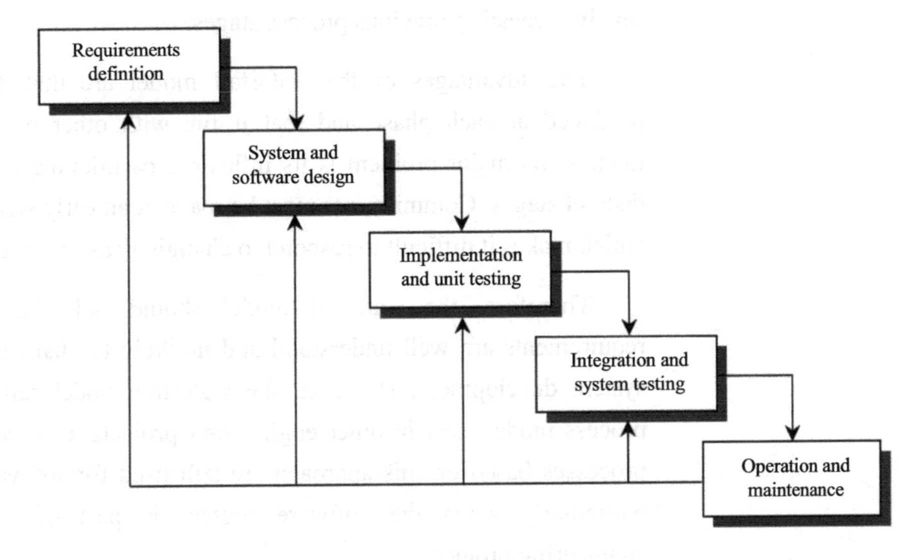


**Figure 4C-1: The Software Life Cycle**

In principle, the result of each phase is one or more documents that are approved. The following phase should not start until the previous phase has finished. In practice, these stages overlap and feed information to each other. During design, problems with requirements are identified; during coding, design problems are found and so on. The software process is not a simple linear model but involves a sequence of **iterations** of the development activities.

**iteration**
/ ˌitəˈreiʃ ən /
n. 迭代（法）；重复

Because of the costs of producing and approving documents, iterations are costly and involve significant rework. Therefore, after a small number of iterations, it is normal to freeze parts of the development, such as the specification, and to continue with the later development stages. Problems are left for later resolution, ignored or programmed around[1]. This **premature** freezing of requirements may mean that the system won't do what the user wants. It may also lead to badly structured systems as design

**premature**
/ ˌpreməˈtjuə; ˌpriː- /
a. 不成熟的；仓促的

[1] *Problems are left for later resolution, ignored or programmed around.*：问题被留待以后解决、置之不顾或者通过编程绕过去。

**circumvent** / ˌsəːkəmˈvent / v. 绕过；规避

problems are **circumvented** by implementation tricks.

During the final life-cycle phase (operation and maintenance), the software is put into use. Errors and omissions in the original software requirements are discovered. Program and design errors emerge and the need for new functionality is identified. The system must therefore evolve to remain useful. Making these changes (software maintenance) may involve repeating previous process stages.

The advantages of the waterfall model are that documentation is produced at each phase and that it fits with other engineering process models. Its major problem is its inflexible partitioning of the project into distinct stages. Commitments must be made at an early stage in the process, which makes it difficult to respond to changing customer requirements.

Therefore, the waterfall model should only be used when the requirements are well understood and unlikely to change radically during system development. However, the waterfall model reflects the type of process model used in other engineering projects. Consequently, software processes based on this approach are still used for software development, particularly when the software project is part of a larger systems engineering project.

### III. Evolutionary Development

**validation** / ˌvæliˈdeiʃən / n. 确认，验证

**interleave** / ˈintəliːv / v. 交错，交叉，交替

**exploratory** / ikˈsplɔrətəri / a. 探索的；勘探的

**throwaway** / ˈθrəuəˌwei / a. 使用后抛弃的，一次性使用的

**prototyping** / ˈprəutəuˌtaipiŋ / n. 原型法；原型开发（或设计）；样机研究

Evolutionary development is based on the idea of developing an initial implementation, exposing this to user comment and refining it through many versions until an adequate system has been developed (Figure 4C-2). Specification, development and **validation** activities are **interleaved** rather than separate, with rapid feedback across activities.

There are two fundamental types of evolutionary development:

- ***Exploratory development*** where the objective of the process is to work with the customers to explore their requirements and deliver a final system. The development starts with the parts of the system that are understood. The system evolves by adding new features proposed by the customer.
- ***Throwaway prototyping*** where the objective of the evolutionary development process is to understand the customer's requirements and hence develop a better requirements definition for the system. The prototype concentrates on experimenting with the customer

requirements that are poorly understood.

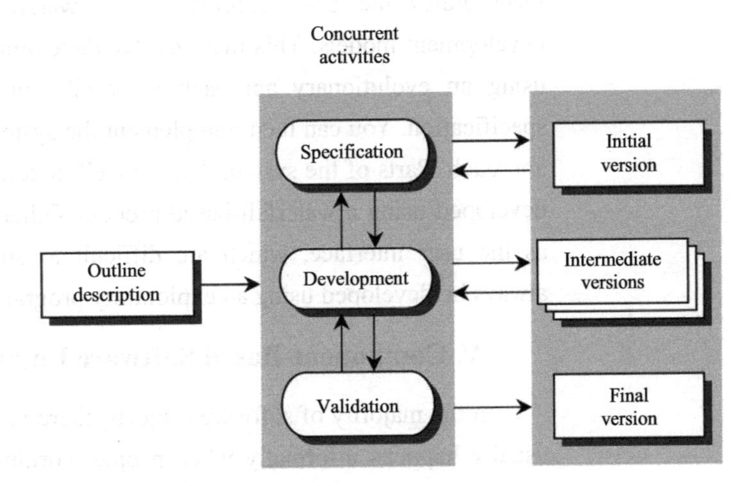


**Figure 4C-2: Evolutionary Development**

An evolutionary approach to software development is often more effective than the waterfall approach in producing systems that meet the immediate needs of customers. The advantage of a software process that is based on an evolutionary approach is that the specification can be developed incrementally. As users develop a better understanding of their problem, this can be reflected in the software system. However, from an engineering and management perspective, the evolutionary approach has two problems:

- *The process is not visible.* Managers need regular deliverables to measure progress. If systems are developed quickly, it is not **cost-effective** to produce documents that reflect every version of the system.
- *Systems are often poorly structured.* Continual change tends to corrupt the software structure. Incorporating software changes becomes increasingly difficult and costly.

**cost-effective**
/ ˈkɔstiˈfektiv /
a. 有成本效益的；合算的

For small and medium-sized systems (up to 500,000 lines of code), the evolutionary approach may be the best approach to development. The problems of evolutionary development become particularly acute for large, complex, long-life-time systems, where different teams develop different parts of the system. It is difficult to establish a stable system architecture using this approach, which makes it hard to integrate contributions from the teams.

For large systems, a mixed process is recommended which incorporates the best features of the waterfall and the evolutionary development models. This may involve developing a throwaway prototype using an evolutionary approach to resolve uncertainties in the system specification. You can then reimplement the system using a more structured approach. Parts of the system that are well understood can be specified and developed using a waterfall-based process. Other parts of the system, such as the user interface, which are difficult to specify in advance, should always be developed using an exploratory programming approach.

## IV. Component-Based Software Engineering

In the majority of software projects, there is some software reuse. This usually happens informally when people working on the project know of designs or code which is similar to that required. They look for these, modify them as needed and incorporate them into their system. In the evolutionary approach, reuse is often essential for rapid system development. This informal reuse takes place **irrespective** of the development process that is used.

**irrespective**
/ ˌiriˈspektiv /
a. 不考虑的；不顾的（*of*）

The component-based software engineering (CBSE[1]), which is a reuse-oriented approach to software development, relies on a large base of reusable software components and some integrating framework for these components. Sometimes, these components are systems in their own right (COTS[2] or commercial off-the-shelf systems) that may provide specific functionality such as text formatting or numeric calculation. The generic process model for CBSE is shown in Figure 4C-3.

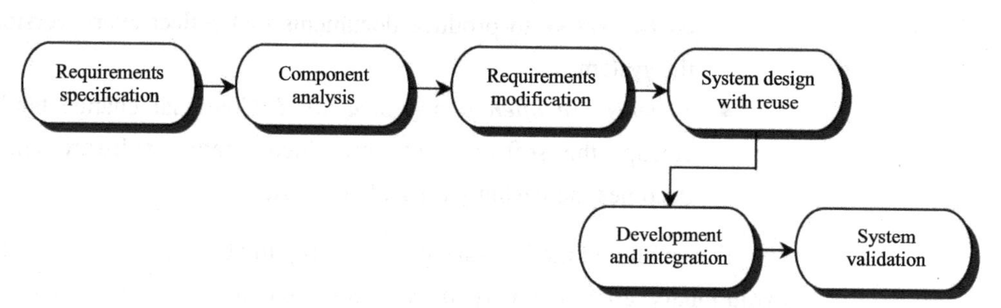


**Figure 4C-3: Component-Based Software Engineering**

While the initial requirements specification stage and the validation

[1] *CBSE*：基于组件的软件工程（component-*b*ased *s*oftware *e*ngineering 的首字母缩略）。
[2] *COTS*：现成商用的，现成民用的（*c*ommercial *o*ff-*t*he-*s*helf 的首字母缩略）。

stage are comparable with other processes, the intermediate stages in a reuse-oriented process are different. These stages are:

- *Component analysis.* Given the requirements specification[1], a search is made for components to implement that specification. Usually, there is no exact match, and the components that may be used only provide some of the functionality required.
- *Requirements* ***modification***. During this stage, the requirements are analyzed using information about the components that have been discovered. They are then modified to reflect the available components. Where modifications are impossible, the component analysis activity may be re-entered to search for alternative solutions.
- *System design with reuse.* During this phase, the framework of the system is designed or an existing framework is reused. The designers take into account the components that are reused and organize the framework to **cater** to this. Some new software may have to be designed if reusable components are not available.
- *Development and integration.* Software that cannot be externally **procured** is developed, and the components and COTS systems are integrated to create the new system. **System integration**, in this model, may be part of the development process rather than a separate activity.

**modification** / ˌmɔdifiˈkeiʃ ən /
n. 修改，更改

**cater** / ˈkeitə /
v. 满足需要；迎合；考虑（***for, to***）

**procure** / prəuˈkjuə /
v. 取得，获得；采办

**system integration**
系统集成

Component-based software engineering has the obvious advantage of reducing the amount of software to be developed and so reducing cost and risks. It usually also leads to faster delivery of the software. However, requirements compromises are inevitable and this may lead to a system that does not meet the real needs of users. Furthermore, some control over the system evolution is lost as new versions of the reusable components are not under the control of the organization using them.

---

1 *Given the requirements specification*：在有需求规格说明的情况下。本句中的 given 作介词用，引起短语作状语，意思相当于“在有…的情况下”“如果有…”。

## Exercises

**I. Fill in the blanks with the information given in the text:**

1. New software may be developed from scratch or through the use of existing systems and ________ software or system components.

2. The fundamental development activities of the waterfall model are ________ analysis and definition, system and software design, implementation and unit testing, ________ and system testing, and operation and maintenance.

3. The two fundamental types of evolutionary development are exploratory development and ________ prototyping.

4. CBSE is a(n) ________ approach to software development, which relies on a large base of reusable software components and an integrating ________ for these components.

**II. Translate the following terms or phrases from English into Chinese and vice versa:**

| | |
|---|---|
| 1. software life cycle | 5. 瀑布模型 |
| 2. evolutionary development process | 6. 系统集成 |
| 3. software reuse | 7. 商用现成软件 |
| 4. system design paradigm | 8. 基于组件的软件工程 |

# Unit 5 Software Engineering
# （软件工程）

## Section A

## Service-Oriented Software Engineering

### I. Introduction

Services may be classified as utility services that provide a general-purpose functionality, business services that implement part of a business process, or coordination services that coordinate the execution of other services. Services can also be thought of as task-oriented or entity-oriented. Task-oriented services are associated with some activity, whereas entity-oriented services are associated with a system resource. Utility or business services may be entity-oriented or task-oriented. Coordination services are always task-oriented.

**composite**
/ 'kɔmpəzit, kəm'pɔz-/
a. 综合成的；合成的；复合的

The underlying principle of service-oriented software engineering is that you compose and configure services to create new, **composite** services. These may be integrated with a user interface implemented in a browser to create a web application, or they may be used as components in some other service composition. The services involved in the composition may be specially developed for the application, business services developed within a company, or services from an external provider.

Adopting a service-oriented approach to software engineering has a number of important benefits:

- Services can be offered by any service provider inside or outside of an organization. **Assuming** these services **conform** to certain standards, organizations can create applications by integrating services from a range of providers.
- The service provider makes information about the service public so that any **authorized** user can use the service. The service provider and the service user do not need to negotiate about what the service does before it can be incorporated in an application program.
- Applications can delay the binding of services until they are **deployed** or until execution. This means that applications can be reactive and adapt their operation to cope with changes to their execution environment.
- **Opportunistic** construction of new services is possible. A service provider may recognize new services that can be created by linking existing services in **innovative** ways.
- Service users can pay for services according to their use rather than their provision. Therefore, instead of buying an expensive component that is rarely used, the application writer can use an external service that will be paid for only when required.
- Applications can be made smaller, which is particularly important for mobile devices with limited processing and memory capabilities. Some computationally intensive processing and exception handling can be **offloaded** to external services.

**assuming** / ə'sju:miŋ; ə'su:- /
conj. 假定，假如

**conform** / kən'fɔ:m /
v. 遵照；一致，符合（*to, with*）

**authorized** / 'ɔ:θəraizd /
a. 经授权的，特许的

**deploy** / di'plɔi /
v. 部署；展开

**opportunistic** / ˌɔpətju:'nistik /
a. 机会主义的

**innovative** / 'inəuveitiv /
a. 革新的，新颖的

**offload** / 'ɔfləud /
v. 卸下，卸载

## II. Service-Oriented Architecture

Service-oriented architecture (SOA[1]) is an approach to software engineering where reusable, standardized services are the basic building blocks for application systems. SOA is an architectural style based on the idea that executable services can be included in applications. Services have well-defined, published interfaces, and applications can choose whether or

[1] *SOA*：面向服务的体系结构（*s*ervice-*o*riented *a*rchitecture 的首字母缩略）。

not these are appropriate. An important idea underlying SOA is that the same service may be available from different providers and that applications could make a runtime decision of which service provider to use.

**registry** / 'redʒistri /
n. 注册表；登记簿

Figure 5A-1 illustrates the structure of a service-oriented architecture. Service providers design and implement services and specify the interface to these services. They also publish information about these services in an accessible **registry**. Service requestors (sometimes called service clients) who wish to make use of a service discover the specification of that service and locate the service provider. They can then bind their application to that specific service and communicate with it, using standard service protocols.

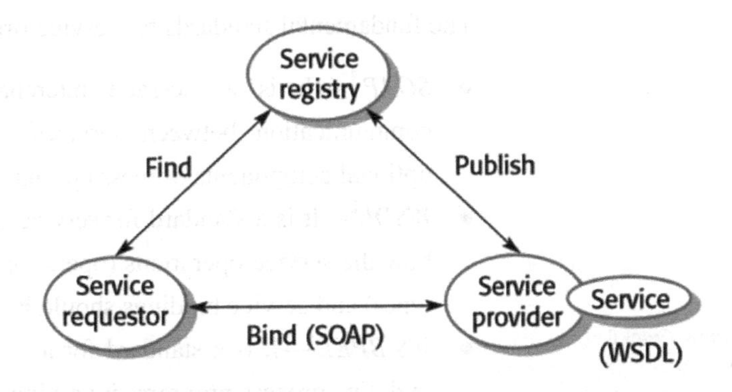


**Figure 5A-1: Service-Oriented Architecture**

**incompatibility** / 'inkəm,pætə'biliti /
n. 不兼容性
**innovation** / ,inəu'veiʃən /
n. 革新，创新
**proprietary** / prəu'praiətəri /
a. 专有的，专用的；专利的
**identifier** / ai'dentifaiə /
n. 标识符

The development and use of internationally agreed standards is fundamental to SOA. As a result, service-oriented architectures have not suffered from the **incompatibilities** that normally arise with technical **innovations**, where different suppliers maintain their **proprietary** version of the technology. Figure 5A-2 shows the stack of key standards that have been established to support web services.

Web service protocols cover all aspects of service-oriented architectures. These standards are all based on XML, a human and machine-readable notation that allows the definition of structured data where text is tagged with a meaningful **identifier**. XML has a range of supporting technologies, such as XSD[1] for schema definition, which are used to extend and manipulate XML descriptions.

---

[1] *XSD*：XML 模式定义（*X*ML *S*chema *D*efinition 的首字母缩略）。

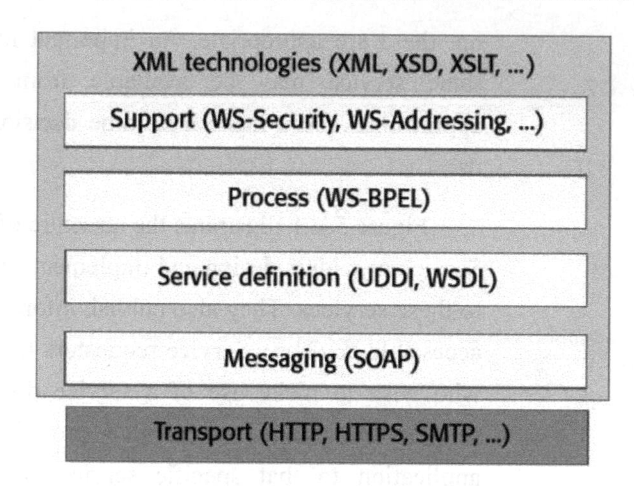


**Figure 5A-2: Web Service Standards**

The fundamental standards for service-oriented architectures are:

- *SOAP*[1]—It is a message interchange standard that supports communication between services. It defines the essential and optional components of messages passed between services.
- *WSDL*[2]—It is a standard for service interface definition. It sets out how the service operations (operation names, parameters, and their types) and service bindings should be defined.
- *WS-BPEL*[3]—It is a standard for a **workflow** language that is used to define process programs involving several different services.

**workflow** /ˈwəːkfləu/
n. 工作流

The UDDI[4] discovery standard defines the components of a service specification intended to help potential users discover the existence of a service. This standard was meant to allow companies to set up registries, with UDDI descriptions defining the services they offered. Some companies set up UDDI registries in the early years of the 21st century, but users preferred standard **search engines** to find services. All public UDDI registries have now closed.

**search engine**
搜索引擎

The principal SOA standards are supported by a range of supporting standards that focus on more specialized aspects of SOA. Some examples of these standards include:

---

1 *SOAP*：简单对象访问协议（*S*imple *O*bject *A*ccess *P*rotocol 的首字母缩略）。
2 *WSDL*：Web 服务描述语言（*W*eb *S*ervice *D*escription *L*anguage 的首字母缩略）。
3 *WS-BPEL*：Web 服务业务流程执行语言（*W*eb *S*ervice *B*usiness *P*rocess *E*xecution *L*anguage 的首字母缩略）。
4 *UDDI*：通用描述、发现与集成（*U*niversal *D*escription, *D*iscovery, and *I*ntegration 的首字母缩略）。

- *WS*[1]*-Reliable Messaging*, a standard for message exchange that ensures messages will be delivered once and once only.
- *WS-Security*, a set of standards supporting web service security, including standards that specify the definition of security policies and standards that cover the use of digital signatures.
- *WS-Addressing*, which defines how address information should be represented in a SOAP message.
- *WS-Transactions*, which defines how transactions across distributed services should be coordinated.

**addressing**
/ ə'dresiŋ /
n. 编址；寻址

## III. RESTful Services[2]

The problem is that web services standards are "heavyweight" standards that are sometimes **overly** general and inefficient. Implementing these standards requires a considerable amount of processing to create, transmit, and interpret the associated XML messages. This slows down communications between services, and, for high-**throughput** systems, additional hardware may be required to deliver the quality of service required. This situation has led to the adoption of an alternative "lightweight" approach, based on so-called RESTful services.

**overly** / 'əuvəli /
ad. 过度地

**throughput**
/ 'θru:put /
n. 吞吐量；吞吐率

REST[3] is an architectural style based on transferring representations of resources from a server to a client. A RESTful architecture is based on resources and standard operations on these resources. In a RESTful architecture, everything is represented as a resource. Essentially, a resource is simply a **data element** such as a catalog and a medical record, or a document. In general, resources may have multiple representations; that is, they can exist in different formats. Resources have a unique identifier, which is their URL[4]. Resources are a bit like objects, with four fundamental **polymorphic** operations associated with them, as shown in Figure 5A-3a:

**data element**
数据元（素）

**polymorphic**
/ ˌpɔli'mɔ:fik /
a. 多（种）形（式）的，多态的

- Create—bring the resource into existence.
- Read—return a representation of the resource.
- Update—change the value of the resource.
- Delete—make the resource inaccessible.

---

[1] *WS*：Web 服务（*W*eb *S*ervice 的首字母缩略）。

[2] *RESTful services*：RESTful 服务。RESTful 系 REST 加形容词后缀-ful 构成。

[3] *REST*：表示层状态转化（*Re*presentational *S*tate *T*ransfer 的缩略）。

[4] *URL*：统一资源定位符，统一资源定位器（*U*niform *R*esource *L*ocator 的首字母缩略），亦称为 *U*niversal *R*esource *L*ocator。

Web page
(万维)网页

The Web is an example of a system that has a RESTful architecture. **Web pages** are resources, and the unique identifier of a web page is its URL. The web protocols http and https[1] are based on four actions, namely, POST, GET, PUT, and DELETE. These map onto the basic resource operations[2], as shown in Figure 5A-3b:

- POST is used to create a resource. It has associated data that defines the resource.
- GET is used to read the value of a resource and return that to the requestor in the specified representation, such as XHTML[3], that can be rendered in a web browser.
- PUT is used to update the value of a resource.
- DELETE is used to delete the resource.

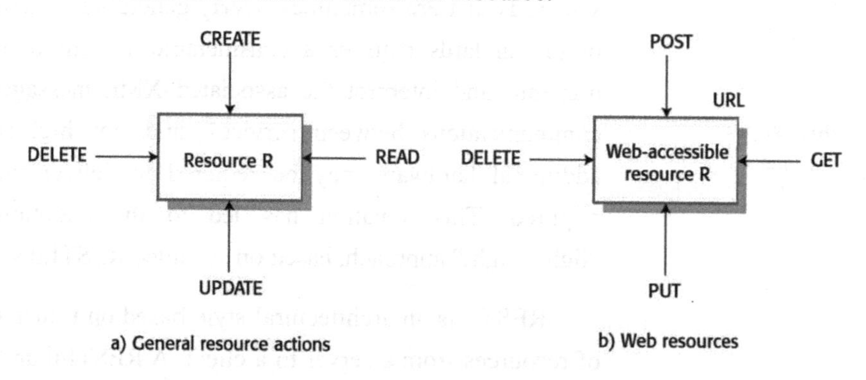


**Figure 5A-3: Resources and Actions**

All services, in some way, operate on data. When a RESTful approach is used, the data is exposed and is accessed using its URL. RESTful services use http or https protocols, with the only allowed actions being POST, GET, PUT, and DELETE.

website /ˈwebsait/
n. 网站,站点

RESTful services have become more widely used over the past few years because of the widespread use of mobile devices. These devices have limited processing capabilities, so the lower overhead of RESTful services allows better system performance. They are also easy to use with existing **websites**—implementing a RESTful API for a website is usually fairly straightforward. However, there are problems with the RESTful approach:

---

1 *https*:亦作 HTTPS,使用安全套接层的超文本传输协议,安全超文本传输协议(*H*yper*t*ext *T*ransfer *P*rotocol over *S*ecure Sockets Layer 的缩略)。

2 *These map onto the basic resource operations*:这些动作映射到基本的资源操作。

3 *XHTML*:可扩展超文本标记语言(e*X*tensible *H*yper*t*ext *M*arkup *L*anguage 的缩略)。

- When a service has a complex interface and is not a simple resource, it can be difficult to design a set of RESTful services to represent this interface.
- There are no standards for RESTful interface description, so service users must rely on informal documentation to understand the interface.
- When you use RESTful services, you have to implement your own **infrastructure** for monitoring and managing the quality of service and the service reliability. SOAP-based services have additional infrastructure support standards.

**infrastructure** / 'infrəˌstrʌktʃə /
n. 基础设施；基础结构

It is often possible to provide both SOAP-based and RESTful interfaces to the same service or resource. This **dual** approach is now common for cloud services from providers such as Microsoft, Google, and Amazon[1]. Service clients can then choose the service access method that is best suited to their applications.

**dual** / 'djuːəl /
a. 双的；双重的

## IV. Service Engineering

Service engineering is the process of developing services for reuse in service-oriented applications. It has much in common with component engineering. Service engineers have to ensure that the service represents a reusable abstraction that could be useful in different systems. They must design and develop generally useful functionality associated with that abstraction and ensure that the service is robust and reliable. They have to document the service so that it can be discovered and understood by potential users.

As shown in Figure 5A-4, there are three logical stages in the service engineering process:

- *Service candidate identification*, where you identify possible services that might be implemented and define the service requirements.
- *Service design*, where you design the logical service interface and its implementation interfaces (SOAP-based and/or RESTful).
- *Service implementation and* ***deployment***, where you implement and test the service and make it available for use.

**deployment** / di'plɔimənt /
n. 部署；展开

[1] *Amazon*：亚马逊公司，一家跨国电子商务公司，创立于1995年，总部设在美国华盛顿州西雅图市。

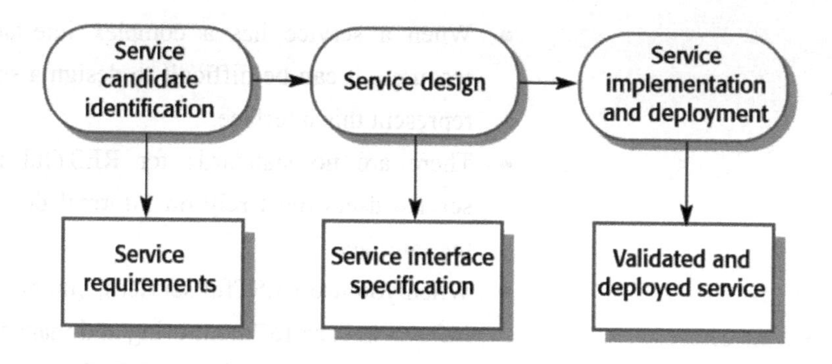


Figure 5A-4: The Service Engineering Process

**generalize** / 'dʒenərəlaiz /
v. 对…进行概括，归纳出
**-specific** / spi'sifik /
comb. form 表示"限定的""特有的"
**generalization** / ,dʒenərəlai'zeiʃən; -li'z- /
n. 概括，归纳

The development of a reusable component may start with an existing component that has already been implemented and used in an application. The same is true for services—the starting point for this process will often be an existing service or a component that is to be converted to a service. In this situation, the design process involves **generalizing** the existing component so that application-**specific** features are removed. Implementation means adapting the component by adding service interfaces and implementing the required **generalizations**.

## Exercises

**I. Fill in the blanks with the information given in the text:**

1. Services may be classified as utility services, ________ services, or coordination services, and can also be thought of as ________ or entity-oriented.

2. The development of software using services is based on the idea that programs are created by composing and ________ services to create new, ________ services and systems.

3. In service-oriented software engineering, reusable, ________ services are the basic building blocks for application systems.

4. Services may be implemented within a service-oriented architecture using a set of web service standards based on ________. These include standards for service ________, service interface definition, and service enactment in workflows.

5. A RESTful architecture is based on resources and standard ________ on these resources.

6. A RESTful approach uses the http and https protocols for service communication and maps operations on the standard http verbs ________, GET, PUT, and ________.

7. The service engineering process involves service ________ identification, service design, service ________ and deployment.

8. Compared with the service-oriented architecture, the ________ approach is simple, but it is less suited to services that offer complex functionality.

**II. Translate the following terms or phrases from English into Chinese and vice versa:**

| | |
|---|---|
| 1. computationally intensive processing | 11. 面向服务的体系结构 |
| 2. reusable component | 12. 面向实体的服务 |
| 3. data element | 13. 简单对象访问协议 |
| 4. service candidate identification | 14. 服务实现与部署 |
| 5. application-specific feature | 15. 数字签名 |
| 6. high-throughput system | 16. 工作流语言 |
| 7. utility service | 17. 业务服务 |
| 8. Representational State Transfer | 18. 模式定义 |
| 9. service binding | 19. 可扩展标记语言 |
| 10. task-oriented service | 20. 移动设备 |

**III. Fill in each of the blanks with one of the words given in the following list, making changes if necessary:**

| | | | |
|---|---|---|---|
| *aspect* | *software* | *environmental* | *most* |
| *different* | *practice* | *maintenance* | *type* |
| *adapt* | *modify* | *way* | *response* |
| *expensive* | *fault* | *necessary* | *system* |

There are three different types of software maintenance. Firstly, there is maintenance to repair software ________. Coding errors are usually relatively cheap to correct; design errors are more ________ as they may involve rewriting several program components. Requirements errors are the ________ expensive to repair because of the extensive system redesign that may be ________. Secondly, there is maintenance to adapt the software to a

________ operating environment. This type of maintenance is required when some ________ of the system's environment such as the hardware, the platform operating ________ or other support software changes. The application system must be ________ to adapt it to cope with these environmental changes. And thirdly, there is ________ to add to or modify the system's functionality. This ________ of maintenance is necessary when the system requirements change in ________ to organizational or business change. The scale of the changes required to the ________ is often much greater than for the other types of maintenance. In ________, there isn't a clear-cut distinction between these types of maintenance. When you ________ the system to a new environment, you may add functionality to take advantage of new ________ features. Software faults are often exposed because users use the system in unanticipated ________. Changing the system to accommodate their way of working is the best way to fix these faults.

**IV. Translate the following passage from English into Chinese:**

Software engineering is the branch of computer science that seeks principles to guide the development of large, complex software systems. The problems faced when developing such systems are more than enlarged versions of those problems faced when writing small programs. For instance, the development of such systems requires the efforts of more than one person over an extended period of time during which the requirements of the proposed system may be altered and the personnel assigned to the project may change. Consequently, software engineering also includes such topics as personnel and project management.

Research in software engineering is currently progressing on two levels: Some researchers, sometimes called *practitioners* (实践者), work toward developing techniques for immediate application, whereas others, called *theoreticians* (理论家), search for underlying principles and theories on which more stable techniques can someday be constructed. Many methodologies developed and promoted by practitioners in the past have been replaced by other approaches that may themselves become *obsolete* (废弃的) with time. Meanwhile, progress by theoreticians continues to be slow.

# Section B

## Software Testing Techniques

There are many possible testing techniques, which can be **intermixed** to provide the best possible testing **coverage** and outcomes. No matter how extensive the testing is, it cannot be **exhaustive** and guarantee program/system correctness. It is never possible to test for all possible data inputs or code **execution paths**.

**intermix**
/ ˌintəˈmiks /
v.（使）混合（***with***）
**coverage**
/ ˈkʌvəridʒ /
n. 覆盖（范围）
**exhaustive**
/ igˈzɔːstiv /
a. 全面而彻底的，详尽无遗的
**execution path**
执行路径

Testing techniques (strategies) can be classified according to various **criteria**. The following discussion is based on five criteria: (1) visibility, (2) automation, (3) partitioning, (4) coverage, and (5) scripting (Figure 5B-1). The criteria are not independent. For example, the two partitioning techniques apply mostly to **black box testing**, but not really to **white box** testing. The scripting techniques (**regression tests** and **exercising tests**) are automatic tests.

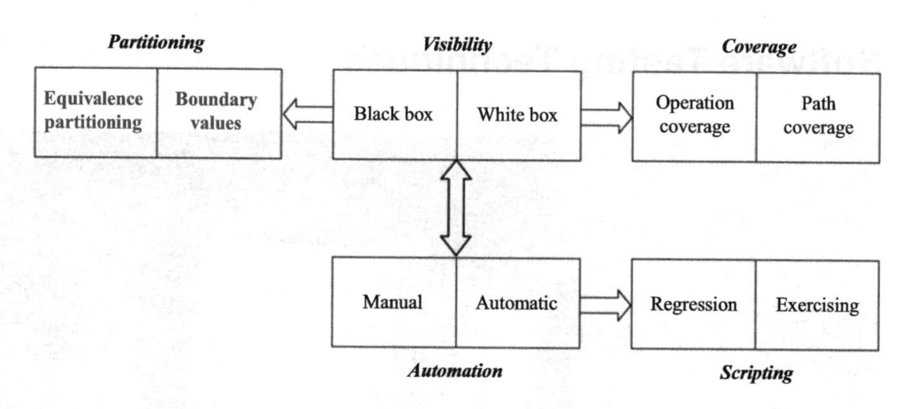


**Figure 5B-1: Testing Techniques**

Black box testing (or testing to **specs**[1]) assumes that the tester does not know or chooses to ignore the internal workings of the program unit under test. The test unit is considered a black box, which takes some input and produces some output. The testing is done by feeding the test unit with data inputs and verifying that the expected output is produced. Because no implementation knowledge is required, black box testing can be conducted by the users. Accordingly, this is the main technique of acceptance testing.

The black box technique tests system functionality, or the system functional outputs, to be precise. It is also applicable for constraint testing, such as system performance or security. A special target of black box testing is missing functionality, i.e. when an expected functional requirement has not been implemented in the software. In such a case, the test unit gets data to trigger such functionality and fails because the code to do the job is not there.

White box testing (or testing to code) is the opposite of black box testing. The test unit is considered a white box or, better, a transparent glass box that reveals its content. In this approach, the tester studies the code and decides on data inputs that are able to "exercise" selected execution paths in the code. Black box testing must normally be followed by white box testing to **pinpoint** the errors discovered when testing to specs.

Unlike black box testing, white box testing is not restricted to code

**criterion** /kraiˈtiəriən/ ([复] **-ria** /-riə/ 或 **-rions**) n. 标准，准则
**black box testing** 黑盒测试（法）
**white box testing** 白盒测试（法）
**regression** /riˈgreʃən/ n. 回归
**regression test** 回归测试
**exercising test** 压力测试
**equivalence** /iˈkwivələns/ n. 等价；相等
**equivalence partitioning** 等价（类）划分
**boundary value** 边界值
**spec** /spek/ n.〈口〉明确说明；规格，规范（= specification）
**pinpoint** /ˈpinpɔint/ v. 精确地确定…的位置；确定

[1] *testing to specs*：按照规格测试。介词 to 表示“按”“按照”。

testing. It is equally suitable for other development **artifacts**, such as design models and specification documents. Testing of this kind uses review techniques, such as **walkthroughs** and inspections. An interesting discovery possible with white box testing, but not with black box, is the existence of "dead code"—statements in the program that cannot be exercised or will never be used no matter what input is given to the code.

**artifact** / 'ɑːtifækt / n. 人工制品，制造物
**walkthrough** / 'wɔːkθruː / n. 走查

The equivalence partitioning and boundary value techniques apply mostly to black box testing, although white box testing can also benefit from them. They are specific approaches to the black box technique to **counteract** the impossibility of conducting exhaustive tests.

**counteract** / ˌkauntə'rækt / v. 对…起反作用；对抗；抵消

Equivalence partitioning groups data inputs (and, implicitly, data outputs) into **partitions** constituting **homogeneous** test targets. The assumption is that testing with any one member of the partition is as good as[1] testing with other members. Therefore, testing with the other members can be **forfeited**. Choosing the homogeneous partitions is the main difficulty. Partitions must be chosen with a good knowledge of data and the application demands for data. In this sense, partitioning **per se** is not really black box (it is not even a testing technique as such). Black box is the testing technique that equivalence partitioning supports.

**partition** / pɑː'tiʃən / n. 分割；划分；（分割而成的）部分
**homogeneous** / ˌhɔməu'dʒiːniəs / a. 同种类的；同性质的；同构的
**forfeit** / 'fɔːfit / v. 丧失，失去
**per se** / ˌpəː'sei, -'siː / ad.〈拉〉自身，本身

Boundary value is merely an additional data analysis technique to assist in equivalence partitioning and, consequently, to assist in black box testing. Boundary values are extreme cases within equivalence partitions. For example, if the partition is a set of **integer** values from 1 to 100, the boundary value analysis will recommend tests to be done on the values on the edges of the partition, i.e. for −1, 0, and +1 as well as for 99, 100, and 101. Naturally, the expected outcome for testing on −1 and 101 would be an error condition.

**integer** / 'intidʒə / n. 整数

Coverage techniques determine how much code is going to be exercised by a white box test. Operation coverage, called also method coverage, ensures that each operation in the code is exercised at least once by the white box test. Operation coverage is a modern object-oriented substitute for statement and branch coverage, which apply in procedural programming languages.

Path coverage aims at numbering possible execution paths in the program and exercising them one by one. Clearly, the number of such paths

[1] *as good as*：与…几乎一样；实际上等于。

**indefinite** /in'definit/ a. 不确定的，不定的

is **indefinite** in large programs. The tester is only able to define the most critical and most frequently used paths, and only these paths will be exercised.

**test script** 测试脚本

**use case** 用例

Testing can be manual or automatic. A human tester who interacts with the application under test conducts manual testing. The tester launches the application, systematically executes its functions, and observes the results. The execution steps are conducted according to a predefined **test script**. The test script defines step-by-step testing actions and expected outcomes. **Use cases** and other requirements specification documents are used to write test scripts.

**baseline** /'beislain/ n. 基线；基准；基础

**manifest** /'mænifest/ v. 显示；使显现

**test harness** 测试框架

The main problem with manual tests is that, in most practical situations, they are performed on "live data." There is no fixed **baseline** data predefined to guarantee the same output for repeating execution of manual tests. For this reason, expected output is not always defined precisely in the test scripts. Frequently, the output is not even presented to the screen, but it is **manifested** in database changes. This forces the tester to write and execute SQL scripts or other **test harnesses** to check the results of test actions[1].

**preparatory** /pri'pærətəri; -tɔːri/ a. 准备性的，预备性的

Manual tests are expensive to prepare, execute, and manage. They are unable to satisfy the "volume" demands of testing. Automated testing employs software testing tools to execute large volumes of tests without human participation. The tools are also able to produce necessary post-test reports to facilitate management of test outcomes. Naturally the **preparatory** tasks, such as deciding what to test, programming some test scripts, and setting up the testing environment, must still be done by human experts.

As indicated in Figure 5B-1, automated testing can be divided into regression testing and exercising testing. Regression testing is a popular term to mean repetitive execution of the same test scripts on the same baseline data to verify that previously accepted functionality of the system has not been broken by successive changes to the code, i.e. changes seemingly unrelated to the tested functionality.

Regression testing is performed by automatic execution of pre-recorded test scripts at scheduled test times. Original test scripts are recorded by a

---

1 *This forces the tester to write and execute SQL scripts or other test harnesses to check the results of test actions.*：这迫使测试员编写并运行 SQL 脚本或其他测试框架，来检查测试行动的结果。

capture/playback tool in the process of capturing the human tester's actions on the application under test. Regression testing is a playback activity of playing back the scripts. In many cases, the tool-captured scripts are modified (improved) by a test programmer to allow regression testing of features, which the tool could not record automatically.

Exercising, for the lack of a better term, is really an automated coverage testing. A tool that implements exercising testing generates automatically and randomly various possible actions that otherwise could be performed by the user on the application under test. This could be compared to a mad user hitting any possible key on the keyboard, selecting any possible menu item, pressing any available command button, etc.

**deem** / 'di:m /
v. 认为，视为

Generated actions are recorded (captured) in a script, sometimes called the best script, which is **deemed** to be able to exercise the application. Any errors or application failures are recorded as well and a separate script is generated for actions that led to the problem. This is called a defect script. The defect script can be played back at any time to reproduce the error and try to determine the reason behind it.

**consecutive**
/ kən'sekjutiv /
a. 连续的
**testbed** / 'testbed /
n. 试验台，测试台

Because both regression and exercising tests use capture/playback tools, which result in programs in some scripting language, they can feed off each other, thus creating a powerful automated testing environment[1]. In practice, the difficulty with automated testing is not in conducting it, but in running **consecutive** tests on a stable testing environment, with an identical **testbed** database, with the same state of open and active applications on the test workstation, with predictable network speed, with network broadcasting to the workstation disabled, with the same desktop appearance (including video resolution, desktop colors, fonts), etc. Finally, all automatic tests should clean up after themselves and re-create the database and application testbed as it was before the tests started.

---

[1] *Because both regression and exercising tests use capture/playback tools, which result in programs in some scripting language, they can feed off each other, thus creating a powerful automated testing environment.*：回归测试和压力测试都使用捕获 / 回放工具，产生以某种脚本语言编成的程序，因此，它们可以互相利用，这样就创建了一个功能强大的自动测试环境。feed off 的基本意思为“以…作为食物(或能量等)的来源”“从…中得到滋养”。

# Exercises

**I. Fill in the blanks with the information given in the text:**

1. Black box testing is the main technique of ________ testing.

2. Black box testing is testing to specs, whereas white box testing is testing to ________.

3. Testing techniques can be classified according to various criteria, including visibility, ________, partitioning, ________, and scripting, on which the discussion in the text is based.

4. Automated testing can be divided into ________ testing and exercising testing.

**II. Translate the following terms or phrases from English into Chinese and vice versa:**

| | |
|---|---|
| 1. black box testing | 6. 捕获/回放工具 |
| 2. acceptance testing | 7. 视频分辨率 |
| 3. code execution path | 8. 白盒测试 |
| 4. test harness | 9. 测试脚本 |
| 5. equivalence partitioning | 10. 用例 |

# Section C

# What Is a Design Pattern[1]

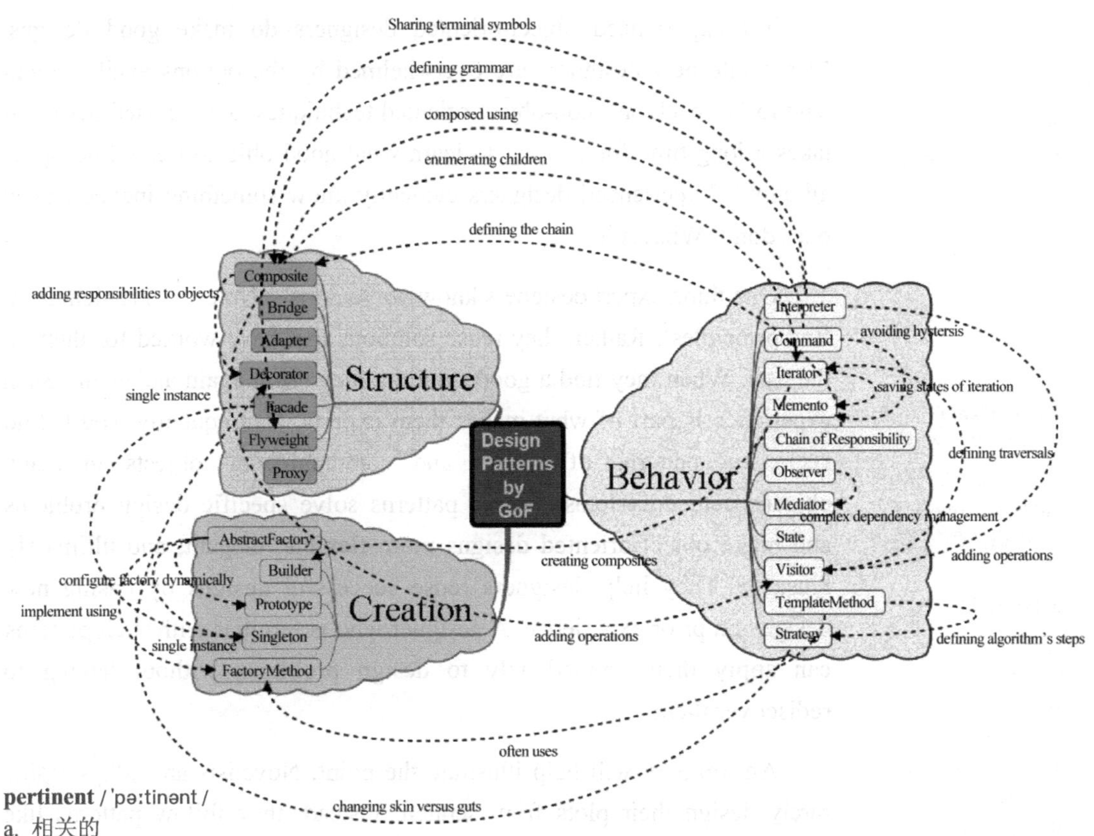


**pertinent** / ˈpəːtinənt /
a. 相关的

**factor** / ˈfæktə /
v. 把…分解成（*into*）

**granularity**
/ ˌɡrænjuˈlæriti /
n.（颗）粒度

**inheritance hierarchy**
继承层次

Designing object-oriented software is hard, and designing *reusable* object-oriented software is even harder. You must find **pertinent** objects, **factor** them into classes at the right **granularity**, define class interfaces and **inheritance hierarchies**, and establish key relationships among them[2]. Your design should be specific to the problem at hand but also general

---

1 "模式"（pattern）的概念最早由建筑大师 Christopher Alexander 于 20 世纪 70 年代提出，应用于建筑领域。20 世纪 80 年代中期，Ward Cunningham 和 Kent Beck 将其思想引入软件领域。软件业界的模式热最早源自 Erich Gamma、Richard Helm、Ralph Johnson 和 John Vlissides（在文献引用中常被称为 GoF，即"四人帮"）的经典著作 *Design Patterns: Elements of Reusable Object-Oriented Software*（本文摘自该书；中译本《设计模式：可复用面向对象软件的基础》已由机械工业出版社出版）。如今，模式已成为软件工程领域内的一个热门话题，在计算机领域的影响超过了在建筑界的影响。

2 *You must find pertinent objects, factor them into classes at the right granularity, define class interfaces and inheritance hierarchies, and establish key relationships among them.*：你必须找到相关的对象，以适当的粒度将它们归类，定义类的接口和继承层次，并建立对象之间的关键关系。

enough to address future problems and requirements. You also want to avoid redesign, or at least minimize it. Experienced object-oriented designers will tell you that a reusable and flexible design is difficult if not impossible to get "right" the first time. Before a design is finished, they usually try to reuse it several times, modifying it each time.

Yet experienced object-oriented designers do make good designs. Meanwhile new designers are overwhelmed by the options available and tend to fall back on[1] non-object-oriented techniques they've used before. It takes a long time for **novices** to learn what good object-oriented design is all about. Experienced designers evidently know something inexperienced ones don't. What is it?

**novice** / ˈnɔvis /
n. 新手，初学者

One thing expert designers know *not* to do is solve every problem from first principles[2]. Rather, they reuse solutions that have worked for them in the past. When they find a good solution, they use it again and again. Such experience is part of what makes them experts. Consequently, you'll find **recurring** patterns of classes and communicating objects in many object-oriented systems[3]. These patterns solve specific design problems and make object-oriented designs more flexible, elegant, and ultimately reusable. They help designers reuse successful designs by basing new designs on prior experience. A designer who is familiar with such patterns can apply them immediately to design problems without having to rediscover them.

**recur** / riˈkəː /
v. 再发生；反复出现

**analogy** / əˈnælədʒi /
n. 比拟，类推，类比

**playwright** / ˈpleirait /
n. 剧作家

**tragic(al)** / ˈtrædʒik(əl) /
a. 悲剧（性）的；悲惨的

**flawed** / flɔːd /
a. 有缺点的，有瑕疵的

**romance** / rəuˈmæns, ˈrəumæns /
n. 浪漫文学，传奇文学

An **analogy** will help illustrate the point. Novelists and **playwrights** rarely design their plots from scratch. Instead, they follow patterns like "**Tragically Flawed** Hero[4]" (Macbeth[5], Hamlet[6], etc.) or "The Romantic Novel" (countless **romance** novels). In the same way, object-oriented designers follow patterns like "represent states with objects" and "decorate objects so you can easily add/remove features." Once you know the pattern, a lot of design decisions follow automatically.

---

[1] *fall back on*：借助于；依赖，依靠。

[2] *One thing expert designers know not to do is solve every problem from first principles.*：内行的设计者知道不要做的一件事是，解决任何问题都从头做起。

[3] *Consequently, you'll find recurring patterns of classes and communicating objects in many object-oriented systems.*：因此，你会在许多面向对象系统中看到类和相互通信的对象的重复模式。

[4] *Tragically Flawed Hero*：带有悲剧性缺陷的主人公（指悲剧主人公存在导致自身毁灭的性格上的缺陷，如骄傲、猜疑等）。

[5] *Macbeth*：麦克白（亦译作麦克佩斯），英国伟大的剧作家、诗人莎士比亚（1564—1616）的同名悲剧中的主人公，其悲剧在于野心战胜了善良的天性。

[6] *Hamlet*：哈姆雷特（亦译作汉姆雷特），莎士比亚的同名悲剧中的主人公，属于性格优柔寡断的思考型人物。

We all know the value of design experience. How many times have you had design **déjà vu**—that feeling that you've solved a problem before but not knowing exactly where or how? If you could remember the details of the previous problem and how you solved it, then you could reuse the experience instead of rediscovering it. However, we don't do a good job of recording experience in software design for others to use.

**déjà vu** /ˌdeɪʒɑːˈvjuː/ n.〈法〉似曾经历的错觉

The purpose of this book is to record experience in designing object-oriented software as design patterns. Each design pattern systematically names, explains, and evaluates an important and recurring design in object-oriented systems. Our goal is to capture design experience in a form that people can use effectively. To this end we have documented some of the most important design patterns and present them as a catalog[1].

Design patterns make it easier to reuse successful designs and architectures. Expressing proven techniques as design patterns makes them more accessible to developers of new systems. Design patterns help you choose design alternatives that make a system reusable and avoid alternatives that compromise reusability. Design patterns can even improve the documentation and maintenance of existing systems by furnishing an explicit specification of class and object interactions and their underlying **intent**. Put simply, design patterns help a designer get a design "right" faster.

**intent** /ɪnˈtent/ n. 意图，目的

Christopher Alexander[2] says, "Each pattern describes a problem which occurs over and over again in our environment, and then describes the core of the solution to that problem, in such a way that you can use this solution a million times over, without ever doing it the same way twice". Even though Alexander was talking about patterns in buildings and towns, what he says is true about object-oriented design patterns. Our solutions are expressed in terms of objects and interfaces instead of walls and doors, but at the core of both kinds of patterns is a solution to a problem in a context.

In general, a pattern has four essential elements:

**handle** /ˈhændl/ n. 句柄；称号

1. The pattern name is a **handle** we can use to describe a design

[1] *To this end we have documented some of the most important design patterns and present them as a catalog.*：为此目的，我们编写了一些最重要的设计模式，并以编目分类的形式将它们展现出来。

[2] *Christopher Alexander*：克里斯托弗·亚历山大（1936—），出生于奥地利首都维也纳，先后毕业于剑桥大学和哈佛大学，获建筑学博士学位。他于20世纪70年代提出"模式"（pattern）的概念，并将其应用于建筑领域，在建筑学界引起强烈反响。但是，他（以及他的著作）对计算机科学的影响甚至远远超过对建筑学的影响。他的"模式语言"理论被应用在面向对象的程序设计上，给软件设计带来了极其深远的影响。

problem, its solutions, and consequences in a word or two. Naming a pattern immediately increases our design vocabulary. It lets us design at a higher level of abstraction. Having a vocabulary for patterns lets us talk about them with our colleagues, in our documentation, and even to ourselves. It makes it easier to think about designs and to communicate them and their **trade-offs** to others. Finding good names has been one of the hardest parts of developing our catalog.

**trade-off** / 'treidɔf /
n. 平衡，权衡（亦作 tradeoff）

2. The problem describes when to apply the pattern. It explains the problem and its context. It might describe specific design problems such as how to represent algorithms as objects. It might describe class or object structures that are **symptomatic** of an inflexible design. Sometimes the problem will include a list of conditions that must be met before it makes sense to apply the pattern.

**symptomatic** / ˌsimptə'mætik /
a. 作为征候（或征兆）的；表明的（*of*）

3. The solution describes the elements that make up the design, their relationships, responsibilities, and collaborations. The solution doesn't describe a particular concrete design or implementation, because a pattern is like a **template** that can be applied in many different situations. Instead, the pattern provides an abstract description of a design problem and how a general arrangement of elements (classes and objects in our case) solves it.

**template** / 'templeit, -plit /
n. 模板，样板

4. The consequences are the results and trade-offs of applying the pattern. Though consequences are often unvoiced when we describe design decisions, they are critical for evaluating design alternatives and for understanding the costs and benefits of applying the pattern. The consequences for software often concern space and time trade-offs. They may address language and implementation issues as well. Since reuse is often a factor in object-oriented design, the consequences of a pattern include its impact on a system's flexibility, **extensibility**, or portability. Listing these consequences explicitly helps you understand and evaluate them.

**extensibility** / ikˌstensə'biliti /
n. 可扩展性，可扩充性

Point of view affects one's interpretation of what is and isn't a pattern. One person's pattern can be another person's primitive building block. For this book we have concentrated on patterns at a certain level of abstraction. The design patterns in this book are *descriptions of communicating objects and classes that are customized to solve a general design problem in a particular context.*

**abstract**
/ æb'strækt /
v. 把…抽象出来；提取，抽取

A design pattern names, **abstracts**, and identifies the key aspects of a common design structure that make it useful for creating a reusable object-oriented design. The design pattern identifies the participating classes and instances, their roles and collaborations, and the distribution of responsibilities. Each design pattern focuses on a particular object-oriented design problem or issue. It describes when it applies, whether it can be applied in view of other design constraints, and the consequences and trade-offs of its use. Since we must eventually implement our designs, a design pattern also provides sample C++ and (sometimes) Smalltalk code to illustrate an implementation.

**mainstream**
/ 'meinstri:m /
a. 主流的

**pragmatic**
/ præg'mætik /
a. 注重实效的；实际的；实用主义的

Although design patterns describe object-oriented designs, they are based on practical solutions that have been implemented in **mainstream** object-oriented programming languages like Smalltalk and C++ rather than procedural languages (Pascal, C, Ada) or more dynamic object-oriented languages (CLOS[1], Dylan[2], Self[3]). We chose Smalltalk and C++ for **pragmatic** reasons: Our day-to-day experience has been in these languages, and they are increasingly popular.

The choice of programming language is important because it influences one's point of view. Our patterns assume Smalltalk/C++-level language features, and that choice determines what can and cannot be implemented easily. If we assumed procedural languages, we might have included design patterns called "Inheritance," "Encapsulation," and "Polymorphism." Similarly, some of our patterns are supported directly by the less common object-oriented languages. CLOS has multi-methods, for example, which lessen the need for a pattern such as Visitor. In fact, there are enough differences between Smalltalk and C++ to mean that some patterns can be expressed more easily in one language than the other.

---

[1] *CLOS*：公共 Lisp 对象系统（*C*ommon *L*isp *O*bject *S*ystem 的首字母缩略）。

[2] *Dylan*：1992 年由苹果公司发布的一种面向对象语言。

[3] *Self*：一种基于原型的面向对象程序设计语言，于 1986 年由施乐帕洛阿尔托研究中心的戴维·昂加尔（David Ungar）和兰迪·史密斯（Randy Smith）给出了最初的设计。

## Exercises

**I. Fill in the blanks with the information given in the text:**

1. A design pattern generally has four essential elements: the pattern name, ________, solution, and consequences.

2. Reusability is often a factor in object-oriented design, so the ________ of a pattern include its impact on a system's flexibility, extensibility, or portability.

3. The solution of a design pattern describes the elements that make up the ________, their relationships, responsibilities, and collaborations.

4. With reusable object-oriented software, your design should be ________ to the problem at hand but also ________ enough to address future problems and requirements.

**II. Translate the following terms or phrases from English into Chinese and vice versa:**

| | |
|---|---|
| 1. procedural language | 5. 设计模式 |
| 2. common design structure | 6. 可复用软件 |
| 3. class and object interaction | 7. 面向对象的系统 |
| 4. design constraint | 8. 继承层次 |

# Unit 6 Database
# （数据库）

## Section A

## Database Overview

### I. Introduction

**flat file**
平面文件，
展开文件

**payroll** /ˈpeirəul/
n. 工资表；在职人员名单；工薪总额

Data storage traditionally used individual, unrelated files, sometimes called **flat files**. In the past, each application program in an organization used its own file. In a university, for example, each department might have its own set of files: the record office kept a file about the student information and their grades, the financial aid office kept its own file about students that needed financial aid to continue their education, the scheduling office kept the names of the professors and the courses they were teaching, the **payroll** department kept its own file about the whole staff (including professors), and so on. Today, however, all of these flat files can be combined in a single entity, the database for the whole university.

Although it is difficult to give a universally agreed definition of a database, we use the following common definition: a database is a collection of related, logically **coherent**, data used by the application programs in an organization.

**coherent**
/ kəu'hiərənt /
a. 相干的；一致的；协调的

## II. Database Management Systems

A database management system (DBMS[1]) defines, creates, and maintains a database. The DBMS also allows controlled access to data in the database. A DBMS is a combination of five components: hardware, software, data, users, and procedures.

### 1. Hardware

The hardware is the physical computer system that allows access to data. For example, the terminals, hard disk, main computer, and workstations are considered part of the hardware in a DBMS.

### 2. Software

The software is the actual program that allows users to access, maintain, and update data. In addition, the software controls which user can access which parts of the data in the database.

### 3. Data

The data in a database is stored physically on the storage devices. In a database, data is a separate entity from the software that accesses it. This separation allows an organization to change the software without having to change the physical data or the way in which it is stored. If an organization decides to use a DBMS, then all the information needed by the organization should be kept together as one entity, to be accessible by the software in the DBMS.

### 4. Users

The term users in a DBMS has a broad meaning. We can divide users into two categories: **end users** and application programs.

End users are those humans who can access the database directly to get information. There are two types of end users: database **administrators**

**end user**
最终用户，终端用户

**administrator**
/ əd'ministreitə /
n.（系统、程序等的）管理员

[1] *DBMS*：数据库管理系统（*d*ata*b*ase *m*anagement *s*ystem 的缩略）。

(DBAs[1]) and normal users. Database administrators have the maximum level of privileges and can control other users and their access to the DBMS, grant some of their privileges to somebody else, but retain the ability to **revoke** them at any time. A normal user, on the other hand, can only use part of the database and has limited access.

**revoke** / ri'vəuk /
v. 撤销；废除

The other users of data in a database are application programs. Applications need to access and process data. For example, a payroll application program needs to access part of the data in a database to create **paychecks** at the end of the month.

**paycheck** / 'peitʃek /
n. 薪金支票；薪水，工资

**5. Procedures**

The last component of a DBMS is a set of procedures or rules that should be clearly defined and followed by the users of the database.

## III. Database Architecture

The American National Standards Institute Standards Planning and Requirements Committee (ANSI/SPARC[2]) has established a three-level architecture for a DBMS: internal, **conceptual**, and external (Figure 6A-1).

**conceptual**
/ kən'septjuəl /
a. 概念的

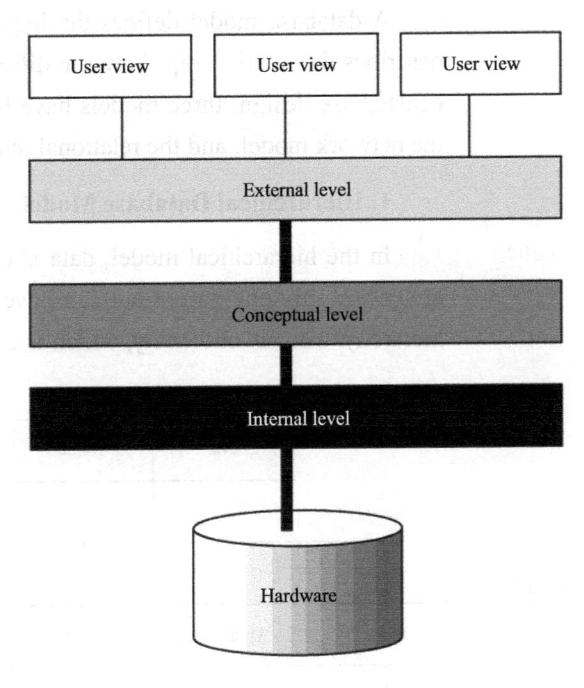


**Figure 6A-1: Database Architecture**

---

1 *DBA*：数据库管理员（*data*base *a*dministrator 的缩略）。

2 *ANSI/SPARC*：美国国家标准协会标准计划与需求委员会（*A*merican *N*ational *S*tandards *I*nstitute *S*tandards *P*lanning *a*nd *R*equirements *C*ommittee 的首字母缩略）。

### 1. Internal Level

The internal level determines where data is actually stored on the storage devices. This level deals with low-level access methods and how bytes are transferred to and from storage devices. In other words, the internal level interacts directly with the hardware.

### 2. Conceptual Level

The conceptual level defines the logical view of the data. The data model is defined on this level, and the main functions of the DBMS, such as queries, are also on this level. The DBMS changes the internal view of data to the external view that users need to see. The conceptual level is an intermediary and frees users from dealing with the internal level.

### 3. External Level

The external level interacts directly with the user (end users or application programs). It changes the data coming from the conceptual level to a format and view that is familiar to the users.

## IV. Database Models

A database model defines the logical design of data. The model also describes the relationships between different parts of the data. In the history of database design, three models have been in use: the hierarchical model, the network model, and the relational model.

### 1. Hierarchical Database Model

**inverted** / in'və:tid /
a. 反向的；倒置的

In the hierarchical model, data is organized as an **inverted** tree. Each entity has only one parent but can have several children. At the top of the hierarchy, there is one entity, which is called the root. Figure 6A-2 shows a

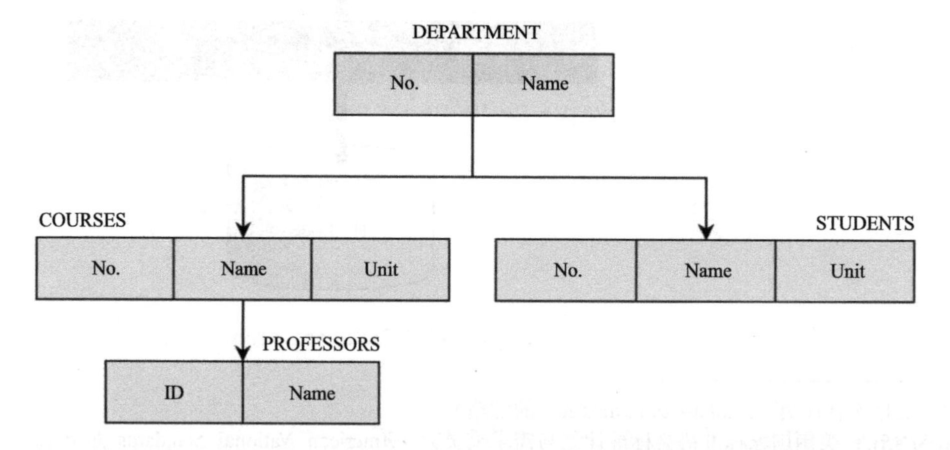


Figure 6A-2: An Example of the Hierarchical Model Representing a University

logical view of an example of the hierarchical model. The hierarchical model is now **obsolete**.

**obsolete**
/ 'ɔbsəli:t, ˌɔbsə'li:t /
a. 废弃的；淘汰的；过时的

### 2. Network Database Model

In the network model, the entities are organized in a graph, in which some entities can be accessed through several paths (Figure 6A-3). There is no hierarchy. This model is now also obsolete.

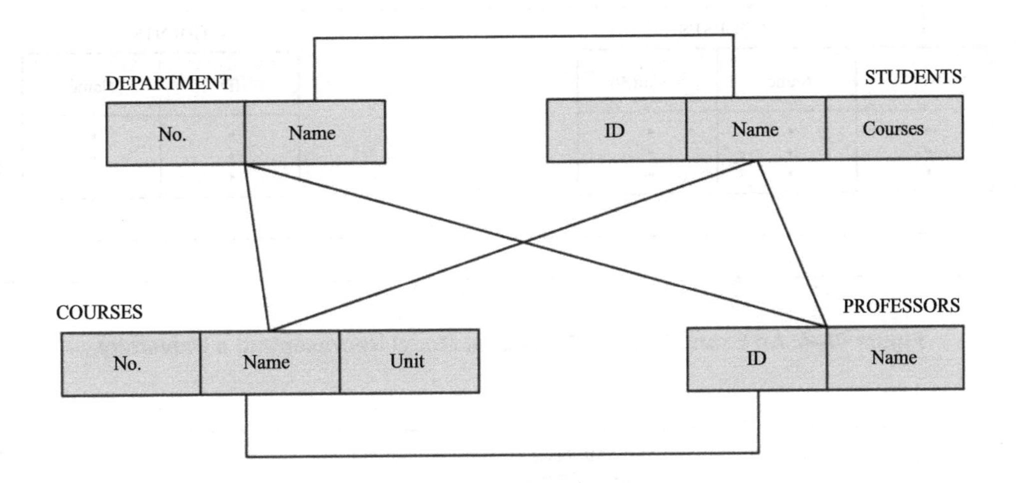


**Figure 6A-3: An Example of the Network Model Representing a University**

### 3. Relational Database Model

**two-dimensional table** 二维表

In the relational model, data is organized in **two-dimensional tables** called relations. There is no hierarchical or network structure imposed on the data. The tables or relations are, however, related to each other (Figure 6A-4). The relational database management system (RDBMS[1]) organizes the data so that its external view is a set of relations or tables. This does not mean that data is stored as tables: the physical storage of the data is independent of the way in which the data is logically organized. Figure 6A-5 shows an example of a relation. A relation in an RDBMS has the following features:

- Name. Each relation in a relational database should have a name that is unique among other relations.

[1] *RDBMS*：关系数据库管理系统（*r*elational *d*ata*b*ase *m*anagement *s*ystem 的缩略）。

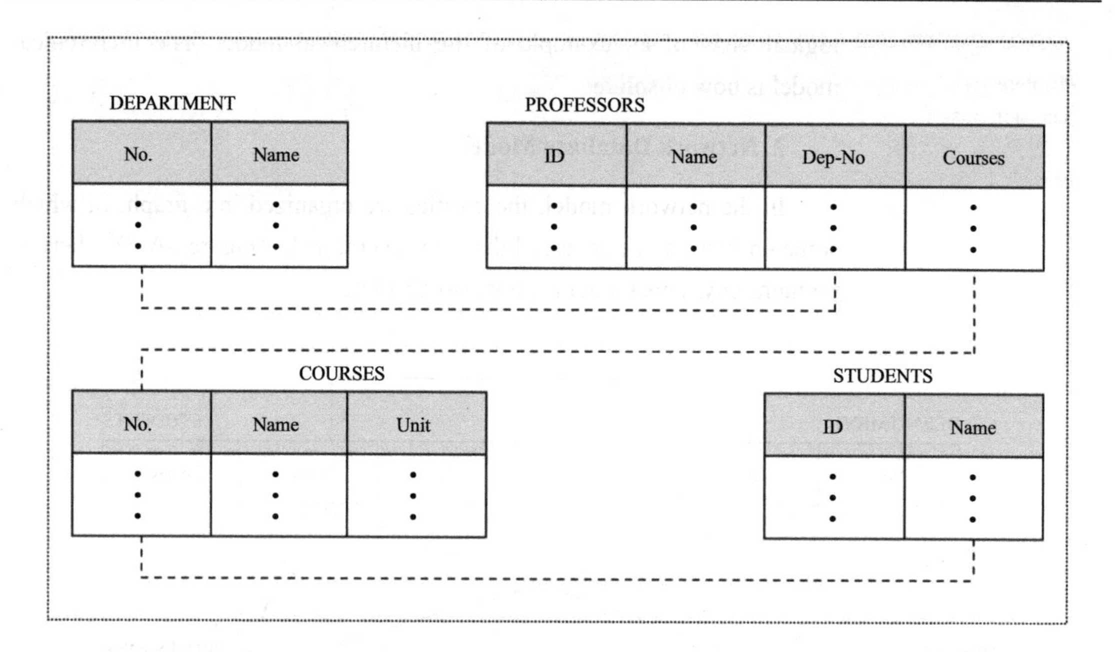


**Figure 6A-4: An Example of the Relational Model Representing a University**

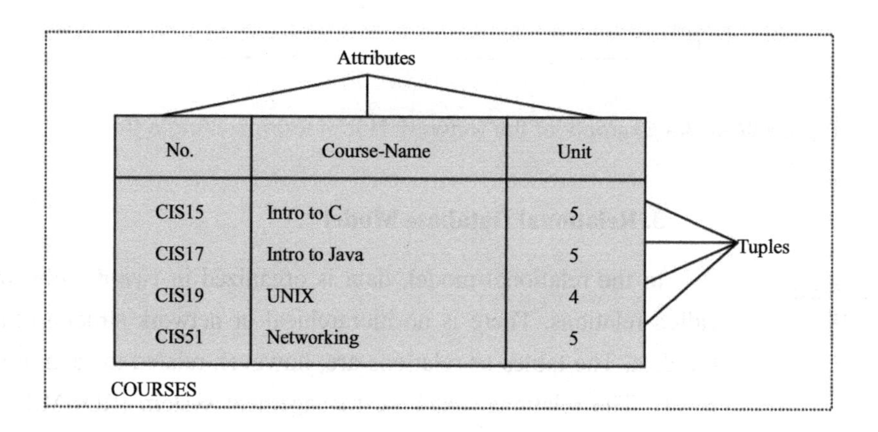


**Figure 6A-5: An Example of a Relation**

- Attributes. Each column in a relation is called an attribute. The attributes are the column headings in the table in Figure 6A-5. Each attribute gives meaning to the data stored under it. Each column in the table must have a name that is unique in the scope of the relation. The total number of attributes for a relation is called the degree of the relation. For example, in Figure 6A-5, the relation has a degree of 3. Note that the attribute names are not stored in the database: the conceptual level uses the attributes to give meaning to each column.

**tuple** / 'tʌpl, 'tju:pl /
n. 元组，字节组
**cardinality** / ˌkɑ:di'næliti /
n. 基数性；（集的）势，（集的）基数

- **Tuples**. Each row in a relation is called a tuple. A tuple defines a collection of attribute values. The total number of rows in a relation is called the **cardinality** of the relation. Note that the cardinality of a relation changes when tuples are added or deleted. This makes the database dynamic.

The relational model is one of the common models in use today. The other two common models that are derived from the relational model are the distributed model and the object-oriented model.

#### 4. Distributed Database Model

**wide area network**
广域网

**replicate** / 'replikeit /
v. 复制，重复

The distributed database model is not a new model, but is based on the relational model. However, the data is stored on several computers that communicate through the Internet or a private **wide area network**. Each computer (or site) maintains either part of the database or the whole database. In other words, data is either fragmented, with each fragment stored at one site, or data is **replicated** at each site.

In a fragmented distributed database, data is localized—locally used data is stored at the corresponding site. However, this does not mean that a site cannot access data stored at another site, but access is mostly local, but occasionally global. Although each site has complete control over its local data, there is global control through the Internet or a wide area network.

**pharmaceutic(al)** / ˌfɑ:mə'sju:tik(əl) /
a. 制药的；药（物）的；药用的

For example, a **pharmaceutical** company may have multiple sites in many countries. Each site has a database with information about its own employees, but a central personnel department could have control of all the databases.

**replica** / 'replikə /
n. 复制品，拷贝

In a replicated distributed database, each site holds an exact **replica** of another site. Any modification to data stored in one site is repeated exactly at every site. The reason for having such a database is security. If the system at one site fails, users at the site can access data at another site.

#### 5. Object-Oriented Database Model

**intersection** / ˌintə'sekʃən /
n. 交，相交，交集；交点
**field** / fi:ld /
n. 字段；域；信息组

The relational database has a specific view of data that is based on the nature of the database's tuples and attributes. The smallest unit of data in a relational database is the **intersection** of a tuple and an attribute. However, some applications need to look at data as other forms, for example, to see data as a structure, such as a record composed of **fields**.

An object-oriented database tries to keep the advantages of the relational model and at the same time allows applications to access structured data. In an object-oriented database, objects and their relations are defined. In addition, each object can have attributes that can be expressed as fields.

**social security number**
社会安全（或保险、保障）号（码）

**denote** / di'nəut /
v. 表示；意思是

For example, in an organization, one could define object types for employee, department, and customer. The employee class could define the attributes of an employee object (first name, last name, **social security number**, salary, and so on) and how they can be accessed. The department object could define the attributes of the department and how they can be accessed. In addition, the database could create a relation between an employee object and a department object to **denote** that the employee works in that department.

## V. Database Design

**involved** / in'vɔlvd /
a.（由于复杂或混乱而）难处理的，棘手的；复杂的

**entity-relationship model**
实体关系模型

The design of any database is a lengthy and **involved** task that can only be done through a step-by-step process. The first step normally involves a lot of interviewing of potential users of the database to collect the information needed to be stored and the access requirement of each department. The second step is to build an **entity-relationship model** (ERM[1]) that defines the entities for which some information must be maintained, the attributes of these entities, and the relationship between these entities.

**normalize** / 'nɔːməlaiz /
v.（使）正常化

**retrieval** / ri'triːvəl /
n. 检索

**anomaly** / ə'nɔməli /
n. 异常（现象）

The next step in design is based on the type of database to be used. In a relational database, the next step is to build relations based on the ERM and **normalize** the relations. Normalization is the process by which a given set of relations are transformed to a new set of relations with a more solid structure. Normalization is needed to allow any relation in the database to be represented, to allow a language like SQL to use powerful **retrieval** operations composed of atomic operations[2], to remove **anomalies** in insertion, deletion, and updating, and to reduce the need for restructuring the database as new data types are added.

---

[1] *ERM*：实体关系模型（*e*ntity-*r*elationship *m*odel的首字母缩略）。

[2] *to use powerful retrieval operations composed of atomic operations*：使用由多个原子操作组成的功能强大的检索操作。atomic operations意思是"原子操作"，指不可分割的最小操作。

# Exercises

**I. Fill in the blanks with the information given in the text:**

1. Data storage traditionally used individual, unrelated files, sometimes called ________ files.

2. According to the text, a DBMS is a combination of five components: hardware, software, ________, users, and procedures.

3. As far as a DBMS is concerned, users can refer to both end users and ________ programs, and end users can refer to both database ________ and normal users.

4. The three-level architecture for a DBMS discussed in the text consists of the internal, ________, and external levels.

5. In the relational database model, data is organized in two-dimensional ________ called relations.

6. A distributed database can be classified either as a ________ distributed database or as a ________ distributed database.

7. While trying to keep the advantages of the relational model, an object-oriented database allows applications to access ________ data.

8. An important step in the design of a database involves the building of a(n) ________ model that defines the entities for which some information must be maintained, the ________ of these entities, and the relationship between these entities.

**II. Translate the following terms or phrases from English into Chinese and vice versa:**

1. end user
2. atomic operation
3. database administrator
4. relational database model
5. local data
6. object-oriented database
7. database management system
8. entity-relationship model
9. distributed database
10. flat file
11. 二维表
12. 数据属性
13. 数据库对象
14. 存储设备
15. 数据类型
16. 数据插入与删除

17. 层次数据库模型
18. 数据库体系结构
19. 关系数据库管理系统
20. 全局控制总线

**III. Fill in each of the blanks with one of the words given in the following list, making changes if necessary:**

| *equipment* | *important* | *access* | *center* |
|---|---|---|---|
| *available* | *text* | *use* | *every* |
| *develop* | *database* | *experimental* | *link* |
| *nonprofit* | *online* | *public* | *limited* |

A database is any collection of data organized for storage in a computer memory and designed for easy ________ by authorized users. The data may be in the form of ________, numbers, or encoded graphics. Small databases were first ________ or funded by the U.S. government for agency or professional ________. In the 1960s, some databases became commercially ________, but their use was *funnelled* (传送) through a few so-called research ________ that collected information inquiries and handled them in *batches* (批，批量). ________ databases—that is, databases available to anyone who could ________ up to them by computer—first appeared in the 1970s. Since their first, ________ appearance in the 1950s, databases have become so ________ that they can be found in almost ________ field of information. Government, military, and industrial ________ are often highly restricted, and professional databases are usually of ________ interest. A wide range of commercial, governmental, and ________ databases are available to the general ________, however, and may be used by anyone who owns or has access to the ________ that they require.

**IV. Translate the following passage from English into Chinese:**

In a relational database, the rows of a table represent records (collections of information about separate items) and the columns represent fields (particular attributes of a record). In conducting searches, a relational database matches information from a field in one table with information in a corresponding field of another table to produce a third table that combines requested data from both tables. For example, if one table contains the fields EMPLOYEE-ID, LAST-NAME, FIRST-NAME, and HIRE-DATE, and another table contains the fields DEPT, EMPLOYEE-ID, and SALARY, a relational database can match the EMPLOYEE-ID fields in the two tables to find such information as the names of all employees earning a certain salary or the departments of all employees hired after a certain date. In other words, a relational database uses matching values in two tables to relate information in one to information in the other. Microcomputer database products typically are relational databases.

# Section B

# Maintaining Database Integrity

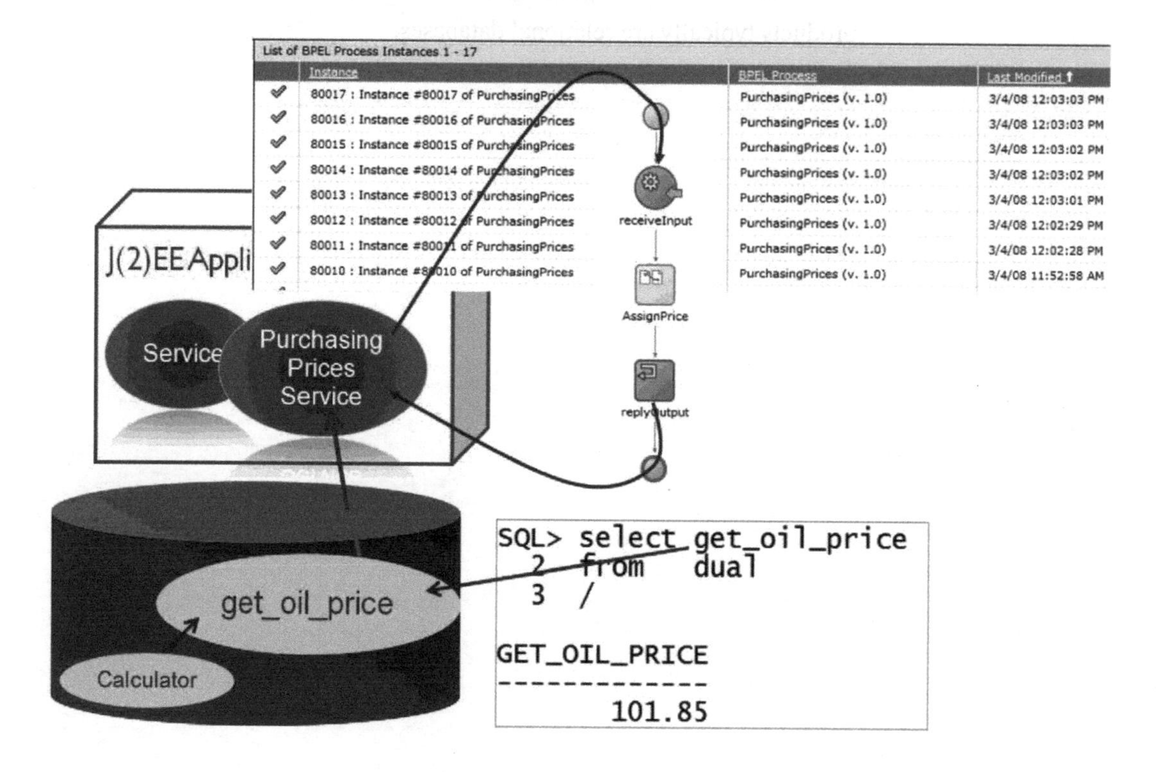


## I. Introduction

Inexpensive database management systems for personal use are relatively simple systems. They tend to have a single objective—to shield the user from the technical details of the database implementation. The databases maintained by these systems are relatively small and generally contain information whose loss or **corruption** would be inconvenient rather than **disastrous**. When a problem does arise, the user can usually correct the **erroneous** items directly or reload the database from a **backup** copy and manually make the modifications required to bring that copy up to date. This process might be inconvenient, but the cost of avoiding the inconvenience tends to be greater than the inconvenience itself. In any case, the inconvenience is restricted to only a few people, and any financial loss is generally limited.

**corruption** /kə'rʌpʃən/
n. 破坏；腐化；讹误

**disastrous** /di'zɑːstrəs/
a. 灾难性的

**erroneous** /i'rəuniəs/
a. 错误的，不正确的

**backup** /'bækʌp/
a. 备份的，后备的

In the case of large, multiuser, commercial database systems, however, the stakes are much higher. The cost of incorrect or lost data can be enormous and can have **devastating** consequences. In these environments, a major role of the DBMS is to maintain the database's integrity by guarding against problems such as operations that for some reason are only partially completed or different operations that might interact **inadvertently** to cause inaccurate information in the database. It is this role of a DBMS that we address in this section.

**devastating** / 'devəsteitiŋ /
a. 破坏性极大的，毁灭性的

**inadvertent** / ˌinəd'vəːtənt /
a. 因疏忽造成的；粗心大意的

**commit** / kə'mit /
v. 提交，委托；承诺

## II. The Commit/Rollback Protocol

A single transaction, such as the transfer of funds from one bank account to another, the **cancellation** of an airline reservation, or the **registration** of a student in a university course, might involve multiple steps at the database level. For example, a transfer of funds between bank accounts requires that the **balance** in one account be **decremented** and the balance in the other be **incremented**. Between such steps the information in the database might be inconsistent. Indeed, funds are missing during the brief period after the first account has been decremented but before the other has been incremented. Likewise, when reassigning a passenger's seat on a flight, there might be an instant when the passenger has no seat or an instant when the passenger list appears to be one passenger greater than it actually is.

**cancellation** / ˌkænsə'leiʃən /
n. 取消；删去

**registration** / ˌredʒi'streiʃən /
n. 登记；注册

**balance** / 'bæləns /
n. 余额；差额；结算

**decrement** / 'dekrimənt /
v. 减少，减缩

**increment** / 'inkrimənt /
v. 增加，增长

In the case of large databases that are subject to heavy transaction loads, it is highly likely that a random **snapshot** will find the database in the middle of some transaction[1]. A request for the execution of a transaction or an equipment malfunction will therefore likely occur at a time when the database is in an inconsistent state.

**snapshot** / 'snæpʃɔt /
n. 快照；瞬象，瞬态图

Let us first consider the problem of a malfunction. The goal of the DBMS is to ensure that such a problem will not freeze the database in an inconsistent state. This is often accomplished by maintaining a **log** containing a record of each transaction's activities in a **nonvolatile** storage system, such as a magnetic disk. Before a transaction is allowed to alter the database, the alteration to be performed is first recorded in the log. Thus the log contains a permanent record of each transaction's actions.

**log** / lɔg /
n.（运行）记录；（系统）日志

**nonvolatile** / nɔn'vɔlətail /
a. 非易失（性）的

The point at which all the steps in a transaction have been recorded in

---

1 *In the case of large databases that are subject to heavy transaction loads, it is highly likely that a random snapshot will find the database in the middle of some transaction.*：在事务量很大的大型数据库的情况下，数据库在任意一个瞬间都极可能有某个事务正在处理当中。

**commit point**
提交点

the log is called the **commit point**. It is at this point that the DBMS has the information it needs to reconstruct the transaction on its own if that should become necessary. At this point the DBMS becomes committed to the transaction in the sense that it accepts the responsibility of guaranteeing that the transaction's activities will be reflected in the database[1]. In the case of an equipment malfunction, the DBMS can use the information in its log to reconstruct the transactions that have been completed (committed) since the last backup was made.

If problems should arise before a transaction has reached its commit point[2], the DBMS might find itself with a partially executed transaction that cannot be completed. In this case the log can be used to roll back (undo) the activities actually performed by the transaction[3]. In the case of a malfunction, for instance, the DBMS could recover by rolling back those transactions that were incomplete (noncommitted) at the time of the malfunction.

**privileged** / ˈprivilidʒd /
a. 享有特权的；特许的

**deadlock** / ˈdedlɔk /
n. 死锁；僵局

**lurk** / lə:k /
v. 潜伏；暗藏；潜行

Rollbacks of transactions are not restricted, however, to the process of recovering from equipment malfunctions. They are often a part of a DBMS's normal operation. For example, a transaction might be terminated before it has completed all its steps because of an attempt to access **privileged** information, or it might be involved in a **deadlock** in which competing transactions find themselves waiting for data being used by each other. In these cases, the DBMS can use the log to roll back a transaction and thus avoid an erroneous database due to incomplete transactions.

To emphasize the delicate nature of DBMS design, we should note that there are subtle problems **lurking** within the rollback process. The rolling back of one transaction might affect database entries that have been used by other transactions. For example, the transaction being rolled back might have updated an account balance, and another transaction might have already based its activities on this updated value. This might mean that these additional transactions must also be rolled back, which might

---

1 *At this point the DBMS becomes committed to the transaction in the sense that it accepts the responsibility of guaranteeing that the transaction's activities will be reflected in the database.*：在这个点上，数据库管理系统在以下意义上向该事务做出承诺：它接受保证该事务的活动将在数据库中得到反映的责任。在本句中，第一个 that 引导的从句系同位语从句，其本位语是 sense；第二个 that 引导的从句属于宾语从句，作 guaranteeing 的宾语。

2 *If problems should arise before a transaction has reached its commit point*：在一个事务达到其提交点之前万一出现问题。本句中的 should 表示语气较强的假设。

3 *to roll back (undo) the activities actually performed by the transaction*：回滚（撤销）该事务实际上已经进行的活动。

adversely affect still other transactions. The result is the problem known as **cascading rollback**.

**cascading** /kæ'skeidiŋ/
a. 级联的；层叠的
**cascading rollback**
级联回滚
**flux** /flʌks/
n.（不断的）变动；波动

## III. Locking

We now consider the problem of a transaction being executed while the database is in a state of **flux** from another transaction[1], a situation that can lead to inadvertent interaction between the transactions and produce erroneous results. For instance, the problem known as the incorrect summary problem can arise if one transaction is in the middle of transferring funds from one account to another when another transaction tries to compute the total deposits in the bank. This could result in a total that is either too large or too small depending on the order in which the transfer steps are performed. Another possibility is known as the lost **update** problem, which is exemplified by two transactions, each of which makes a deduction from the same account. If one transaction reads the account's current balance at the point when the other has just read the balance but has not yet calculated the new balance, then both transactions will base their deductions on the same initial balance. In turn, the effect of one of the deductions will not be reflected in the database.

**update** /'ʌpdeit/
n. 更新；修改

To solve such problems, a DBMS could force transactions to execute in their **entirety** on a one-at-a-time basis[2] by holding each new transaction in a queue until those preceding it have completed. But a transaction often spends a lot of time waiting for mass storage operations to be performed. By **interweaving** the execution of transactions, the time during which one transaction is waiting can be used by another transaction to process data it has already retrieved. Most large database management systems therefore contain a scheduler to coordinate time-sharing among transactions in much the same way that a multiprogramming operating system coordinates interweaving of processes.

**entirety** /in'taiəti/
n. 全部；整体
**interweave** /ˌintə'wi:v/
v.（使）交织

To guard against such anomalies as the incorrect summary problem and the lost update problem, these schedulers incorporate a locking protocol in which the items within a database that are currently being used by some transaction are marked as such[3]. These marks are called locks; marked

---

1 *the database is in a state of flux from another transaction*：数据库因另一事物而处于变迁状态。

2 *a DBMS could force transactions to execute in their entirety on a one-at-a-time basis*：数据库管理系统可以强制事务一次完整地执行一个。

3 *these schedulers incorporate a locking protocol in which the items within a database that are currently being used by some transaction are marked as such*：这些调度程序都包含了一个锁定协议。在该协议中，数据库中当前正在被某个事务使用的项目都要加上说明这种状态的标记。

**shared lock**
共享锁
**exclusive lock**
排他锁，
互斥（型）锁
**data item**
数据项

items are said to be locked. Two types of locks are common—**shared locks** and **exclusive locks**. They correspond to the two types of access to data that a transaction might require—shared access and exclusive access[1]. If a transaction is not going to alter a **data item**, then it requires shared access, meaning that other transactions are also allowed to view the data. However, if the transaction is going to alter the item, it must have exclusive access, meaning that it must be the only transaction with access to that data.

In a locking protocol, each time a transaction requests access to a data item, it must also tell the DBMS the type of access it requires. If a transaction requests shared access to an item that is either unlocked or locked with a shared lock, that access is granted and the item is marked with a shared lock. If, however, the requested item is already marked with an exclusive lock, the additional access is denied. If a transaction requests exclusive access to an item, that request is granted only if the item has no lock associated with it. In this manner, a transaction that is going to alter data protects that data from other transactions by obtaining exclusive access, whereas several transactions can share access to an item if none of them are going to change it. Of course, once a transaction is finished with an item, it notifies the DBMS, and the associated lock is removed.

Various algorithms are used to handle the case in which a transaction's access request is rejected. One algorithm is that the transaction is merely forced to wait until the requested item becomes available. This approach, however, can lead to deadlock, since two transactions that require exclusive access to the same two data items could block each other's progress if each obtains exclusive access to one of the items and then insists on waiting for the other. To avoid such deadlocks, some database management systems give priority to older transactions. That is, if an older transaction requires access to an item that is locked by a younger transaction, the younger transaction is forced to release all of its data items, and its activities are rolled back (based on the log). Then, the older transaction is given access to the item it required, and the younger transaction is forced to start again. If a younger transaction is repeatedly preempted, it will grow older in the process and ultimately become one of the older transactions with high priority[2]. This

---

[1] *They correspond to the two types of access to data that a transaction might require—shared access and exclusive access.*：它们对应于事务可能需要的两种数据访问类型——共享访问和互斥访问。

[2] *If a younger transaction is repeatedly preempted, it will grow older in the process and ultimately become one of the older transactions with high priority.*：如果一个较年轻的事务一再被抢先，它将在这一过程中变得较老，并最终成为一个具有高优先级的较老事务。

protocol, known as the wound-wait protocol (old transactions wound young transactions, young transactions wait for old ones), ensures that every transaction will ultimately be allowed to complete its task.

---

## Exercises

**I. Fill in the blanks with the information given in the text:**

1. The ________ point is the point at which all the steps in a transaction have been recorded in the log of the DBMS.

2. A large DBMS often contains a ________ to coordinate time-sharing among transactions in much the same way that a multiprogramming operating system coordinates interweaving of processes.

3. The items within a database that are currently being used by some transaction are marked with locks. There are two common types of locks: ________ locks and ________ locks.

4. With the ________ protocol, if an older transaction requires access to an item that is locked by a younger transaction, the younger transaction is forced to release all of its data items, and its activities are rolled back.

**II. Translate the following terms or phrases from English into Chinese and vice versa:**

| | |
|---|---|
| 1. nonvolatile storage system | 6. 共享锁 |
| 2. equipment malfunction | 7. 数据库实现 |
| 3. wound-wait protocol | 8. 级联回滚 |
| 4. multiprogramming operating system | 9. 数据项 |
| 5. database integrity | 10. 排他锁 |

# Section C

## What Is Data Mining

### I. Introduction

**data mining** 数据挖掘
**disseminate** / di'semineit / v. 散布；传播
**big data** 大数据
**data set** 数据集
**biotechnology** / ˌbaiəutek'nɔlədʒi / n. 生物技术
**velocity** / vi'lɔsiti / n. 速度；速率
**extract** / ik'strækt / v. 提取，抽取，析取

Rapid advances in data collection and storage technology, coupled with the ease with which data can be generated and **disseminated**, have triggered the explosive growth of data, leading to the current age of **big data**. Deriving actionable insights [1] from these large **data sets** is increasingly important in decision making across almost all areas of society, including business and industry; science and engineering; medicine and **biotechnology**; and government and individuals. However, the amount of data (volume), its complexity (variety), and the rate at which it is being collected and processed (**velocity**) have simply become too great for humans to analyze unaided[2]. Thus, there is a great need for automated tools for **extracting** useful information from the big data despite the challenges

[1] *actionable insights*：可行的见解（Actionable insights are conclusions drawn from data that can be turned directly into an action or a response.）。

[2] *have simply become too great for humans to analyze unaided*：对于人类来说已变得简直太大了，无法进行独立分析。analyze 的逻辑宾语是本句的主语，unaided 是形容词，表示“未受协助的”“无外援的”“独立的”，可作定语，也可放在动词或动宾结构后面（If you do something unaided, you do it without help from anyone or anything else.）。

posed by its **enormity** and diversity. Data mining blends traditional data analysis methods with sophisticated algorithms for processing this **abundance** of data.

**enormity** / iˈnɔːmiti /
n. 巨大，广大
**abundance** / əˈbʌndəns /
n. 大量，丰富，充足

## II. Definition and Scope

Data mining is the process of automatically discovering useful information in large data repositories. Data mining techniques are deployed to **scour** large data sets in order to find novel and useful patterns that might otherwise remain unknown. They also provide the capability to predict the outcome of a future observation, such as the amount a customer will spend at an online or a **brick-and-mortar** store.

**scour** / skauə /
v. 四处搜索；细查
**mortar** / ˈmɔːtə /
n. 砂浆，灰浆
**brick-and-mortar** / ˌbrikəndˈmɔːtə /
a. 实体的，具体的

Not all information discovery tasks are considered to be data mining. Examples include queries, e.g., looking up individual records in a database or finding web pages that contain a particular set of **keywords**. This is because such tasks can be accomplished through simple interactions with a database management system or an information retrieval system. These systems rely on traditional computer science techniques, which include sophisticated indexing structures and query processing algorithms, for efficiently organizing and retrieving information from large data repositories. **Nonetheless**, data mining techniques have been used to enhance the performance of such systems by improving the quality of the search results based on their **relevance** to the input queries.

**keyword** / ˈkiːwəːd /
n. 关键词，关键字
**nonetheless** / ˌnʌnðəˈles /
ad. 尽管如此，然而
**relevance** / ˈreləvəns /
n. 相关(性)，关联；重要性

## III. Motivating Challenges[1]

Traditional data analysis techniques have often encountered practical difficulties in meeting the challenges posed by big data applications. The following are some of the specific challenges that motivated the development of data mining.

### 1. Scalability

Because of advances in data generation and collection, data sets with sizes of **terabytes**, **petabytes**, or even **exabytes** are becoming common. If data mining algorithms are to handle these massive data sets, they must be **scalable**. Many data mining algorithms employ special search strategies to handle **exponential** search problems. Scalability may also require the implementation of novel data structures to access individual records in an

**scalability** / ˌskeiləˈbiliti /
n. 可缩放性，可伸缩性
**terabyte** / ˈterəbait /
n. 太字节，万亿字节
**petabyte** / ˈpetəbait /
n. 拍字节，千万亿字节
**exabyte** / ˈeksəbait /
n. 艾字节，百亿亿字节
**scalable** / ˈskeiləbl /
a. 可缩放的，可伸缩的
**exponential** / ˌekspəˈnenʃəl /
a. 指数的；迅速增长的

[1] *Motivating Challenges*：起到激励作用的挑战，给予动力的挑战，即下文提到的 challenges that motivated the development of data mining。

**out-of-core**
/ 'autəv'kɔː /
a. 核外的，基于外存的
**sampling**
/ 'sɑːmpliŋ; 'sæ- /
n. 取样，抽样，采样
**dimensionality**
/ diˌmenʃə'næliti, dai- /
n. 维度，维数
**bioinformatics**
/ ˌbaiəuˌinfə'mætiks /
n. 生物信息学
**microarray**
/ ˌmaikrəuə'rei /
n. 微阵列
**temporal**
/ 'tempərəl /
a. 时间的
**spatial** / 'speiʃəl /
a. 空间的

efficient manner. For instance, **out-of-core** algorithms may be necessary when processing data sets that cannot fit into main memory. Scalability can also be improved by using **sampling** or developing parallel and distributed algorithms.

**2. High Dimensionality**

It is now common to encounter data sets with hundreds or thousands of attributes instead of the handful common a few decades ago[1]. In **bioinformatics**, progress in **microarray** technology has produced gene expression data involving thousands of features. Data sets with **temporal** or **spatial** components also tend to have high dimensionality. For example, consider a data set that contains measurements of temperature at various locations. If the temperature measurements are taken repeatedly for an extended period, the number of dimensions (features) increases in proportion to the number of measurements taken. Traditional data analysis techniques that were developed for low-dimensional data often do not work well for such high-dimensional data due to issues such as curse of dimensionality[2]. Also, for some data analysis algorithms, the computational complexity increases rapidly as the dimensionality (the number of features) increases.

**heterogeneous**
/ ˌhetərəu'dʒiːniəs /
a. 各种各样的；异构的
**categorical**
/ ˌkæti'gɔrikəl /
a. 类的，分类的
**hyperlink**
/ 'haipəliŋk /
n. 超（级）链接
**sequential**
/ si'kwenʃəl /
a. 顺序的，序列的
**autocorrelation**
/ 'ɔːtəuˌkɔri'leiʃən /
n. 自相关（性）
**connectivity**
/ kənek'tiviti /
n. 连通（性），连接（性）

**3. Heterogeneous and Complex Data**

Traditional data analysis methods often deal with data sets containing attributes of the same type, either continuous or **categorical**. As the role of data mining in business, science, medicine, and other fields has grown, so has the need for techniques that can handle heterogeneous attributes. Recent years have also seen the emergence of more complex data objects. Examples of such non-traditional types of data include web and social media data containing text, **hyperlinks**, images, audio, and videos; DNA data with **sequential** and three-dimensional structure; and climate data that consists of measurements (temperature, pressure, etc.) at various times and locations on the Earth's surface. Techniques developed for mining such complex objects should take into consideration relationships in the data, such as temporal and spatial **autocorrelation**, graph **connectivity**, and parent-child relationships between the elements in semi-structured text and XML documents.

---

1 *with hundreds or thousands of attributes instead of the handful common a few decades ago*：具有成百上千属性，而不是几十年前常见的少量属性，common a few decades ago 是 handful 的后置定语。

2 *curse of dimensionality*：维灾难，维度（或维数）灾难，维度（或维数）的诅咒。

### 4. Data Ownership and Distribution

Sometimes, the data needed for an analysis is not stored in one location or owned by one organization. Instead, the data is geographically distributed among resources belonging to multiple entities. This requires the development of distributed data mining techniques. The key challenges faced by distributed data mining algorithms include the following: (1) how to reduce the amount of communication needed to perform the distributed computation, (2) how to effectively consolidate the data mining results obtained from multiple sources, and (3) how to address data security and privacy issues.

### 5. Non-traditional Analysis

**hypothesize**
/ hai'pɔθisaiz /
v. 假设，假定
**hypothesis**
/ hai'pɔθisis /
（[复] **-ses** / -si:z /）
n. 假设，假说

The traditional statistical approach is based on a **hypothesize**-and-test paradigm. In other words, a **hypothesis** is proposed, an experiment is designed to gather the data, and then the data is analyzed with respect to the hypothesis. Unfortunately, this process is extremely labor-intensive. Current data analysis tasks often require the generation and evaluation of thousands of hypotheses, and consequently, the development of some data mining techniques has been motivated by the desire to automate the process of hypothesis generation and evaluation. Furthermore, the data sets analyzed in data mining are typically not the result of a carefully designed experiment and often represent opportunistic samples of the data, rather than random samples[1].

## IV. The Origins of Data Mining

While data mining has traditionally been viewed as an intermediate process within the KDD[2] (knowledge discovery in databases) framework, as shown in Figure 6C-1, it has emerged over the years as an academic field within computer science, focusing on all aspects of KDD, including data preprocessing, mining, and postprocessing. Its origin can be traced back to the late 1980s, following a series of workshops organized on the topic of knowledge discovery in databases. The workshops brought together researchers from different disciplines to discuss the challenges and opportunities in applying computational techniques to extract actionable knowledge from large databases. The workshops quickly grew into hugely popular conferences that

---

1 *often represent opportunistic samples of the data, rather than random samples*：经常代表数据的时机性样本，而不是随机样本。

2 *KDD*：数据库知识发现（*k*nowledge *d*iscovery in *d*atabases 的首字母缩略）。

**practitioner**
/ præk'tiʃənə /
n. 实践者；开业者
**academia**
/ ˌækə'di:miə /
n. 学术界
**hire** / haiə /
n. 新雇员；新员工

were attended by researchers and **practitioners** from both the **academia** and industry. The success of these conferences, along with the interest shown by businesses and industry in recruiting new **hires** with data mining background, have fueled the tremendous growth of this field.

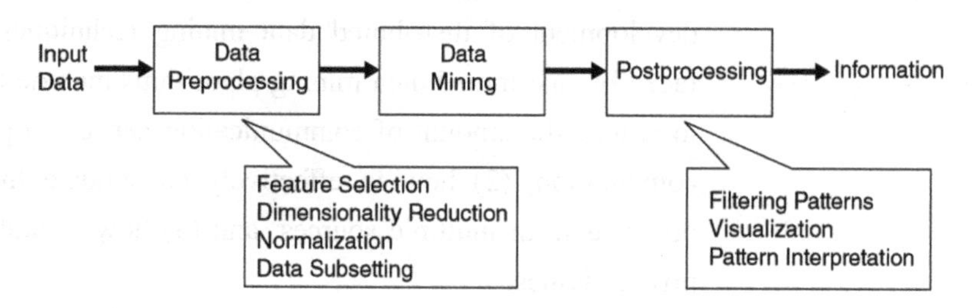


Figure 6C-1: The Process of Knowledge Discovery in Databases (KDD)

The field was initially built upon the methodology and algorithms that researchers had previously used. In particular, data mining researchers draw upon ideas, such as (1) sampling, estimation, and hypothesis testing from statistics and (2) search algorithms, modeling techniques, and learning theories from artificial intelligence, pattern recognition, and machine learning. Data mining has also been quick to adopt ideas from other areas, including optimization, evolutionary computing, information theory, signal processing, visualization, and information retrieval, and extending them to solve the challenges of mining big data.

A number of other areas also play key supporting roles. In particular, database systems are needed to provide support for efficient storage, indexing, and query processing. Techniques from high performance (parallel) computing are often important in addressing the massive size of some data sets. Distributed techniques can also help address the issue of size and are essential when the data cannot be gathered in one location. Figure 6C-2 shows the relationship of data mining to other areas.

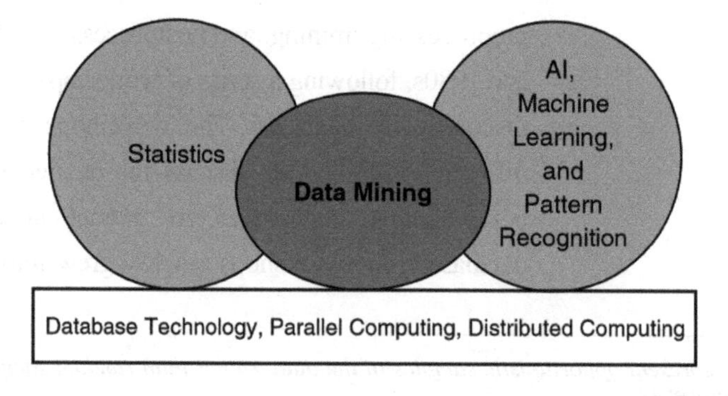


Figure 6C-2: Data Mining as a Confluence of Many Disciplines

**confluence**
/ 'kɔnfluəns /
n. 汇合；聚集

## V. Data Mining Tasks

Data mining tasks are generally divided into two major categories:

- **Predictive tasks:** The objective of these tasks is to predict the value of a particular attribute based on the values of other attributes. The attribute to be predicted is commonly known as the target or **dependent variable**, while the attributes used for making the **prediction** are known as the **explanatory** or **independent variables.**
- **Descriptive tasks:** Here, the objective is to derive patterns (**correlations**, trends, **clusters, trajectories**, and anomalies) that summarize the underlying relationships in data. Descriptive data mining tasks are often exploratory in nature and frequently require postprocessing techniques to validate and explain the results.

Figure 6C-3 illustrates four of the core data mining tasks.

**predictive**
/ pri'diktiv /
a. 预言性的，预测性的
**dependent variable**
因变量
**prediction**
/ pri'dikʃən /
n. 预言，预测
**explanatory**
/ ik'splænətəri /
a. 解释的，说明的
**independent variable** 自变量
**correlation**
/ ˌkɔri'leiʃən /
n. 相关（性），相互关联
**cluster** / 'klʌstə /
n. 一群，一串，一簇；聚类
**trajectory**
/ trə'dʒektəri /
n. 轨迹，轨道

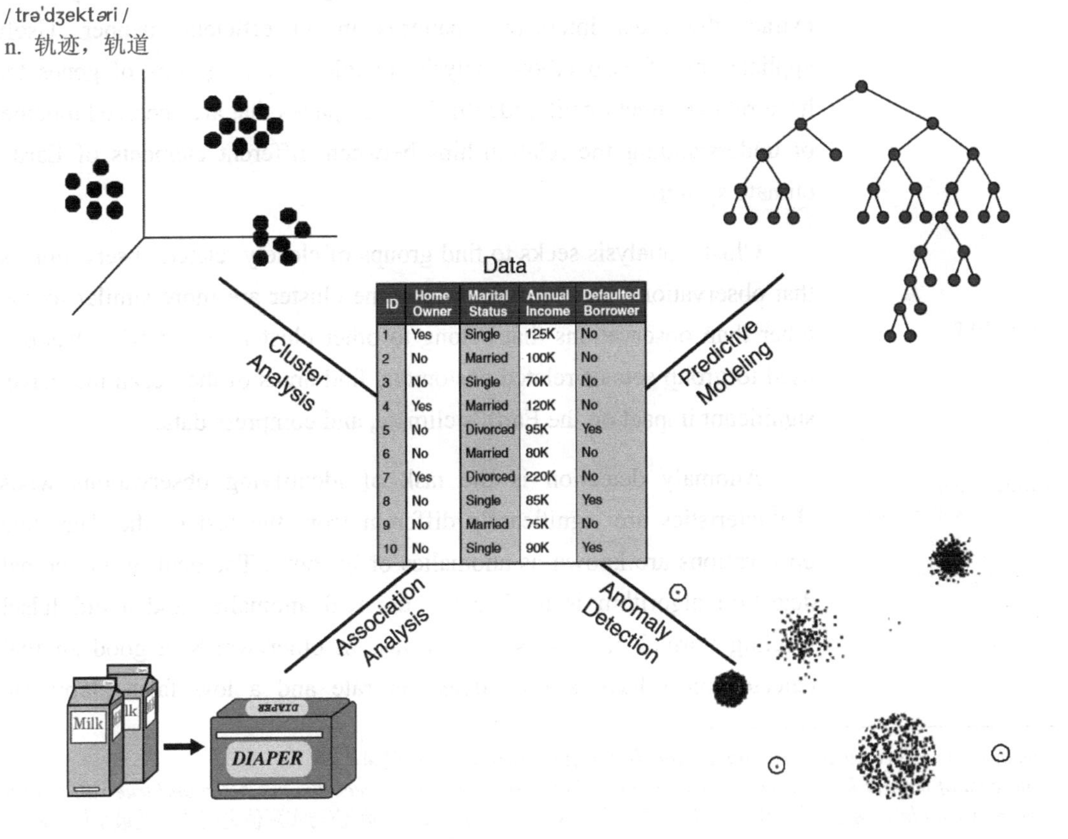


| ID | Home Owner | Marital Status | Annual Income | Defaulted Borrower |
|---|---|---|---|---|
| 1 | Yes | Single | 125K | No |
| 2 | No | Married | 100K | No |
| 3 | No | Single | 70K | No |
| 4 | Yes | Married | 120K | No |
| 5 | No | Divorced | 95K | Yes |
| 6 | No | Married | 80K | No |
| 7 | Yes | Divorced | 220K | No |
| 8 | No | Single | 85K | Yes |
| 9 | No | Married | 75K | No |
| 10 | No | Single | 90K | Yes |

**Figure 6C-3: Four of the Core Data Mining Tasks**

Predictive modeling refers to the task of building a model for the target variable as a function of the explanatory variables. There are two types of predictive modeling tasks: classification, which is used for **discrete** target variables, and regression, which is used for continuous target variables. For example, predicting whether a web user will make a purchase at an online bookstore is a classification task because the target variable is binary-valued[1]. On the other hand, forecasting the future price of a stock is a regression task because price is a continuous-valued attribute. The goal of both tasks is to learn a model that minimizes the error between the predicted and true values of the target variable[2]. Predictive modeling can be used to identify customers who will respond to a marketing campaign, predict **disturbances** in the Earth's **ecosystem**, or judge whether a patient has a particular disease based on the results of medical tests.

**discrete** / 'diskri:t /
a. 分离的；离散的

**disturbance** / di'stə:bəns /
n. 打扰；干扰；扰动；骚乱

**ecosystem** / 'i:kəuˌsistəm /
n. 生态系统

**subset** / 'sʌbset /
n. 子集

Association analysis is used to discover patterns that describe strongly associated features in the data. The discovered patterns are typically represented in the form of implication rules or feature **subsets**[3]. Because of the exponential size of its search space, the goal of association analysis is to extract the most interesting patterns in an efficient manner. Useful applications of association analysis include finding groups of genes that have related functionality, identifying web pages that are accessed together, or understanding the relationships between different elements of Earth's climate system.

Cluster analysis seeks to find groups of closely related observations so that observations that belong to the same cluster are more similar to each other than observations that belong to other clusters. **Clustering** has been used to group sets of related customers, find areas of the ocean that have a significant impact on the Earth's climate, and compress data.

**clustering** / 'klʌstəriŋ /
n. 聚类

Anomaly detection is the task of identifying observations whose characteristics are significantly different from the rest of the data. Such observations are known as anomalies or **outliers**. The goal of an anomaly detection algorithm is to discover the real anomalies and avoid falsely labeling normal objects as **anomalous**. In other words, a good anomaly detector must have a high detection rate and a low false alarm rate.

**outlier** / 'autˌlaiə /
n. 与主体分开的人（或物）；离群值；离群点；异常值

**anomalous** / ə'nɔmələs /
a. 异常的

---

[1] *because the target variable is binary-valued*：因为该目标变量是二值变量。

[2] *The goal of both tasks is to learn a model that minimizes the error between the predicted and true values of the target variable.*：两项任务的目标都是训练一个模型，使目标变量的预测值与实际值之间的误差最小化。learn a model 相当于 make a model learn，that 引导的从句是 model 的定语从句。

[3] *in the form of implication rules or feature subsets*：以蕴涵规则或特征子集的形式。

**intrusion**
/ in'tru:ʒən /
n. 侵入；打扰
**drought** / draut /
n. 干旱；旱灾
**hurricane**
/ 'hʌrikən /
n. 飓风

Applications of anomaly detection include the detection of fraud, network **intrusions**, unusual patterns of disease, and ecosystem disturbances, such as **droughts**, floods, fires, **hurricanes**, etc.

---

## Exercises

**I. Fill in the blanks with the information given in the text:**

1. Scalability, high ________, heterogeneous and complex data, data ownership and distribution, and ________ analysis are among the specific challenges that motivated the development of data mining.

2. Traditionally viewed as a(n) ________ process within the KDD framework, data mining now focuses on all aspects of KDD, including data preprocessing, mining, and ________.

3. Data mining tasks are generally divided into two major categories, namely predictive tasks and ________ tasks.

4. Predictive modeling, ________ analysis, cluster analysis, and ________ detection are four of the core data mining tasks.

**II. Translate the following terms or phrases from English into Chinese and vice versa:**

1. data set
2. independent variable
3. predictive modeling
4. cluster analysis
5. 数据挖掘
6. 因变量
7. 大数据
8. 数据预处理

# Unit 7 Computer Network

# （计算机网络）

## Section A

## Network Fundamentals

The need to share information and resources among different computers has led to linked computer systems, called networks, in which computers are connected so that data can be transferred from machine to machine. In these networks, computer users can exchange messages and share resources—such as printing capabilities, software packages, and data storage facilities—that are scattered throughout the system. The underlying software required to support such applications has grown from simple **utility packages** into an expanding system of network software that provides a sophisticated network-wide infrastructure. In a sense, network software is evolving into a network-wide operating system.

**utility package** 实用软件包，公用程序包

### I. Network Classifications

A computer network is often classified as being either a **personal area**

**personal area network**
个人域网
**local area network**
局域网
**metropolitan**
/ ˌmetrəˈpɔlitən /
a. 大城市的，大都会的
**metropolitan area network**
城域网
**headset** / ˈhedset /
n.（头戴式）耳机

**network** (PAN[1]), a **local area network** (LAN[2]), a **metropolitan area network** (MAN[3]), or a wide area network (WAN[4]). A PAN is normally used for short-range communications—typically less than a few meters—such as between a wireless **headset** and a smartphone or between a wireless mouse and its PC. In contrast, a LAN normally consists of a collection of computers in a single building or building complex. For example, the computers on a university campus or those in a manufacturing plant might be connected by a LAN. A MAN is a network of intermediate size, such as one spanning a local community. Finally, a WAN links machines over a greater distance—perhaps in neighboring cities or on opposite sides of the world.

**domain** / dəuˈmein /
n. 领域，域

Another means of classifying networks is based on whether the network's internal operation is based on designs that are in the public **domain** or on innovations owned and controlled by a particular entity[5] such as an individual or a corporation. A network of the former type is called an open network; a network of the latter type is called a closed, or sometimes a proprietary, network.

**protocol suite**
协议簇，协议组

The Internet is an open system. In particular, communication throughout the Internet is governed by an open collection of standards known as the TCP/IP[6] **protocol suite**. Anyone is free to use these standards without paying fees or signing license agreements. In contrast, a company such as Novell Inc.[7] might develop proprietary systems for which it chooses to maintain ownership rights, allowing the company to draw income from selling or leasing these products.

**topology**
/ təuˈpɔlədʒi, tɔ- /
n. 拓扑（结构），布局

Still another way of classifying networks is based on the **topology** of the network, which refers to the pattern in which the machines are connected. Two of the more popular topologies are the bus, in which the

---

1 *PAN*：个人域网（*p*ersonal *a*rea *n*etwork 的首字母缩略）。
2 *LAN*：局域网（*l*ocal *a*rea *n*etwork 的首字母缩略）。
3 *MAN*：城域网（*m*etropolitan *a*rea *n*etwork 的首字母缩略）。
4 *WAN*：广域网（*w*ide *a*rea *n*etwork 的首字母缩略）。
5 *whether the network's internal operation is based on designs that are in the public domain or on innovations owned and controlled by a particular entity*：网络的内部操作是基于公共领域的设计，还是基于特定实体所拥有并控制的革新。句中的 public domain 有“不受专利权限制（或保护）”“不受版权限制（或保护）”等意思。
6 *TCP/IP*：TCP/IP 协议，传输控制（协议）/ 网际（或网间）协议（*T*ransmission *C*ontrol *P*rotocol/*I*nternet *P*rotocol 的首字母缩略）。
7 *Novell Inc.*：Novell股份有限公司，网威公司，一家网络系统公司，创立于1979年，总部设在美国犹他州普罗沃市（Provo），2014 年被 Micro Focus（英国微福斯有限公司）收购。

machines are all connected to a common communication line called a bus, and the star, in which one machine serves as a central **focal point** to which all the others are connected. The **bus topology** was popularized in the 1990s when it was implemented under a set of standards known as **Ethernet**, and Ethernet networks remain one of the most popular **networking** systems in use today. The **star topology** has roots as far back as the 1970s. It evolved from the paradigm of a large central computer serving many users. As the simple terminals employed by these users grew into small computers themselves, a star network emerged. Today, the star configuration is popular in wireless networks where communication is conducted by means of radio broadcast and the central machine, called the access point (AP[1]), serves as a focal point around which all communication is coordinated.

**focal** /ˈfəukəl/
a. 焦点的
**focal point**
焦点，活动（或注意、兴趣等）的中心
**bus topology**
总线拓扑（结构）
**Ethernet** /ˈiːθəˌnet/
n. 以太网（标准）（亦作 ethernet）
**networking**
/ˈnetˌwəːkiŋ/
n. 连网，联网，建网
**star topology**
星型拓扑（结构）

## II. Protocols

For a network to function reliably, it is important to establish rules by which activities are conducted. Such rules are called protocols. By developing and adopting protocol standards, vendors are able to build products for network applications that are compatible with products from other vendors. Thus, the development of protocol standards is an indispensable process in the development of networking technologies.

As an introduction to the protocol concept, let us consider the problem of coordinating the transmission of messages among computers in a network. Without rules governing this communication, all the computers might insist on transmitting messages at the same time or fail to assist other machines when that assistance is required.

In a bus network based on the Ethernet standards, the right to transmit messages is controlled by the protocol known as **Carrier Sense**, Multiple Access with Collision Detection (CSMA/CD[2]). This protocol dictates that each message be broadcast to all the machines on the bus (Figure 7A-1). Each machine monitors all the messages but keeps only those addressed to itself. To transmit a message, a machine waits until the bus is silent, and at this time it begins transmitting while continuing to monitor the bus. If another machine also begins transmitting, both machines detect the clash and pause for a brief, independently random period of time before trying to

**carrier** /ˈkæriə/
n. 载波
**carrier sense**
载波侦听，载波监听

---

[1] *AP*：（访问）接入点（*a*ccess *p*oint 的首字母缩略）。

[2] *CSMA/CD*：带冲突检测的载波侦听多址（或多路）访问（协议）(*C*arrier *S*ense, *M*ultiple *A*ccess with *C*ollision *D*etection 的首字母缩略）。

transmit again.

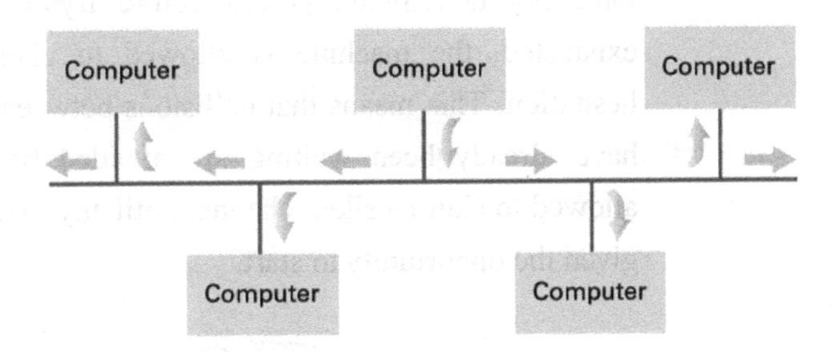


**Figure 7A-1: Communication over a Bus Network**

**collide** / kə'laid /
v. 冲突；碰撞

**avoidance**
/ ə'vɔidəns /
n. 避免；避开

Note that CSMA/CD is not compatible with wireless star networks in which all machines communicate through a central AP. This is because a machine may be unable to detect that its transmissions are **colliding** with those of another. For example, the machine may not hear the other because its own signal drowns out that of the other machine. Another cause might be that the signals from the different machines are blocked from each other by objects or distance even though they can all communicate with the central AP (a condition known as the hidden terminal problem, Figure 7A-2). The result is that wireless networks adopt the policy of trying to avoid collisions rather than trying to detect them. Such policies are classified as Carrier Sense, Multiple Access with Collision **Avoidance** (CSMA/CA[1]), many of which are standardized by IEEE[2] within the protocols defined in IEEE 802.11 and commonly referred to as WiFi[3]. We emphasize that collision avoidance protocols are designed to avoid collisions and may not eliminate them completely. When collisions do occur, messages must be retransmitted.

The most common approach to collision avoidance is based on giving advantage to machines that have already been waiting for an opportunity to transmit. The protocol used is similar to Ethernet's CSMA/CD. The basic difference is that when a machine first needs to transmit a message and finds the communication channel silent, it does not start transmitting immediately. Instead, it waits for a short period of time and then starts transmitting only if the channel has remained silent throughout that period.

---

1 *CSMA/CA*：带冲突避免的载波侦听多址（或多路）访问（协议）（*C*arrier *S*ense, *M*ultiple *A*ccess with *C*ollision *A*voidance 的首字母缩略）。

2 *IEEE*：（美国）电气和电子工程师协会（*I*nstitute of *E*lectrical and *E*lectronics *E*ngineers 的首字母缩略）。

3 *WiFi*：无线保真（*Wi*reless *Fi*delity 的缩略），亦作 Wi-Fi。WiFi 是一种无线传输规范。WiFi 网络使用 IEEE 802.11b 或 802.11a 无线电技术，提供安全、可靠、快速的无线连通性。

If a busy channel is experienced during this process, the machine waits for a randomly determined period before trying again. Once this period is exhausted, the machine is allowed to claim a silent channel without hesitation. This means that collisions between "newcomers" and those that have already been waiting are avoided because a "newcomer" is not allowed to claim a silent channel until any machine that has been waiting is given the opportunity to start.

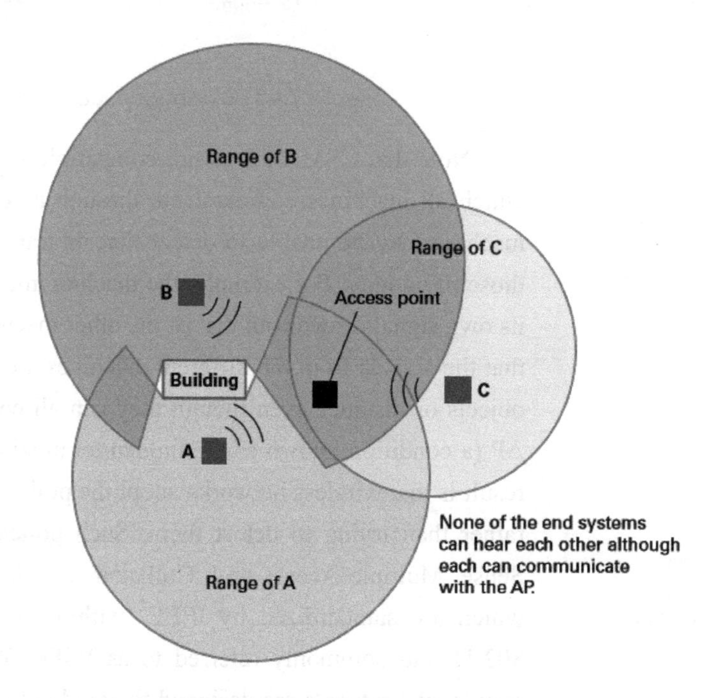


**Figure 7A-2: The Hidden Terminal Problem**

This protocol, however, does not solve the hidden terminal problem. After all, any protocol based on distinguishing between a silent or busy channel requires that each individual station be able to hear all the others. To solve this problem, some WiFi networks require that each machine send a short "request" message to the AP and wait until the AP acknowledges that request before transmitting an entire message. If the AP is busy because it is dealing with a "hidden terminal," it will ignore the request, and the requesting machine will know to wait. Otherwise, the AP will acknowledge the request, and the machine will know that it is safe to transmit. Note that all the machines in the network will hear all acknowledgments sent from the AP and thus have a good idea of whether the AP is busy at any given time, even though they may not be able to hear the transmissions taking place.

## III. Methods of Process Communication

The various activities (or processes) executing on the different computers within a network (or even executing on the same machine via time-sharing/multitasking) must often communicate with each other to coordinate their actions and to perform their designated tasks. Such communication between processes is called **interprocess communication**.

**interprocess communication**
进程间通信

A popular convention used for interprocess communication is the client/server model. This model defines the basic roles played by the processes as either a client, which makes requests of other processes, or a server, which satisfies the requests made by clients.

An early application of the client/server model appeared in networks connecting all the computers in a cluster of offices. In this situation, a single, high-quality printer was attached to the network where it was available to all the machines in the network. In this case the printer played the role of a server (often called a **print server**), and the other machines were programmed to play the role of clients that sent print requests to the print server.

**print server**
打印服务器

Today the client/server model is used extensively in network applications. However, the client/server model is not the only means of interprocess communication. Another model is the **peer-to-peer** (often abbreviated P2P[1]) model. Whereas the client/server model involves one process (the server) providing a service to numerous others (clients), the peer-to-peer model involves processes that provide service to and receive service from each other (Figure 7A-3). Moreover, whereas a server must execute continuously so that it is prepared to serve its clients at any time, the peer-to-peer model usually involves processes that execute on a temporary basis. For example, applications of the peer-to-peer model include **instant messaging** in which people carry on a written conversation over the Internet as well as situations in which people play competitive **interactive** games.

**peer-to-peer**
/ ˈpiətəˈpiə /
a. 对等的

**messaging**
/ ˈmesidʒiŋ /
n. 消息接发，通信

**instant messaging**
即时通信，
即时消息

**interactive**
/ ˌintərˈæktiv /
a. 交互（式）的

The peer-to-peer model is also a popular means of distributing files such as music recordings and **motion pictures** via the Internet. In this case, one peer may receive a file from another and then provide that file to other peers. The collection of peers participating in such a distribution is sometimes called a **swarm**. The swarm approach to file distribution is in

**motion picture**
〈美〉影片，电影

**swarm** / swɔːm /
n. 一大群；蜂群

[1] *P2P*：对等的（*peer-to-peer* 的缩略）。

contrast to earlier approaches that applied the client/server model by establishing a central distribution center (the server) from which clients downloaded files (or at least found sources for those files).

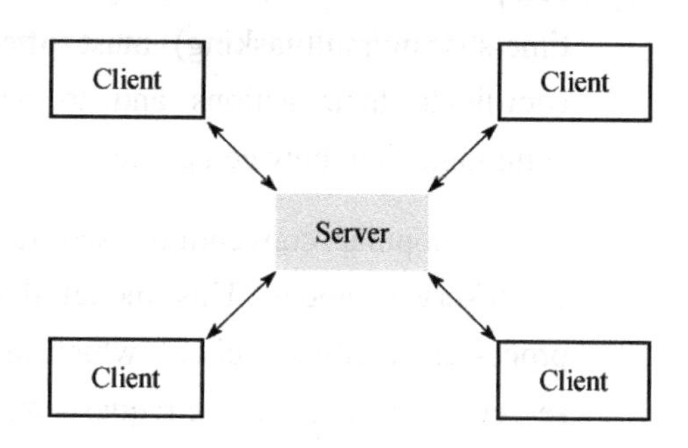


a) Server must be prepared to serve multiple clients at any time

b) Peers communicate as equals on a one-to-one basis

**Figure 7A-3: The Client/Server Model Compared to the Peer-to-Peer Model**

**peer** / piə /
n. 同级设备；同层，对等层

One reason that the P2P model is replacing the client/server model for file sharing is that it distributes the service task over many peers rather than concentrating it at one server. This lack of a centralized base of operation leads to a more efficient system. Unfortunately, another reason for the popularity of file distribution systems based on the P2P model is that, in cases of questionable **legality**, the lack of a central server makes legal efforts to enforce copyright laws more difficult.

**legality** / li'gæliti /
n. 合法（性）

You might often read or hear the term *peer-to-peer network*, which is an example of how misuse of **terminology** can evolve when technical terms are adopted by the nontechnical community[1]. The term *peer-to-peer* refers to a system by which two processes communicate over a network (or **internet**[2]). It is not a property of the network (or internet). A process might use the peer-to-peer model to communicate with another process and later use the client/server model to communicate with another process over the same network. Thus, it would be more accurate to speak of communicating

**terminology** / ˌtəːmiˈnɔlədʒi /
n. [总称] 术语

**internet** / 'intənet /
n. 互联网，互连网

1 *an example of how misuse of terminology can evolve when technical terms are adopted by the nontechnical community*：当技术术语被非技术界采用时可能发生术语误用的一个例子。

2 *internet*：常译作“互联网（络）”，系 internetwork 的简写形式，指的是一组通过网关联结在一起的可能互不相同的计算机网络。注意 internet 与 Internet 的不同，后者常译作“因特网”“国际互联网”。

by means of the peer-to-peer model rather than communicating over a peer-to-peer network.

---

## Exercises

**I. Fill in the blanks with the information given in the text:**

1. According to the geographical area covered, a computer network is often classified as either a(n) ________, a LAN, a(n) ________, or a WAN.

2. A network can also be classified as either a(n) ________ network or a(n) ________ network according to whether its internal operation is based on designs in the public domain or on innovations owned and controlled by a particular entity.

3. A network can also be classified according to its topology. Two of the more popular topologies are the ________ topology and ________ topology.

4. Many policies classified as CSMA/CA are standardized by IEEE within the protocols defined in IEEE 802.11 and commonly referred to as ________.

5. In a bus network based on the ________ standards, the right to transmit messages is controlled by the protocol known as CSMA/CD.

6. The ________ model for interprocess communication defines the basic roles played by the processes as either a client or a server.

7. An application example of the peer-to-peer model is instant ________ in which people carry on a written conversation over the Internet.

8. According to the text, although it is often used, the term peer-to-peer ________ is not very appropriate.

**II. Translate the following terms or phrases from English into Chinese and vice versa:**

1. personal area network
2. carrier sense
3. protocol suite
4. peer-to-peer model
5. star topology
6. communication channel
7. access point
8. proprietary network

| | |
|---|---|
| 9. utility package | 15. 总线拓扑 |
| 10. wireless headset | 16. 客户机 / 服务器模型 |
| 11. 局域网 | 17. 以太网标准 |
| 12. 无线鼠标 | 18. 进程间通信 |
| 13. 城域网 | 19. 打印服务器 |
| 14. 封闭式网络 | 20. 广域网 |

**III. Fill in each of the blanks with one of the words given in the following list, making changes if necessary:**

| | | | |
|---|---|---|---|
| *space* | *memory* | *access* | *manage* |
| *maintenance* | *advantage* | *workstation* | *bulk* |
| *reduce* | *server* | *networked* | *expensive* |
| *remote* | *within* | *exchange* | *without* |

Computers can communicate with other computers through a series of connections and associated hardware called a network. The ________ of a network is that data can be ________ rapidly, and software and hardware resources, such as hard-disk ________ or printers, can be shared. Networks also allow ________ use of a computer by a user who cannot physically ________ the computer.

One type of network, a local area network (LAN), consists of several PCs or ________ connected to a special computer called a server, often ________ the same building or office complex. The server stores and ________ programs and data. A server often contains all of a ________ group's data and enables LAN workstations or PCs to be set up ________ large storage capabilities. In this *scenario* (方案), each PC may have "local" ________ (for example, a hard drive) specific to itself, but the ________ of storage resides on the server. This ________ the cost of the workstation or PC because less ________ computers can be purchased, and it simplifies the ________ of software because the software resides only on the ________ rather than on each individual workstation or PC.

**IV. Translate the following passage from English into Chinese:**

A network, in computer science, is a group of computers and associated devices that are connected by communications facilities. A network can involve permanent connections, such as cables, or temporary connections made through telephone or other communications links. A network can be as small as a local area network consisting of a few computers, printers, and other devices, or it can consist of many small and large computers distributed over a vast geographic area. Small or large, a computer network exists to provide computer users with the means of communicating and transferring information electronically. Some types of communication are simple user-to-user messages; others, of the type known as distributed processes, can involve several computers and the sharing of workloads or *cooperative* (合作的) efforts in performing a task.

# Section B

# A Guide to Network Topology

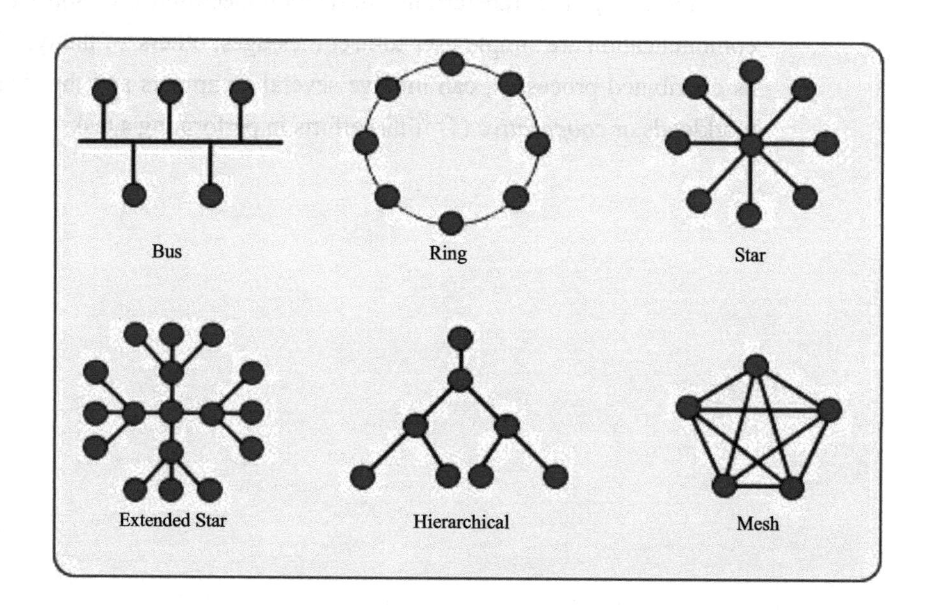


**routing** / 'ru:tiŋ /
n. 路由选择
**routing path**
路由路径，路由选择通路
**familiarize**
/ fə'miliəraiz /
v. 使熟悉；使通晓
**mesh** / meʃ /
n. 网；网格；网状结构；网状网络

A network topology is how computers, printers, and other devices are connected over a network. It describes the layout of wires, devices, and **routing paths**. Essentially there are six different common topologies you should **familiarize** yourself with: bus, ring, star, extended star, hierarchical, and **mesh**.

## I. Bus Topology

The bus topology was fairly popular in the early years of networking. It's easy to set up—not to mention inexpensive. All devices on the bus topology are connected using a single cable. If you need help remembering how the bus topology operates, think of it as the route a bus takes throughout a city.

**terminator**
/ 'tə:mineitə /
n. 终结器；端子；终结符
**disrupt** / dis'rʌpt /
v. 扰乱；使中断

It is extremely important to note that both ends of the main cable need to be terminated. If there is no **terminator**, the signal will bounce back when it reaches the end. The result: a bunch of collisions and noise that will **disrupt** the entire network.

The bus topology is less common these days. In fact, this topology is

**coaxial** / ˌkəu'æksiəl /
a. 同轴的
**coaxial cable**
同轴电缆

commonly used to network computers via **coaxial cable**—when's the last time you can say you've done that?

## II. Ring Topology

**ring topology**
环型拓扑（结构）

**redundant** / ri'dʌndənt /
a. 冗余的

The **ring topology** is a very interesting topology indeed. It is a lot more complex than it may seem—it looks like just a bunch of computers connected in a circle! But behind the scenes, the ring topology is providing a collision-free and **redundant** networking environment[1].

**node** / nəud /
n. 节点

Note that since there is no end on a ring topology, no terminators are necessary. A frame travels along the circle, stopping at each **node**. If that node wants to transmit data, it adds destination address and data information to the frame. The frame then travels around the ring, searching for the destination node. When it's found, the data is taken out of the frame and the cycle continues.

**slack** / slæk /
n. 松弛（部分）
**down** / daun /
v. 击倒；击落；倒下

But wait—it gets better! We have two types of ring topologies in networking: the one we just reviewed, and dual-ring topology. In a dual-ring topology, we use two rings instead of one. This creates a sense of redundancy so that if any point in the network fails, the second ring will (hopefully) be able to pick up the **slack**[2]. If both rings were to fail at separate locations, we can even use the opposite ring at each point to "patch" the **downed** node[3].

In the diagram on the right in Figure 7B-1, you can see that although

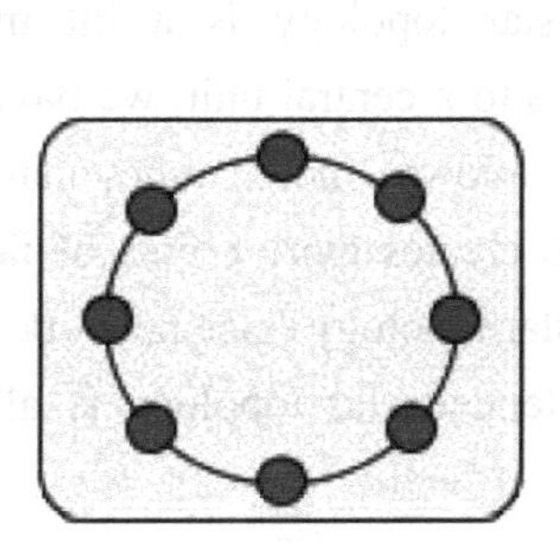

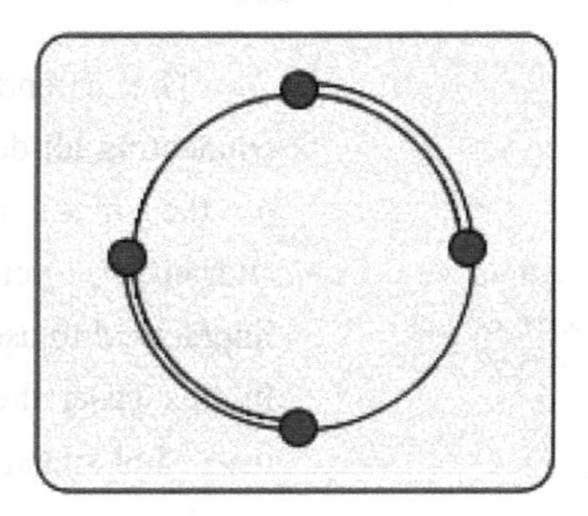

Figure 7B-1: The Ring and Dual-Ring Topology

1 *But behind the scenes, the ring topology is providing a collision-free and redundant networking environment.*：但是，在幕后，环型拓扑结构却提供着一个无冲突的冗余联网环境。

2 *This creates a sense of redundancy so that if any point in the network fails, the second ring will (hopefully) be able to pick up the slack.*：这在一定意义上产生了冗余，因此，如果网络中的任何点失灵，第二个环将（有望）能够予以弥补。

3 *If both rings were to fail at separate locations, we can even use the opposite ring at each point to "patch" the downed node.*：如果两个环分别在不同的点失灵，我们甚至可以在每个点使用对面的环来“修补”发生故障的节点。本句前半部分是一个虚拟条件状语从句，表示所述的情况发生的可能性不大。

the outer ring and inner ring failed at separate parts of the network, thanks to redundancy, the network is still fully operational. This is generally more expensive to implement than other topologies—so it isn't as common as the star or extended star topology.

### III. Star/Extended Star Topology

One of the most popular topologies for Ethernet LANs is the star and extended star topology. It's easy to set up, it's relatively cheap, and it creates more redundancy than the bus topology.

The star topology works by connecting each node to a central device. This central connection allows us to have a fully functioning network even when other devices fail. The only real threat to this topology is that if the central device goes down, so does the entire network.

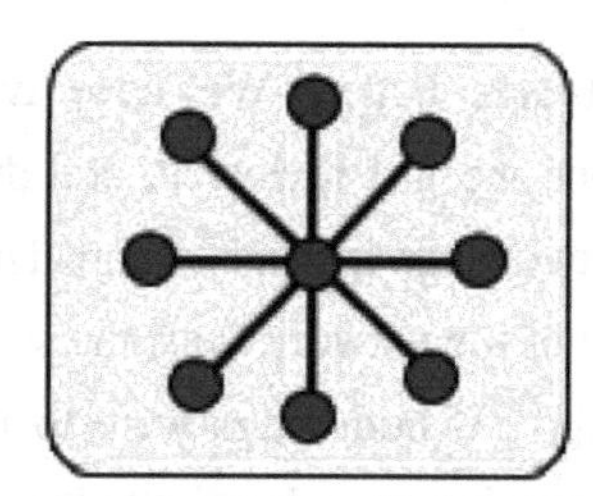

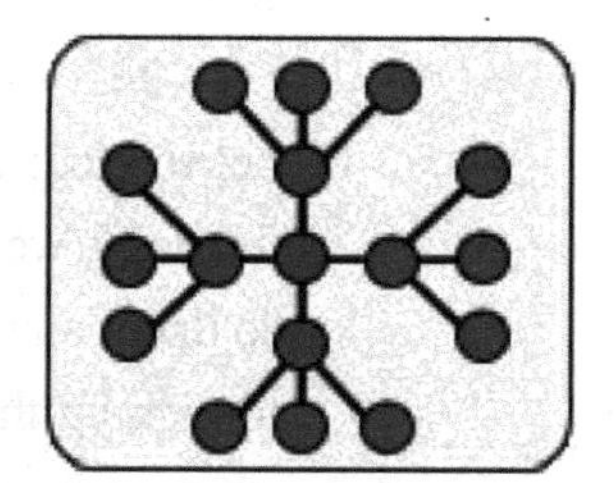

**Figure 7B-2: The Star and Extended-Star Topology**

The extended star topology is a bit more advanced. Instead of connecting all devices to a central unit, we have sub-central devices added to the mix. This allows more functionality for organization and **subnetting**—yet also creates more points of failure[1]. In many cases, it is impractical to use a star topology since networks can span an entire building. In this case, the extended star topology is all but necessary to prevent **degraded** signals[2].

**subnetting**
/ 'sʌbˌnetiŋ /
n. 子网划分，子网组建

**degraded**
/ di'greidid /
a. 降级的，退化的

Whereas the star topology is better suited for small networks, the extended star topology is generally better for the larger ones.

1 *This allows more functionality for organization and subnetting—yet also creates more points of failure.*：这就在网络的组织与子网划分方面提供了更多功能——但也造成了更多的故障点。

2 *In this case, the extended star topology is all but necessary to prevent degraded signals.*：在这种情况下，为防止信号衰减，扩展星型拓扑结构几乎是必需的。all but 意为“几乎”“差不多”。

## IV. Hierarchical Topology

The hierarchical topology is much like the star topology, except that it doesn't use a central node. Although Cisco[1] prefers to call this hierarchical, you may see it as instead referred to as the tree topology.

This type of topology suffers from the same centralization flaw as the star topology. If the device that is on top of the chain fails, consider the entire network down. Obviously this is impractical and not used a great deal in real applications.

## V. Mesh Topology

If you haven't noticed, we've had a little problem with a fully redundant network. The dual-ring topology helped, but it wasn't perfect. If you are looking for a truly redundant network, look no further than the mesh topology. You will see two main types of mesh topology: full-mesh and partial-mesh.

The full-mesh topology connects every single node together. This will create the most redundant and reliable network around—especially for large networks. If any link fails, we (should) always have another link to send data through. So why don't we use it more often? Simple: how many wires would it take to link a computer to every device on a network of over 100 devices? Now multiply that for every device on the network—not a pleasant number is it? Obviously you should only use this in smaller networks. **Alternatively**, you could try a partial-mesh topology.

**alternatively**
/ ɔːl'təːnətivli /
ad. 或者，非此即彼

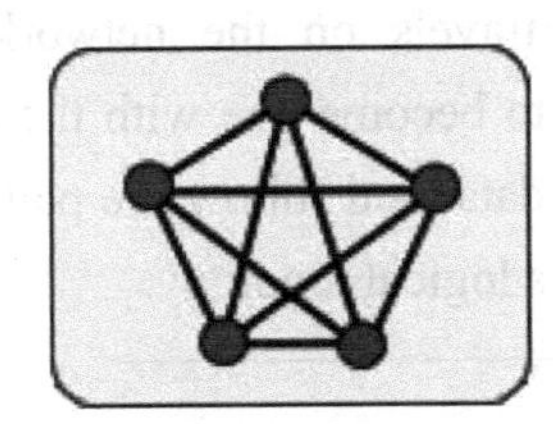

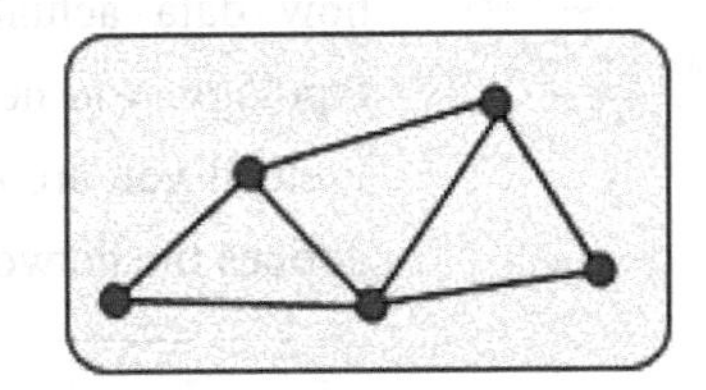

**Figure 7B-3: The Full-Mesh and Partial-Mesh Topology**

The partial-mesh topology is much like the full-mesh, only we don't connect each device to every other device on the network. Instead, we only implement a few **alternate** routes. After all, what are the odds a network will fail in multiple times near the same device[2]?

**alternate**
/ ɔːl'təːnit, 'ɔːltə- /
a. 供选择的，供替换的；备用的

[1] *Cisco*：思科公司（Cisco Systems, Inc.，思科系统有限公司），一家网络解决方案供应公司，创立于 1984 年，总部位于美国加利福尼亚州圣何塞市（San Jose）。Cisco 取自 San Fran*cisco*（旧金山）后 5 个字母。

[2] *After all, what are the odds a network will fail in multiple times near the same device?*：毕竟，一个网络在同一个设备附近多次出故障的可能性有多大呢？言外之意，这种可能性不大。

**backbone**
/ 'bækbəun /
n. 骨干（网），主干（网），基干（网）
**Internet Service Provider**
因特网服务提供商
**router** / 'ru:tə /
n. 路由器

You'll see the partial-mesh topology in **backbone** environments, since these are often vital networks that depend on redundancy to keep services running (such as an **Internet Service Provider**). The full-mesh topology is commonly seen in WANs between **routers**, yet also on smaller networks that depend on a redundant connection.

## VI. Closing Comments

**hybrid** / 'haibrid /
n. 混合物
**categorize**
/ 'kætigəraiz /
v. 将…分类
**endpoint**
/ 'endpɔint /
n. 端点

Keep in mind that network topology isn't limited to the above examples. There are **hybrids** and variations of the topologies mentioned above. Oddly enough, Cisco fails to **categorize** point-to-point topology in their course material—but don't worry, it's just a simple connection between two **endpoints**. Perhaps it was considered too simple to include in the course material—either way, make sure you commit the above topologies to memory[1]. You'll be expected to know them when exam day comes—not to mention it could save you from a disorganized mess of a network!

**hub** / hʌb /
n.（网络）集线器；（轮）毂；（兴趣、活动等的）中心

Another point to emphasize is that each network has a logical topology as well as a physical topology. The physical and logical topology of the network can be the same, but often they aren't. Think of the physical topology as the network layout you'd see if you looked at the physical network from outside. To understand the physical topology, you completely ignore how the data travels, and focus instead on the appearance of cables, **hubs**, and nodes that make up the network. The logical topology describes how data actually travels on the network. To understand the logical topology, you need to become one with the data traveling on the network. Pretend you are the data, and follow the path it takes. The path you follow defines the network's logical topology.

---

# Exercises

**I. Fill in the blanks with the information given in the text:**

1. In the bus topology, a(n) ________ should be installed at each end of the bus or main cable.

2. As compared with the star topology, the extended star topology adds

---

1 *commit the above topologies to memory*：记住上面的拓扑结构。commit sth to memory 表示“把…记住”“把…背熟”。

________ devices instead of connecting all devices to a central unit.

3. The hierarchical topology, also referred to as the ________ topology, is much like the star topology, except that it doesn't use a(n) ________ node.

4. A truly redundant network can be created with the ________ topology.

**II. Translate the following terms or phrases from English into Chinese and vice versa:**

| | |
|---|---|
| 1. routing path | 6. 同轴电缆 |
| 2. dual-ring topology | 7. 逻辑拓扑结构 |
| 3. extended star topology | 8. 无冲突联网环境 |
| 4. backbone network | 9. 树型拓扑结构 |
| 5. mesh topology | 10. 目的地节点 |

# Section C

## Network Connecting Devices

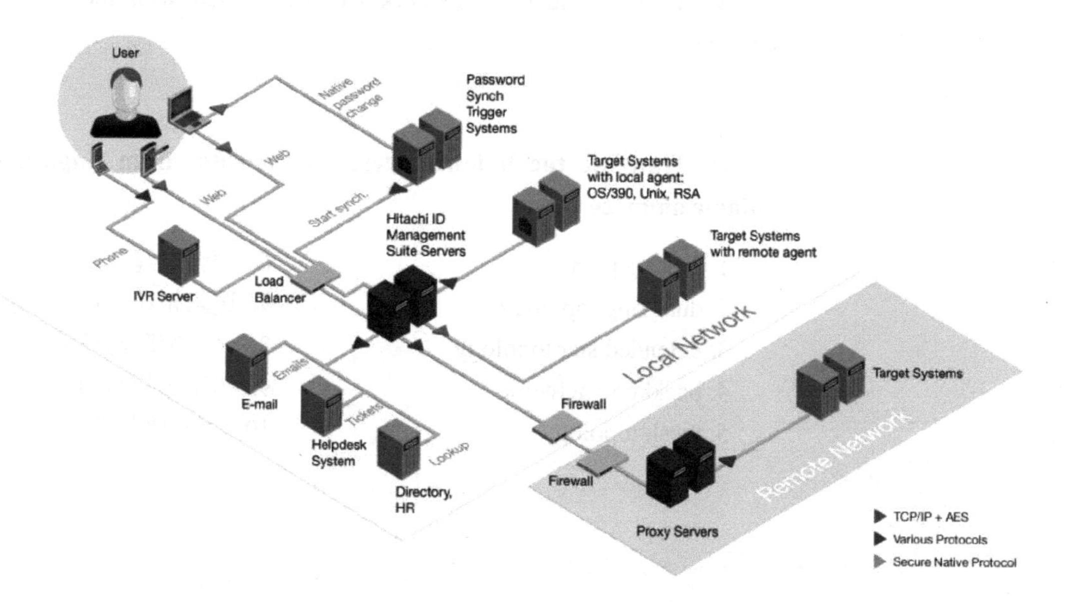


**repeater** / ri'pi:tə /
n. 中继器，转发器
**bridge** / bridʒ /
n. 网桥，桥接器
**gateway** / 'geitwei /
n. 网关
**internetwork** / ˌintə'netwə:k /
n. 互联网，互连网

We can divide computer networks into three broad categories: local area networks (LANs), metropolitan area networks (MANs), and wide area networks (WANs). The three network types can be connected using connecting devices. The interconnection of networks makes global communication from one side of the world to the other possible. Connecting devices can be divided into four types based on their functionality as related to the layers in the OSI model: **repeaters**, **bridges**, routers, and **gateways**. Repeaters and bridges typically connect devices in a network. Routers and gateways typically connect networks into **internetworks** (Figure 7C-1).

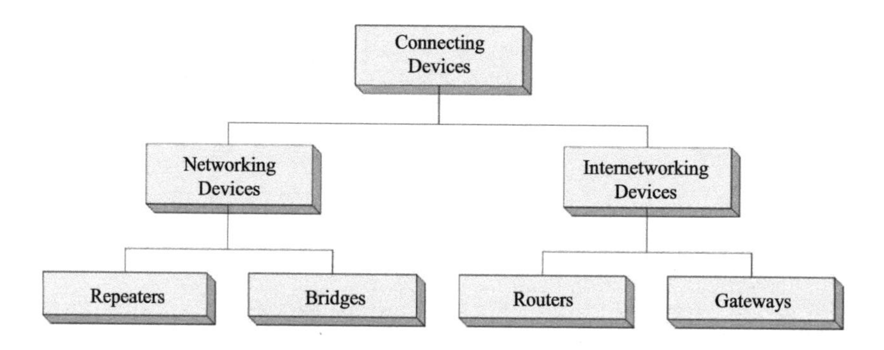


Figure 7C-1: Network Connecting Devices

A repeater is an electronic device that regenerates data. It extends the physical length of a network. As a signal is transmitted, it may lose strength, and a weak signal may be interpreted erroneously by a receiver. A repeater can regenerate the signal and send it to the rest of the network. Figure 7C-2 shows a network with and without a repeater.

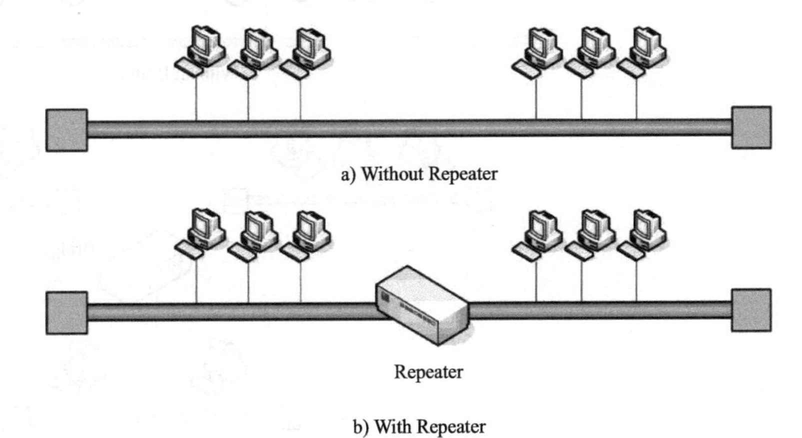


**Figure 7C-2: Repeater**

Repeaters operate only in the physical layer of the OSI model. They do not recognize physical or logical addresses. They simply regenerate every signal they receive. Repeaters, popular when the dominant topology was the bus topology, often connected two buses to increase the length of the network.

When a network uses a bus topology, the medium is shared between all stations. In other words, when a station sends a frame, the common bus is occupied by this one station, and no other station is allowed to send a frame (if it does, the two frames collide). This implies a **degradation** of performance. Stations need to wait until the bus is free. This is similar to an airport that has only one **runway**. When the runway is used by one aircraft, the other aircraft ready for takeoff must wait.

**degradation** /ˌdegrəˈdeiʃən/ n. 降级，退化
**runway** /ˈrʌnwei/ n.（机场的）跑道

A bridge is a traffic controller. It can divide a long bus into smaller segments so that each segment is independent **trafficwise**[1]. A bridge installed between two segments can pass or block frames based on the destination address in the frame. If a frame **originates** in one segment and the destination is in the same segment, there is no reason for the frame to pass the bridge and go to the other segments. The bridge uses a table to

**trafficwise** /ˈtræfikwaiz/ ad. 在交通方面
**originate** /əˈridʒineit/ v. 发源；产生

[1] *A bridge is a traffic controller. It can divide a long bus into smaller segments so that each segment is independent trafficwise.*：网桥相当于交通管制员。它可以把一条长总线分成若干较小的段，这样每一段就交通（这里指通信）而言都是独立的。

decide if the frame needs to be forwarded to another segment. With a bridge, two or more pairs of stations can communicate at the same time (Figure 7C-3).

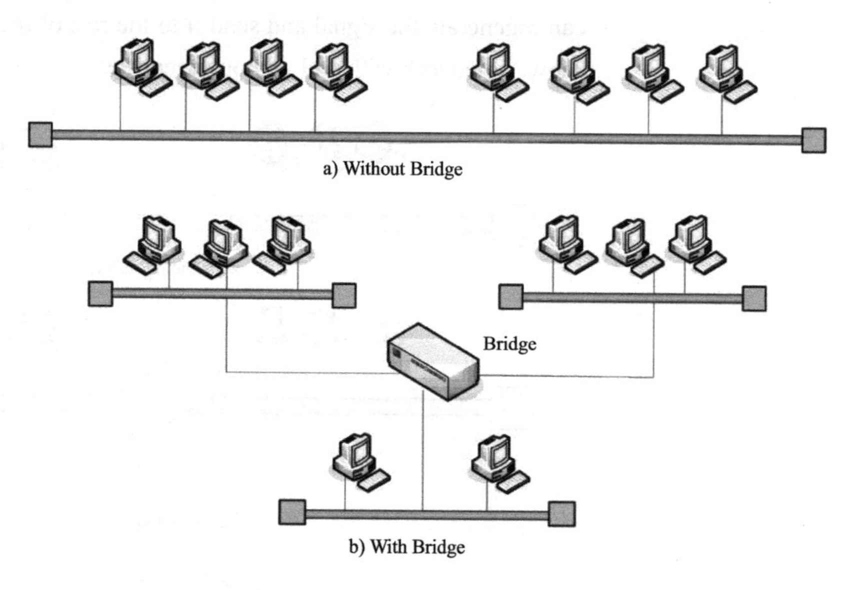


**Figure 7C-3: Bridge**

In addition to its traffic controlling duties, a bridge also functions as a repeater by regenerating the frame. As discussed previously, this means that a bridge operates at the physical layer. But because a bridge needs to interpret the address embedded in the frame to make filtering decisions, it also operates at the **data-link** layer of the OSI model.

**data link**
数据链路

The need for better performance has led to the design of a device referred to as a second-layer switch, which is simply a sophisticated bridge with multiple interfaces. For example, a network with 20 stations can be divided into four segments using a four-interface bridge. Or the same network can be divided into 20 segments (with one station per segment) using a 20-interface switch. A switch in this case increases performance; a station that needs to send a frame sends it directly to the switch. The media are not shared; each station is directly connected to the switch (Figure 7C-4).

Routers are devices that connect LANs, MANs, and WANs. A router operates at the third layer of the OSI model. Whereas a bridge filters a frame based on the physical (data-link layer) address of the frame, a router **routes** a packet based on the logical (network layer) address of the packet.

**route** /ru:t/
v.(按特定路径)发送，传递

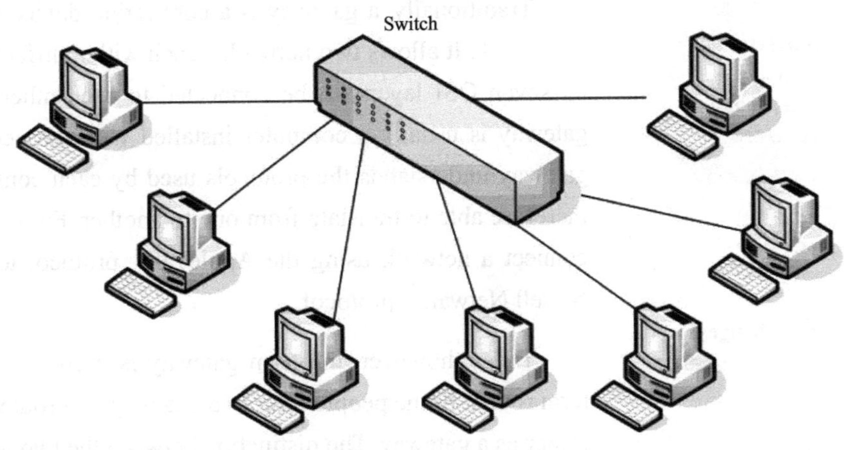


Figure 7C-4: Switch

Whereas a bridge may connect two segments of a LAN or two LANs belonging to the same organization, a router can connect two independent networks: a LAN to a WAN, a LAN to a MAN, a WAN to another WAN, and so on. The result is an internetwork (or an internet). The Internet (the unique global internet) that connects the whole world together is an example of an internetwork where many networks are connected together through routers. Figure 7C-5 shows an example of an internetwork.

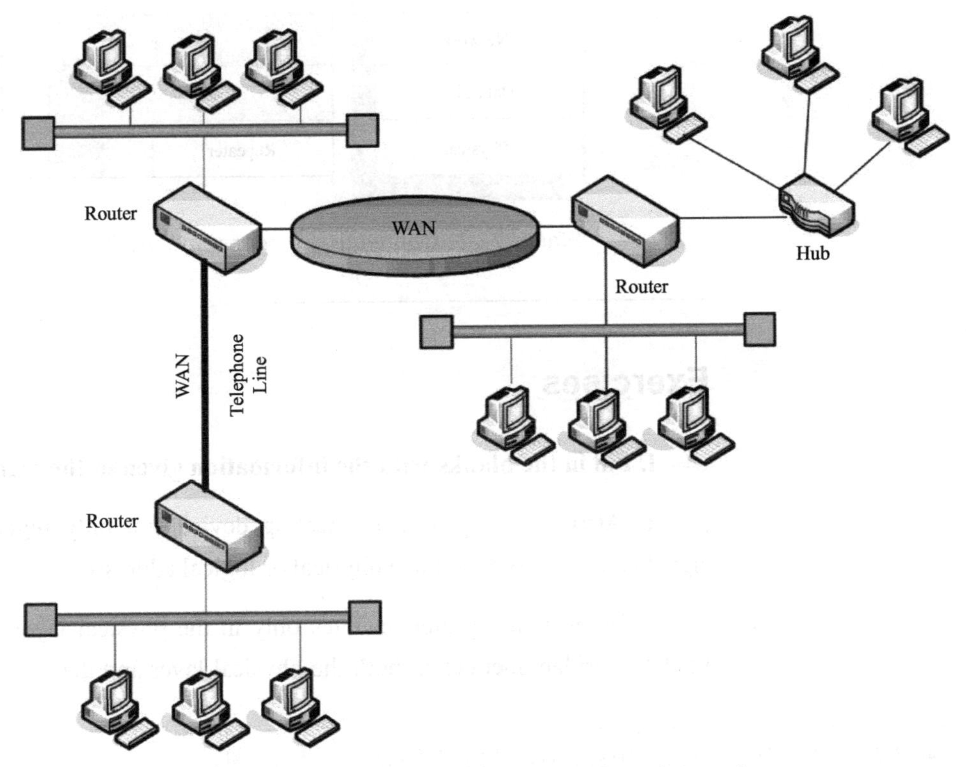


Figure 7C-5: Routers in an Internet

**converter** /kən'və:tə/ n. 转换器，转换程序

Traditionally, a gateway is a connecting device that acts as a protocol **converter**. It allows two networks, each with a different set of protocols for all seven OSI layers, to be connected to each other and communicate. A gateway is usually a computer installed with the necessary software. The gateway understands the protocols used by each connected network and is therefore able to translate from one to another. For example, a gateway can connect a network using the AppleTalk[1] protocol to a network using the Novell Netware[2] protocol.

**interchangeable** /ˌintə'tʃeindʒəbəl/ a. 可交换的，可互换的

Today, however, the term gateway is used **interchangeably** with the term router. Some people refer to a gateway as a router, and others refer to a router as a gateway. The distinction between the two terms is disappearing.

Figure 7C-6 presents the relationship between the connecting devices and the OSI model.

**session** /'seʃən/ n. 对话（期），会话

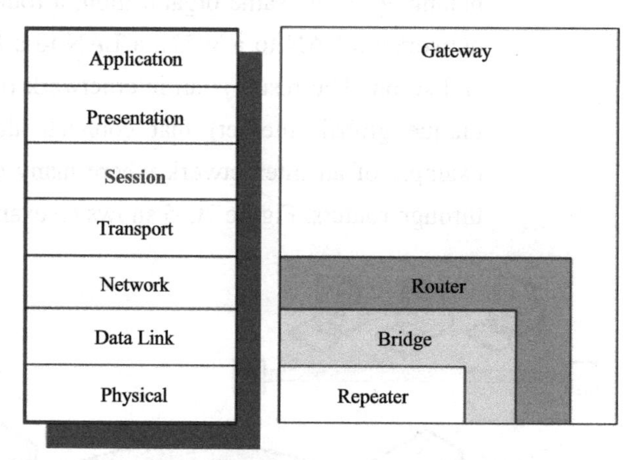


**Figure 7C-6: Connecting Devices and the OSI Model**

---

## Exercises

**I. Fill in the blanks with the information given in the text:**

1. A(n) ________ is a connecting device that only regenerates the signal and does not recognize physical or logical addresses.

2. Whereas a repeater operates only in the physical layer of the OSI model, a bridge operates in both the physical layer and the ________ layer

---

1 *AppleTalk*：苹果公司创建的一组网络协议，用于苹果系列的个人计算机。

2 *Novell Netware*：美国网威（Novell）公司推出的网络操作系统。

of the OSI model.

3. A(n) ________ is usually a computer installed with the necessary software and allows two networks, each with a completely different set of protocols, to communicate.

4. The distinction between the two terms gateway and ________ is disappearing. They are now used interchangeably.

**II. Translate the following terms or phrases from English into Chinese and vice versa:**

| | |
|---|---|
| 1. destination address | 5. 数据链路层 |
| 2. performance degradation | 6. 协议转换器 |
| 3. four-interface bridge | 7. 开放式系统互联 |
| 4. common bus | 8. 物理地址 |

# Unit 8 The Internet

# （因特网）

## Section A

## The Internet

**uppercase**
/ ˈʌpəˈkeis /
a.（字母）大写的

The most notable example of an internet is the Internet (note the **uppercase** I), which originated from research projects going back to the early 1960s. The goal was to develop the ability to link a variety of computer networks so that they could function as a connected system that would not be disrupted by local disasters. Much of this work was sponsored by the U.S. government through the Defense Advanced Research Projects

**undertaking** /ˌʌndəˈteikiŋ, ˈʌndət-/ n. 任务；事业

Agency (DARPA[1] —pronounced "DAR–pa"). Over the years, the development of the Internet shifted from a government-sponsored project to an academic research project, and today it is largely a commercial **undertaking** that links a worldwide combination of PANs, LANs, MANs, and WANs involving millions of computers.

## I. Internet Architecture

**customary** /ˈkʌstəməri/ a. 习惯的；按惯例的

The Internet is a collection of connected networks. In general, these networks are constructed and maintained by organizations called Internet Service Providers (ISPs[2]). It is also **customary** to use the term ISP in reference to[3] the networks themselves. Thus, we will speak of connecting to an ISP, when what we really mean is connecting to the network provided by an ISP.

The system of networks operated by the ISPs can be classified in a hierarchy according to the role they play in the overall Internet structure (Figure 8A-1). At the top of this hierarchy are relatively few tier-1 ISPs that consist of very high-speed, high-capacity, international WANs. These networks are thought of as the backbone of the Internet. They are typically operated by large companies that are in the communications business.

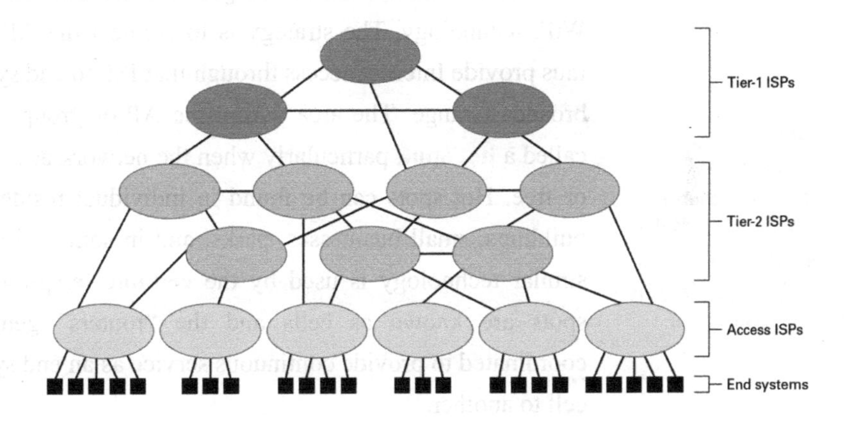


**Figure 8A-1: Internet Composition**

**potent** /ˈpəutənt/ a. 有（效）力的，效力大的

Connecting to the tier-1 ISPs are the tier-2 ISPs that tend to be more regional in scope and less **potent** in their capabilities. Again, these networks tend to be operated by companies in the communications business.

---

1 *DARPA*：（美国）国防部高级研究计划局（*D*efense *A*dvanced *R*esearch *P*rojects *A*gency 的首字母缩略）。
2 *ISP*：因特网服务提供商（*I*nternet *S*ervice *P*rovider 的首字母缩略）。
3 *in reference to*：关于，就…而论。

Tier-1 and tier-2 ISPs are essentially networks of routers that collectively provide the Internet's communication infrastructure. As such, they can be thought of as the core of the Internet. Access to this core is usually provided by an intermediary called an access or tier-3 ISP. An access ISP is essentially an independent internet, sometimes called an **intranet**, operated by a single authority that is in the business of supplying Internet access to individual homes and businesses. Examples include cable and telephone companies that charge for their service as well as organizations such as universities or corporations that take it upon themselves to provide Internet access to individuals within their organizations.

**intranet** / 'intrənet /
n. 内联网，
内部网络

The devices that individual users connect to the access ISPs are known as end systems or **hosts**. These end systems may be laptops or PCs, but increasingly range over a **multitude** of other devices including telephones, **video cameras**, automobiles, and **home appliances**. After all, the Internet is essentially a communications system, and thus any device that would benefit from communicating with other devices is a potential end system.

**host** / həust /
n. 主机
**multitude**
/ 'mʌltitjuːd /
n. 大量，许多
**video camera**
摄像机
**home appliance**
家用电器

The technology by which end systems connect to larger networks is also varied. Perhaps the fastest growing are wireless connections based on WiFi technology. The strategy is to connect the AP to an access ISP and thus provide Internet access through that ISP to end systems within the AP's broadcast range. The area within the AP or group of APs' range is often called a **hot spot**, particularly when the network access is publicly available or free. Hot spots can be found in individual residences, hotel and office buildings, small businesses, parks, and in some cases span entire cities. A similar technology is used by the **cellular telephone** industry where hot spots are known as cells and the "routers" generating the cells are coordinated to provide continuous service as an end system moves from one cell to another.

**hot spot**
（网络）热点
**cellular** / 'seljulə /
a. 蜂窝状的，多孔的
**cellular (tele)phone**
蜂窝电话，移动电话，手机

Other popular techniques for connecting to access ISPs use telephone lines or cable/satellite systems. These technologies may be used to provide direct connection to an end system or to a customer's router to which multiple end systems are connected. This latter **tactic** is popular for individual residences where a local hot spot is created by a router/AP connected to an access ISP by means of existing cable or telephone lines.

**tactic** / 'tæktik /
n. 战术；策略，
手段

Telephone, **cable television**, and satellite wide area networks of the twentieth century were designed to carry analog communications, such as

**cable television**
有线电视

the human voice or pre-digital television signals. Modern networks can be designed to carry digital data directly between computers, but older, analog network infrastructure still comprises a significant portion of the Internet. Issues arising from these **legacy** analog **linkages** are often referred to collectively as the last mile problem. The main **arteries** of WANs, MANs, and many LANs are relatively easy to modernize with high-speed digital technology such as **fiber optics**, but it can be far costlier to replace the existing copper telephone lines and coaxial cables that connect these arteries to each individual home or office. Several clever schemes have been developed to extend these legacy analog links to **accommodate** transmission of digital data. DSL[1] modems, cable modems, satellite **uplinks**, and even direct **fiber-optic** connections to the home are used to bring **broadband** Internet access to end users.

**legacy** / ˈlegəsi /
a. 遗留的，旧版本的
**linkage** / ˈliŋkidʒ /
n. 联系；连接；链接
**artery** / ˈɑːtəri /
n. 动脉；干线
**optics** / ˈɔptiks /
n. 光学
**fiber optics**
光学纤维，光纤；纤维光学
**accommodate**
/ əˈkɔmədeit /
v. 容纳；使适应
**uplink** / ˈʌpliŋk /
n. 上行链路
**fiber-optic**
/ ˈfaibəˈɔptik /
a. 光学纤维的，光纤的；纤维光学的
**broadband**
/ ˈbrɔːdbænd /
n. & a. 宽带（的）
**nonprofit**
/ nɔnˈprɔfit /
a. 非营利的
**decimal notation**
十进制记数法
**dotted decimal notation**
点分十进制记数法
**base 10 notation**
以 10 为底的记数法

## II. Internet Addressing

An internet needs an internet-wide addressing system that assigns a unique identifying address to each computer in the system. In the Internet these addresses are known as IP[2] addresses. Originally, each IP address was a pattern of 32 bits, but to provide a larger set of addresses, the process of converting to 128-bit addresses is currently underway. Blocks of consecutively numbered IP addresses[3] are awarded to ISPs by the Internet Corporation for Assigned Names and Numbers (ICANN[4]), which is a **nonprofit** corporation established to coordinate the Internet's operation. The ISPs are then allowed to allocate the addresses within their awarded blocks to machines within their region of authority. Thus, machines throughout the Internet are assigned unique IP addresses.

IP addresses are traditionally written in **dotted decimal notation** in which the bytes of the address are separated by periods and each byte is expressed as an integer represented in traditional **base 10 notation**[5]. For example, using dotted decimal notation, the pattern 5.2 would represent the two-byte bit pattern 0000010100000010, which consists of the byte

[1] *DSL*：数字用户线路（*D*igital *S*ubscriber *L*ine 的首字母缩略）。

[2] *IP*：IP 协议，网际协议，网间协议（*I*nternet *P*rotocol 的首字母缩略）。

[3] *Blocks of consecutively numbered IP addresses*：连续编号的 IP 地址块。

[4] *ICANN*：因特网名称与数字地址分配机构（*I*nternet *C*orporation for *A*ssigned *N*ames and *N*umbers 的首字母缩略），读作 / ˈaikæn /，成立于 1998 年，总部设在美国加利福尼亚州洛杉矶县马里纳德尔雷（Marina del Rey）。

[5] *the bytes of the address are separated by periods and each byte is expressed as an integer represented in traditional base 10 notation*：地址的字节用句号分隔，每个字节用一个整数来表示，而该整数是用传统的以 10 为底的记数法来表示的。

**binary notation** 二进制记数法
**binary** /ˈbainəri/ n. 二进制(数)

00000101 (represented by 5) followed by the byte 00000010 (represented by 2), and the pattern 17.12.25 would represent the three-byte bit pattern consisting of the byte 00010001 (which is 17 written in **binary notation**), followed by the byte 00001100 (12 written in **binary**), followed by the byte 00011001 (25 written in binary). In summary, a 32-bit IP address might appear as 192.207.177.133 when expressed in dotted decimal notation.

**conducive** /kənˈdjuːsiv/ a. 有助的,有益的(*to*)
**mnemonic** /niːˈmɔnik/ a. 助记的;记忆的
**mnemonic name** 助记名,缩写名
**registrar** /ˌredʒiˈstrɑː, ˈredʒis-/ n. 登记员;(域名)注册(服务)商
**domain name** 域名

Addresses in bit-pattern form (even when compressed using dotted decimal notation) are rarely **conducive** to human consumption[1]. For this reason the Internet has an alternative addressing system in which machines are identified by **mnemonic names**. This addressing system is based on the concept of a domain, which can be thought of as a "region" of the Internet operated by a single authority such as a university, club, company, or government agency. (The word region is in quotations here because such a region may not correspond to a physical area of the Internet.) Each domain must be registered with ICANN—a process handled by companies, called **registrars**, that have been assigned this role by ICANN. As a part of this registration process, the domain is assigned a mnemonic **domain name**, which is unique among all the domain names throughout the Internet. Domain names are often descriptive of the organization registering the domain, which enhances their utility for humans.

**suffix** /ˈsʌfiks/ n. 后缀
**top-level domain** 顶级域名
**country code** 国家代码

As an example, the domain name of Marquette University[2] is `mu.edu`. Note the **suffix** following the period. It is used to reflect the domain's classification, which in this case is "educational" as indicated by the `edu` suffix. These suffixes are called **top-level domains** (TLDs[3]). Other TLDs include `com` for commercial institutions, `gov` for U.S. government institutions, `org` for nonprofit organizations, `museum` for museums, `info` for unrestricted use, and `net`, which was originally intended for ISPs but is now used on a much broader scale. In addition to these general TLDs, there are also two-letter TLDs for specific countries (called **country-code** TLDs) such as `au` for Australia and `ca` for Canada.

Once a domain's mnemonic name is registered, the organization that registered the name is free to extend the name to obtain mnemonic identifiers for individual items within the domain. For example, an

---

1 *Addresses in bit-pattern form (even when compressed using dotted decimal notation) are rarely conducive to human consumption.*:用位模式形式表示的地址(即使采用点分十进制记数法压缩)很难为人们所用。
2 *Marquette University*:马凯特大学,位于美国威斯康星州密尔沃基市(Milwaukee)。
3 *TLD*:顶级域名(*top-level domain* 的缩略)。

**subdomain**
/ ˌsʌbdəʊ'mein /
n. 子域
**jurisdiction**
/ ˌdʒuəris'dikʃən /
n. 司法权；管辖权；管辖范围

**mnemonic address**
助记地址

**conversion**
/ kən'vəːʃən /
n. 转换；转变
**name server**
名称服务器，名字服务器

individual host within Marquette University may be identified as `eagle.mu.edu`. Note that domain names are extended to the left and separated by a period. In some cases multiple extensions, called **subdomains**, are used as a means of organizing the names within a domain. These subdomains often represent different networks within the domain's **jurisdiction**. For example, if Yoyodyne Corporation was assigned the domain name `yoyodyne.com`, then an individual computer at Yoyodyne might have a name such as `overthruster.propulsion.yoyodyne.com`, meaning that the computer `overthruster` is in the subdomain `propulsion` within the domain `yoyodyne` within the TLD `com`. (We should emphasize that the dotted notation used in **mnemonic addresses** is not related to the dotted decimal notation used to represent addresses in bit pattern form[1].)

Although mnemonic addresses are convenient for humans, messages are always transferred over the Internet by means of IP addresses. Thus, if a human wants to send a message to a distant machine and identifies the destination by means of a mnemonic address, the software being used must be able to convert that address into an IP address before transmitting the message. This **conversion** is performed with the aid of numerous servers, called **name servers**, that are essentially directories that provide address translation services to clients. Collectively, these name servers are used as an Internet-wide directory system known as the domain name system (DNS[2]). The process of using DNS to perform a translation is called a DNS lookup.

Thus, for a machine to be accessible by means of a mnemonic domain name, that name must be represented in a name server within the DNS. In those cases in which the entity establishing the domain has the resources, it can establish and maintain its own name server containing all the names within that domain. However, many individuals or small organizations want to establish a domain presence on the Internet without committing the resources necessary to support it. In this case, they can contract with an access ISP to create the appearance of a registered domain using the resources already established by the ISP. Typically, perhaps with the assistance of the ISP, they register the name chosen and contract with the

1 *the dotted notation used in mnemonic addresses is not related to the dotted decimal notation used to represent addresses in bit pattern form*：助记地址中使用的点分表示法与用于以位模式形式表示地址的点分十进制记数法没有关系。

2 *DNS*：域名系统（*d*omain *n*ame *s*ystem 的首字母缩略）。

ISP to have that name included in the ISP's name server. This means that all DNS lookups regarding the new domain name will be directed to the ISP's name server, from which the proper translation will be obtained. In this way, many registered domains can reside within a single ISP, each often occupying only a small portion of a single computer.

---

# Exercises

**I. Fill in the blanks with the information given in the text:**

1. The Internet originated from ________ projects going back to the early 1960s, and much of the initial work in its development was sponsored by the U.S. government through ________.

2. The backbone of the Internet is composed of very high-speed, high-capacity, international ________.

3. An access ISP is essentially an independent ________ operated by a single authority that is in the business of supplying ________ access to individual homes and businesses.

4. ISPs are awarded blocks of consecutively numbered IP addresses by a nonprofit corporation called ________.

5. Each domain of the Internet must be registered with ICANN. This registration process is handled by companies, called ________, that have been assigned this role by ICANN.

6. The process of converting IP addresses from 32-bit addresses to ________ addresses is currently underway.

7. Although ________ addresses are convenient for humans, messages are always transferred over the Internet by means of ________ addresses.

8. The conversion of mnemonic addresses into IP addresses is performed with the aid of name ________.

**II. Translate the following terms or phrases from English into Chinese and vice versa:**

1. cellular telephone
2. IP address
3. analog communications
4. access ISP

5. fiber-optic cable
6. binary notation
7. mnemonic name
8. Internet-wide directory system
9. name server
10. broadband Internet access
11. 助记标识符
12. 电缆调制解调器
13. 网络基础设施
14. 顶级域名
15. 因特网编址
16. 点分十进制记数法
17. 摄像机
18. 域名系统
19. 卫星上行链路
20. 有线电视

**III. Fill in each of the blanks with one of the words given in the following list, making changes if necessary:**

| | | | |
|---|---|---|---|
| *origin* | *risk* | *channel* | *connection* |
| *sender* | *Internet* | *own* | *academic* |
| *network* | *computer* | *technological* | *data* |
| *develop* | *establish* | *researcher* | *packet* |

Early computer networks used leased telephone company lines for their connections. Telephone company systems of that time established a single connection between ________ and receiver for each telephone call, and that connection carried all ________ along a single path. When a company wanted to connect computers it ________ at two different locations, the company placed a telephone call to ________ the connection, and then connected one computer to each end of that single ________.

The U.S. Defense Department was concerned about the inherent ________ of this single-channel method for connecting computers, and its researchers ________ a different method of sending information through multiple ________. In this method, files and messages are broken into ________ that are labeled electronically with codes for their ________, sequences, and destinations. In 1969, Defense Department ________ in the Advanced Research Projects Agency (ARPA) used this network model to connect four ________ into a network called the ARPANET. The ARPANET was the earliest of the ________ that eventually combined to become what we now call the ________. Throughout the 1970s and 1980s, many researchers in the ________ community connected to the ARPANET and contributed to the ________ developments that increased its speed and efficiency.

**IV. Translate the following passage from English into Chinese:**

The Internet provides only the physical and logical infrastructure that connects millions of computers together. Many believe that the World Wide Web (WWW, or simply the Web) provides the *killer application* (制胜法宝) for this global network. The Web is considered the content of the Internet, providing all sorts of information by using a rich set of tools that manage and link text, graphics, sound, and video. Providing and viewing information on the Web is accomplished using server applications and client applications.

If you've already explored the Web, you'll recognize the client-side application as the Web browser. A Web browser receives, interprets, and displays pages of information from the Web. The user can navigate within pages, jump to other pages by clicking *hypertext* (超文本) links, and point to just about any page on the Web.

# Section B

# The Layered Approach to Internet Software

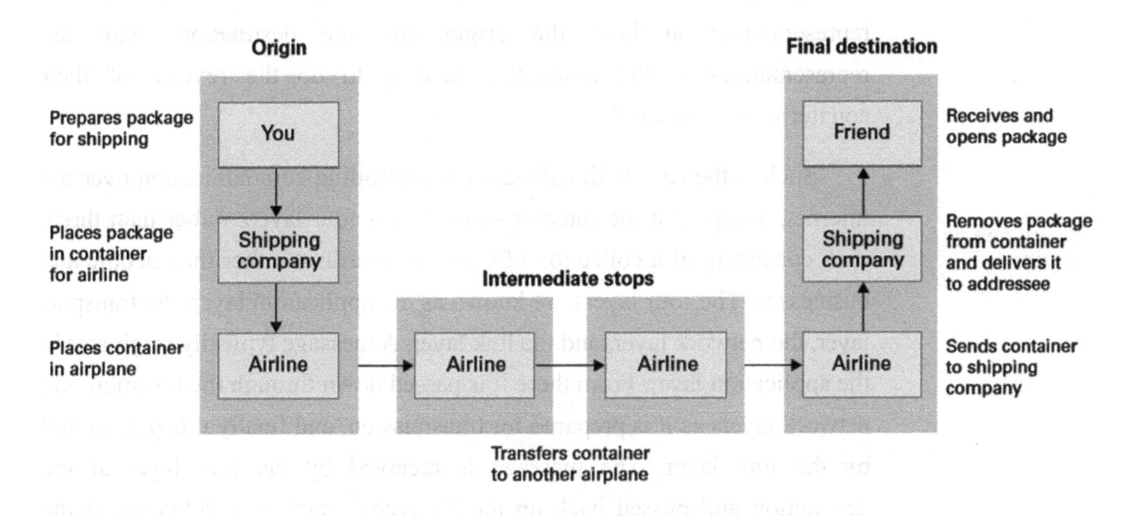


**Figure 8B-1: Package-Shipping Example**

A principal task of networking software is to provide the infrastructure required for transferring messages from one machine to another. In the Internet, this message-passing activity is accomplished by means of a hierarchy of software units, which perform tasks **analogous** to those that would be performed if you were to send a gift in a package from the West Coast of the United States to a friend on the East Coast (Figure 8B-1). You would first wrap the gift as a package and write the appropriate address on the outside of the package. Then, you would take the package to a **shipping** company such as the U.S. Postal Service[1]. The shipping company might place the package along with others in a large container and deliver the container to an airline, whose services it has contracted[2]. The airline would place the container in an aircraft and transfer it to the destination city, perhaps with intermediate stops along the way. At the final destination, the airline would remove the container from the aircraft and give it to the shipping company's office at the destination. In turn, the shipping company would take your package out of the container and deliver it to the **addressee**.

**analogous**
/ ə'næləgəs /
a. 相似的；可比拟的（***to/with***）

**shipping** / 'ʃipiŋ /
n. 运送，运输；航运业

**addressee**
/ ˌædre'siː /
n. 收信人；收件人

[1] *the U.S. Postal Service*：美国邮政管理局，美国邮政服务公司。

[2] *deliver the container to an airline, whose services it has contracted*：将集装箱送往与其签有服务合同的航空公司。

In short, the transportation of the gift would be carried out by a three-level hierarchy: (1) the user level (consisting of you and your friend), (2) the shipping company, and (3) the airline. Each level uses the next lower level as an abstract tool. (You are not concerned with the details of the shipping company, and the shipping company is not concerned with the internal operations of the airline.) Each level in the hierarchy has representatives at both the origin and the destination, with the representatives at the destination tending to do the reverse of their counterparts at the origin[1].

**software routine**
软件例程

Such is the case with software for controlling communication over the Internet, except that the Internet software has four layers rather than three, each consisting of a collection of **software routines** rather than people and businesses. The four layers are known as the application layer, the transport layer, the network layer, and the link layer. A message typically originates in the application layer. From there it is passed down through the transport and network layers as it is prepared for transmission, and finally it is transmitted by the link layer. The message is received by the link layer at the destination and passed back up the hierarchy[2] until it is delivered to the application layer at the message's destination.

Let us investigate this process more thoroughly by tracing a message as it finds its way through the system (Figure 8B-2). We begin our journey with the application layer.

**login** / ˈlɒgin; ˈlɔːg- /
n. 注册，登录，进入系统

The application layer consists of those software units such as clients and servers that use Internet communication to carry out their tasks[3]. Although the names are similar[4], this layer is not restricted to software in the application classification, but also includes many utility packages. For example, software for transferring files using FTP[5] or for providing remote **login** capabilities using SSH[6] have become so common that they are

---

1 *Each level in the hierarchy has representatives at both the origin and the destination, with the representatives at the destination tending to do the reverse of their counterparts at the origin.*：层次体系中的每一层在起点和目的地都有代理，目的地的代理所做的事往往与起点相应代理所做的相反。

2 *passed back up the hierarchy*：逆着层次体系向上传递。

3 *those software units such as clients and servers that use Internet communication to carry out their tasks*：客户机软件和服务器软件等使用因特网通信来完成其任务的软件单元。本句中的 clients 和 servers 指客户机与服务器上安装的相关软件。

4 *the names are similar*：这里所说的“名称相似”是指 application layer（应用层）与 application software（应用软件）都包含 application。

5 *FTP*：文件传送协议，文件传输协议（File Transfer Protocol 的首字母缩略）。

6 *SSH*：安全外壳（Secure Shell 的缩略）。

utility software
实用软件

normally considered **utility software**.

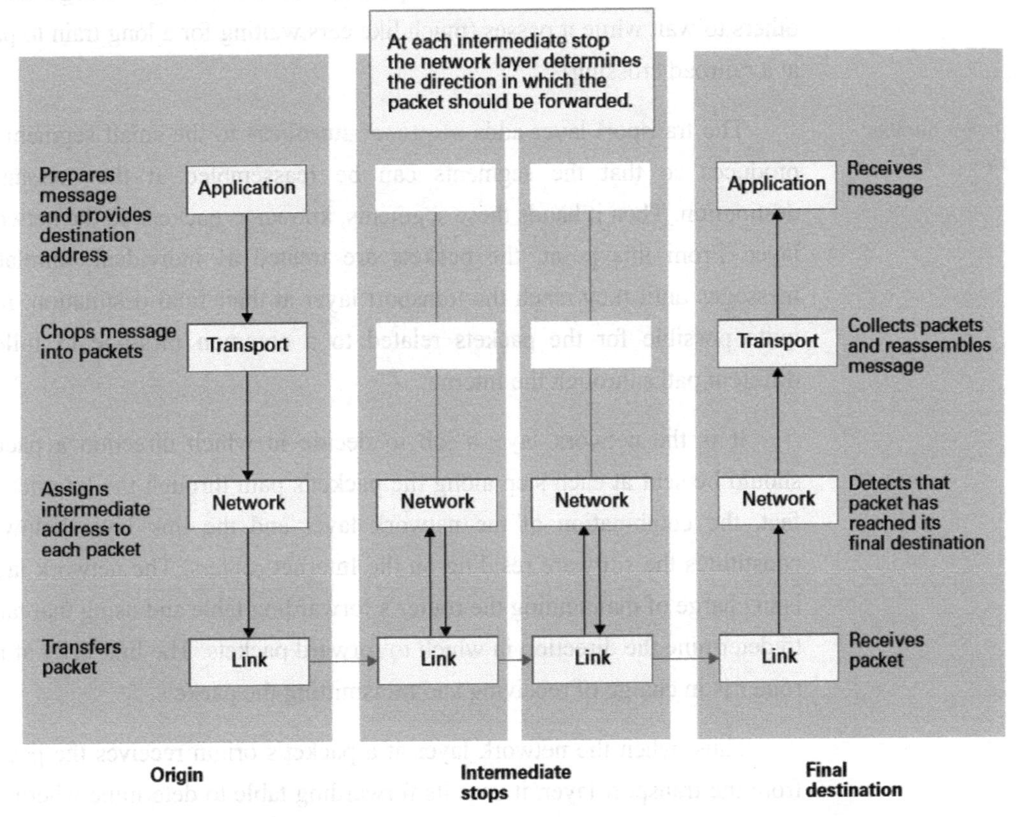


Figure 8B-2: Following a Message through the Internet

The application layer uses the transport layer to send and receive messages over the Internet in much the same way that you would use a shipping company to send and receive packages. Just as it is your responsibility to provide an address compatible with the specifications of the shipping company, it is the application layer's responsibility to provide an address that is compatible with the Internet infrastructure. To fulfill this need, the application layer may use the services of the name servers within the Internet to translate mnemonic addresses used by humans into Internet-compatible IP addresses.

obstruct
/ əb'strʌkt, ɔb- /
v. 阻塞；阻碍

An important task of the transport layer is to accept messages from the application layer and to ensure that the messages are properly formatted for transmission over the Internet. Toward this latter goal, the transport layer divides long messages into small segments, which are transmitted over the Internet as individual units. This division is necessary because a single long message can **obstruct** the flow of other messages at the Internet routers where numerous messages cross paths[1]. Indeed, small segments of messages can interweave at these points, whereas a long message forces others to wait while it passes (much like cars waiting for a long train to pass at a railroad crossing).

**sequence number**
序列号，序号

The transport layer adds **sequence numbers** to the small segments it produces so that the segments can be reassembled at the message's destination. Then it hands these segments, known as packets, to the network layer. From this point, the packets are treated as individual, unrelated messages until they reach the transport layer at their final destination. It is quite possible for the packets related to a common message to follow different paths through the Internet.

It is the network layer's job to decide in which direction a packet should be sent at each step along the packet's path through the Internet. In fact, the combination of the network layer and the link layer below it constitutes the software residing on the Internet routers. The network layer is in charge of maintaining the router's forwarding table and using that table to determine the direction in which to forward packets. The link layer at the router is in charge of receiving and transmitting the packets.

Thus, when the network layer at a packet's origin receives the packet from the transport layer, it uses its forwarding table to determine where the packet should be sent to get it started on its journey. Having determined the proper direction, the network layer hands the packet to the link layer for actual transmission.

The link layer has the responsibility of transferring the packet. Thus the link layer must deal with the communication details particular to the individual network in which the computer resides. For instance, if that network is an Ethernet, the link layer applies CSMA/CD. If the network is a WiFi network, the link layer applies CSMA/CA.

When a packet is transmitted, it is received by the link layer at the other end of the connection. There, the link layer hands the packet up to its network layer where the packet's final destination is compared to the network layer's forwarding table to determine the direction of the packet's next step. With this decision made, the network layer returns the packet to the link layer to be forwarded along its way. In this manner each packet **hops** from machine to machine on its way to its final destination.

**hop** / hɔp /
v. 跳跃

[1] *numerous messages must cross paths*：许多报文必须路径相交。

Note that only the link and network layers are involved at the intermediate stops during this journey (see again Figure 8B-2), and thus these are the only layers present on routers, as previously noted. Moreover, to minimize the delay at each of these intermediate "stops," the forwarding role of the network layer within a router is closely integrated with the link layer. In turn, the time required for a modern router to forward a packet is measured in millionths of a second.

At a packet's final destination, it is the network layer that recognizes that the packet's journey is complete. In that case the network layer hands the packet to its transport layer rather than forwarding it. As the transport layer receives packets from the network layer, it extracts the underlying message segments and reconstructs the original message according to the sequence numbers that were provided by the transport layer at the message's origin. Once the message is assembled, the transport layer hands it to the appropriate unit within the application layer—thus completing the message transmission process.

**incoming** / ˈinˌkʌmiŋ / a. 进来的；输入的
**port** / pɔːt / n. 端口，通信口
**port number** 端口号
**append** / əˈpend / v. 附加

Determining which unit within the application layer should receive an **incoming** message is an important task of the transport layer. This is handled by assigning unique **port numbers** to the various units and requiring that the appropriate port number be **appended** to a message's address before starting the message on its journey. Then, once the message is received by the transport layer at the destination, the transport layer merely hands the message to the application layer software at the designated port number.

**mail server** 邮件服务器

Users of the Internet rarely need to be concerned with port numbers because the common applications have universally accepted port numbers. For example, if a Web browser is asked to retrieve the document whose URL is `http://www.zoo.org/animals/frog.html`[1], the browser assumes that it should contact the HTTP server at `www.zoo.org` via port number 80. Likewise, when sending email, an SMTP[2] client assumes that it should communicate with the SMTP **mail server** through port number 25.

In summary, communication over the Internet involves the interaction of four layers of software. The application layer deals with messages from the application's point of view. The transport layer converts these messages

[1] *html*：超文本标记语言（*H*yper*t*ext *M*arkup *L*anguage 的缩略），亦作 HTML。
[2] *SMTP*：简单邮件传送协议，简单邮件传输协议（*S*imple *M*ail *T*ransfer *P*rotocol 的首字母缩略）。

into segments that are compatible with the Internet and reassembles messages that are received before delivering them to the appropriate application. The network layer deals with directing the segments through the Internet. The link layer handles the actual transmission of segments from one machine to another. With all this activity, it is somewhat amazing that the response time of the Internet is measured in milliseconds, so that many transactions appear to take place instantaneously.

---

## Exercises

**I. Fill in the blanks with the information given in the text:**

1. For an outgoing message, the four layers of software involved in Internet communication work in the order of the ________ layer, the ________ layer, the ________ layer, and the ________ layer; for an incoming message, it is just the opposite.

2. The ________ layer converts long messages into segments compatible with the Internet before their transmission and reassembles them at the destination.

3. The packets related to a common message may follow ________ paths through the Internet.

4. Only the ________ and ________ layers are involved at the intermediate stops during a packet's journey to its final destination.

**II. Translate the following terms or phrases from English into Chinese and vice versa:**

| | |
|---|---|
| 1. incoming message | 6. 端口号 |
| 2. application layer | 7. 软件例程 |
| 3. utility software | 8. 转发表 |
| 4. sequence number | 9. 文件传送协议 |
| 5. remote login capabilities | 10. 万维网浏览器 |

# Section C

# Web Basics

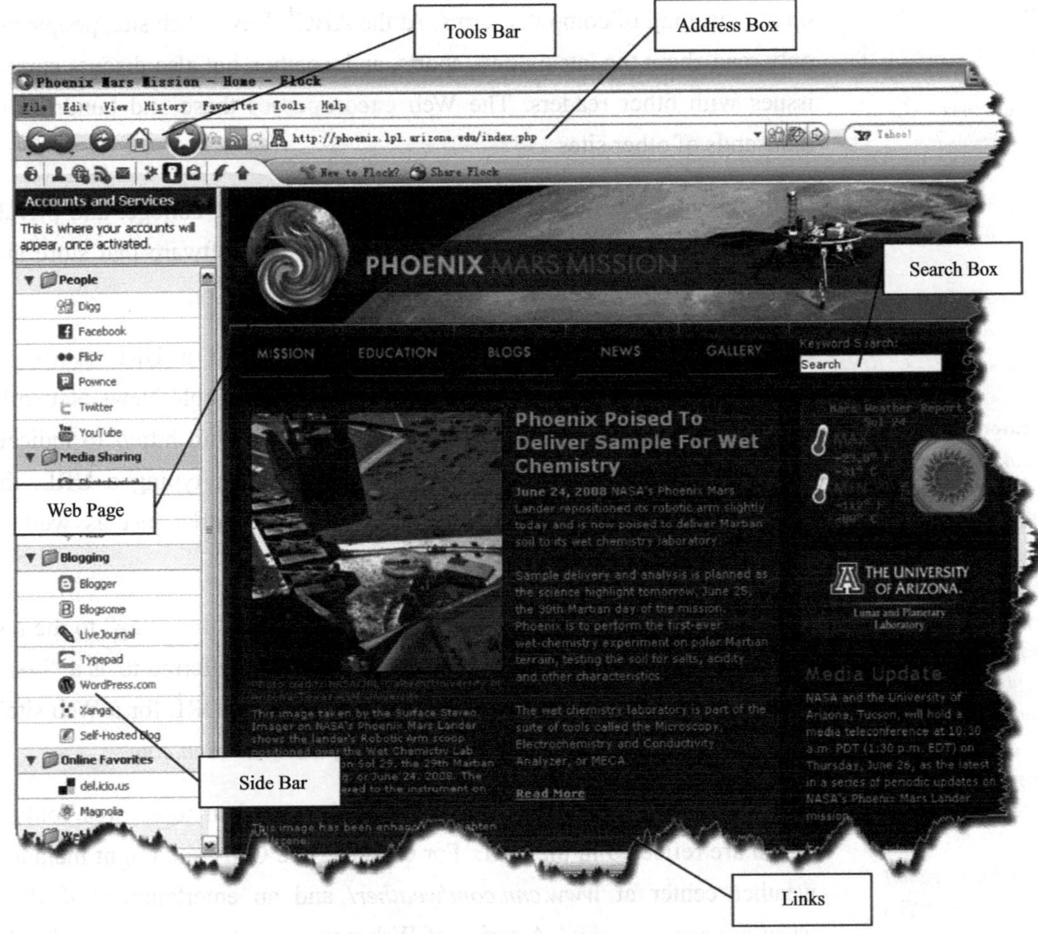


**captivating**
/ ˈkæptiveitiŋ /
a. 迷人的，可爱的

**instrumental**
/ ˌinstruˈmentəl, ˌinstrə- /
a. 起作用的，有帮助的（***in, to***）

**ferry** / ˈferi /
v. 渡运；运送

**animation**
/ ˌæniˈmeiʃən /
n. 动画（制作）

## I. The World Wide Web

One of the Internet's most **captivating** attractions, the World Wide Web (also referred to as WWW, W3, or the Web) is a collection of files that can be linked and accessed using HTTP. HTTP (Hypertext Transfer Protocol) is the communications standard that's **instrumental** in **ferrying** Web documents to all corners of the Internet.

Many Web-based files produce documents called Web pages. Other files contain photos, videos, **animations**, and sounds that can be incorporated into specific Web pages. Most Web pages contain links (sometimes called "hyperlinks") to related documents and media files.

A series of Web pages can be grouped into a Web site—a sort of virtual "place" in **cyberspace**. Every day, thousands of people shop at Nordstrom's[1] Web site, an online department store featuring clothing, shoes, and **jewelry**[2]. Thousands of people visit the Webopedia[3] Web site to look up the meaning of computer terms. At the ABC[4] News Web site, people not only read about the latest news, sports, and weather, but also discuss current issues with other readers. The Web **encompasses** these and hundreds of thousands of other sites.

Web sites are **hosted** by **corporate**, government, college, and private computers all over the world. The computers and software that store and distribute Web pages are called **Web servers**.

Every Web page has a unique address called a URL (**Uniform Resource Locator**). For example, the URL for the Cable News Network[5] Web site is *http://www.cnn.com*. Most URLs begin with http:// to indicate the Web's standard communications protocol. When typing a URL, the http:// can usually be omitted, so *www.cnn.com* works just as well as *http://www.cnn.com*.

Most Web sites have a **main page** that acts as a "**doorway**" to the rest of the pages at the site. This main page is sometimes referred to as a "**home page**[6]", although this term has another meaning. The URL for a Web site's main page is typically short and to the point, like *www.cnn.com*.

The pages for a Web site are typically stored in topic area folders, which are reflected in the URL. For example, the CNN site might include a weather center at *www.cnn.com/weather/* and an entertainment **desk** at *www.cnn.com/showbiz/*. A series of Web pages are then grouped under the appropriate topic. For example, you might find a page about hurricanes at the URL *www.cnn.com/weather/hurricanes.html*, and you could find a page

**cyberspace** / 'saibəspeis /
n. 计算机空间，网络空间
**jewelry** / 'dʒu:əlri /
n. 〈美〉[总称] 珠宝，首饰
**encompass** / in'kʌmpəs /
v. 包含，包括
**host** / həust /
v. 作…的主机
**corporate** / 'kɔ:pərit /
a. 公司的；社团的
**Web server**
万维网服务器
**locator** / ləu'keitə; 'ləukeitə /
n. 定位器，定位符
**Uniform Resource Locator**
统一资源定位符，统一资源定位器
**main page**
主页
**doorway** / 'dɔ:wei /
n. 出入口，门口
**home page** 主页
**desk** / desk /
n. 服务台；部门
**showbiz** / 'ʃəubiz /
n.〈口〉娱乐性行业；娱乐界（= show business）

1 *Nordstrom*：诺德斯特龙公司，高档连锁百货店，创立于1901年，总部位于美国华盛顿州西雅图市。
2 *an online department store featuring clothing, shoes, and jewelry*：一家主要销售衣服、鞋子和珠宝的在线百货公司。
3 *Webopedia*：网络百科全书，与计算机和互联网技术有关的在线词典和搜索引擎。
4 *ABC*：美国广播公司（*A*merican *B*roadcasting *C*ompany 的首字母缩略）。
5 *Cable News Network*：（美国）有线电视新闻网，首字母缩略词为 CNN。
6 *home page*：既可指一家网站的主页（网站的首页），也可指为浏览器设的主页（每次打开浏览器时所显示的网页）。

about El Niño[1] at *www.cnn.com/weather/elnino.html*. The file name of a specific Web page always appears last in the URL—*hurricanes.html* and *elnino.html* are the names of two Web pages. Web page file names usually have an .htm or .html extension, indicating that the page was created with HTML (**Hypertext Markup Language**), a standard format for Web documents.

A URL never contains a **space**, even after a **punctuation mark**, so do not type any spaces within a URL. An **underline** character is sometimes used to give the appearance of a space between words, as in the URL *www.detroit.com/top_10.html*. Be sure to use the correct type of slash—always a **forward slash** (/)—and **duplicate** the URL's **capitalization** exactly. The servers that run some Web sites are case sensitive[2], which means that an uppercase letter is not the same as a **lowercase** letter. On these servers, typing *www.cmu.edu/Overview.html* (with an uppercase "O") will not locate the page that's actually stored as *www.cmu.edu/overview.html* (with a lowercase "o").

## II. Browsers

A Web browser—usually referred to simply as a browser—is a software program that runs on your computer and helps you access Web pages. Two of today's most popular browsers are Microsoft Internet Explorer[3] (IE) and Netscape Navigator (Navigator). A browser provides a sort of "window" in which it displays a Web page. The borders of the window contain a set of menus and controls to help you navigate from one Web page to another. Despite small **cosmetic** differences and some variations in terminology, Web browsers offer a remarkably similar set of features and capabilities.

A browser fetches and displays Web pages. Suppose that you want to view the Web page located at *www.e-course.com/boxer.html*. You enter the URL into a special **Address box** that's provided by your browser. When you press the **Enter key**, the browser contacts the Web server at *www.e-course.com* and requests the *boxer.html* page. The server sends your computer the data stored in *boxer.html*. This data includes two things: the information you want to view, and **embedded** codes, called **HTML tags**,

**Hypertext Markup Language**
超文本标记语言
**space** / speis /
n. 空格，空白
**punctuation mark**
标点符号
**underline**
/ ˈʌndəˌlain /
n. 下划线
**forward slash**
正斜杠
**duplicate**
/ ˈdjuːplikeit /
v. 复制；重复
**capitalization**
/ ˌkæpitəlaiˈzeiʃən; -liˈz /
n. 大写字母的使用
**lowercase**
/ ˈləuəˈkeis /
a.（字母）小写的
**cosmetic**
/ kɔzˈmetik /
a. 化妆用的；装饰性的；非实质性的
**boxer** / ˈbɔksə /
n. 拳击运动员，拳师
**Address box**
地址框
**Enter key**
回车键
**embedded**
/ imˈbedid /
a. 嵌入（式）的
**HTML tag**
HTML 标记

[1] *El Niño*：厄尔尼诺现象。El Niño 来自西班牙语，意为“圣婴”，即上帝之子。厄尔尼诺现象的基本特征是，赤道太平洋中东部海域大范围海水温度异常升高，海水水位上涨。该现象发生时，往往出现全球范围的灾害天气。

[2] *The servers that run some Web sites are case sensitive*：运行有些网站的服务器区分大小写。这里的 case 指“（字母的）大小写”，sensitive 表示“敏感的”，case sensitive 意指“对大小写敏感的”，即“区分大小写的”。

[3] *Internet Explorer*：IE 网页浏览器，IE 浏览器，微软公司推出的图形用户界面网页浏览器。

that tell your browser how to display it. The tags specify details such as the background color, the text color and size, and the placement of graphics.

**Copy command**
复制命令

Most browsers provide a **Copy command** that allows you to copy a section of text from a Web page, which you can then paste into one of your own documents. To keep track of the source for each text section, you can use the Copy command to record the Web page's URL from the Address box, and then paste the URL into your document.

## III. Search Engines

The term search engine popularly refers to a Web site that provides a variety of tools to help you find information. Search engines such as Google and Yahoo![1] are indispensable tools when it comes to finding information on the Web. Depending on the search engine you use, you can find information by entering a description, filling out a form, or clicking a series of links to drill down through a list of topics and subtopics. Based on your input, the search engine displays a list of Web pages like the one shown in Figure 8C-1.

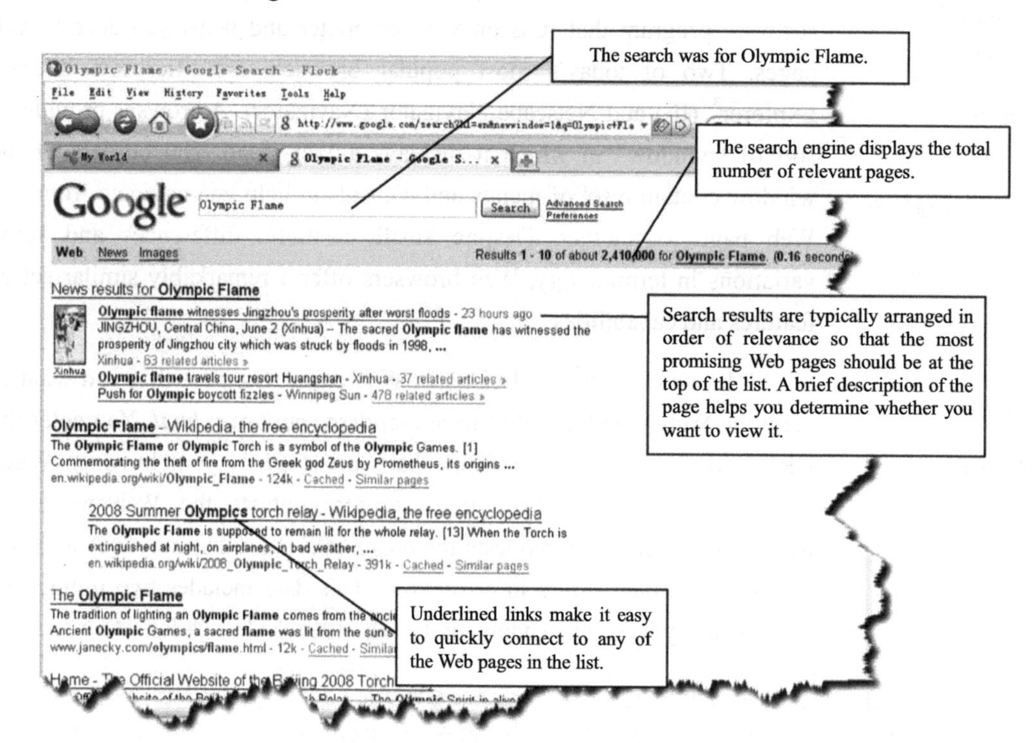


Figure 8C-1: In response to a query, a search engine produces a list of relevant Web pages, along with a brief description of each page and a link to it

[1] *Yahoo!*: 雅虎，网上搜索引擎。同名公司创立于 1995 年，总部位于美国加利福尼亚州森尼韦尔(Sunnyvale)。

A query describes the information you want to find. It includes one or more keywords and can also include search **operators**. A keyword (sometimes called a "search term") is any word that describes the information you're trying to find. You can enter more than one search term. Separate each term with a space or a search operator.

**operator** / 'ɔpəreitə /
n.（运）算子，算符

Search engines have a tendency to **inundate** you with possibilities—often finding thousands of potentially relevant Web pages. To receive a more manageable list of results, you need to **formulate** a more specific search. A search operator is a word or symbol that describes a relationship between keywords and thereby helps you create a more focused query. The search operators that you can use with each search engine vary slightly. To discover exactly how to formulate a query for a particular search engine, refer to its Help pages. Most search engines allow you to formulate queries with the search operators described in Figure 8C-2.

**inundate** / 'inəndeit /
v. 淹没；（似洪水般）布满
**formulate** / 'fɔːmjuleit /
v. 构想出；系统地阐述

| | |
|---|---|
| AND | When two search terms are joined by AND, both terms must appear on a Web page before it can be included in the search results. The query *railroad AND cars* will locate pages that contain both the words "railroad" and "cars." Your search results might include pages containing information about old **railroad cars**, about railroad car construction, and even about railroads that haul automobiles ("cars"). Some search engines use the plus symbol (+) instead of the word AND. |
| OR | When two search terms are joined by OR, either one or both of the search words could appear on a page. Entering the query *railroad OR cars* produces information about railroad fares, railroad routes, railroad cars, automobile safety records, and even car **ferries**. |
| NOT | The keyword following NOT must not appear on any of the pages found by the search engine. Entering *railroad NOT cars* would tell the search engine to look for pages that include "railroad" but not the keyword "cars." In some search engines, the minus sign (−) can be used instead of the word NOT. |
| Quotation Marks | Surrounding a series of keywords with **quotation marks** indicates that the search engine must treat the words as a phrase. The complete phrase must exist on a Web page for it to be included in the list of results. Entering *"green card"* would indicate that you are looking for information on **immigration**, not information on the color green, golf **greens**, or **greeting cards**. |
| NEAR | The NEAR operator tells a search engine that you want documents in which one of the keywords is located close to but not necessarily next to the other keyword. The query *library NEAR/15 congress* means that the words "library" and "congress" must appear within 15 words of each other. Successful searches could include documents containing phrases such as "Library of Congress" or "Congress funds special library research." |

**railroad car**
〈美〉火车车厢

**ferry** / 'feri /
n. 渡船；摆渡；渡口

**quotation marks**
[复]（一对）引号
**immigration**
/ ˌimi'greiʃən /
n. 移民，移居
**green** / griːn /
n.（高尔夫）球穴区；高尔夫球场
**greeting card**
贺卡

**wildcard**
/ 'waildkaːd /
n. 通配符，万能符
**wildcard character**
通配符
**derivation**
/ ˌderi'veiʃən /
n. 派生（物）
**medic** / 'medik /
n.〈口〉医生；医科学生
**medication**
/ ˌmedi'keiʃən /
n. 药物治疗；药物
**medicinal**
/ me'disinəl /
a.（医）药的；药用的；有疗效的
**backcountry**
/ 'bækˌkʌntri /
a. 偏僻乡村的
**recipe** / 'resipi /
n. 烹饪法，食谱；处方
**colon** / 'kəulən /
n. 冒号

| | |
|---|---|
| Wildcards | The asterisk (*) is sometimes referred to as a "**wildcard character**." It allows a search engine to find pages with any **derivation** of a basic word. For example, the query *medic** would not only produce pages containing the word "**medic**," but also "medical," "**medication**," "medicine," and "**medicinal**." |
| Field Searches | Some search engines allow you to search for a Web page by its title or by any part of its URL. The query *T:Backcountry Recipe Book* indicates that you want to find a specific Web page titled "**Backcountry Recipe** Book." In this search, the *T:* tells the search engine to look at Web page titles, and the information following the **colon** identifies the name of the title. |

**Figure 8C-2: Search Operators**

A topic directory is a list of topics and subtopics, such as Arts, Business, Computers, and so on, which are arranged in a hierarchy (Figure 8C-3). The top level of the hierarchy contains general topics. Each successive level of the hierarchy contains increasingly specific subtopics[1]. A topic directory might also be referred to as a "category list," an "index," or a "directory." To use a topic directory, simply click a general topic. When a list of subtopics appears, click the one that's most relevant to the information you are trying to locate. If your selection results in another list of subtopics, continue to select the most relevant one until the search engine presents a list of Web pages. You can then link to these pages just as though you had used a keyword query.

**targeted** / 'tɑːgitid /
a. 有目标的；指向目标的

Many search engines provide an advanced search form that helps you formulate a very **targeted** search. A search form helps you enter complex queries. It might also allow you to search for pages that are written in a particular language, located on a specific Web server, or created within a limited range of dates. It is usually accessible by clicking an Advanced Search link, which often is located on the main page of the search engine Web site.

Instead of entering a cryptic query such as *movie+review+"The Producers"*, wouldn't it be nice to enter a more straightforward question like *Where can I find a review of The Producers?* A few search engines specialize in natural language queries, which accept questions written in plain English.

[1] *Each successive level of the hierarchy contains increasingly specific subtopics.*：分层结构顶层下面依次相继的各层含有越来越具体的子主题。

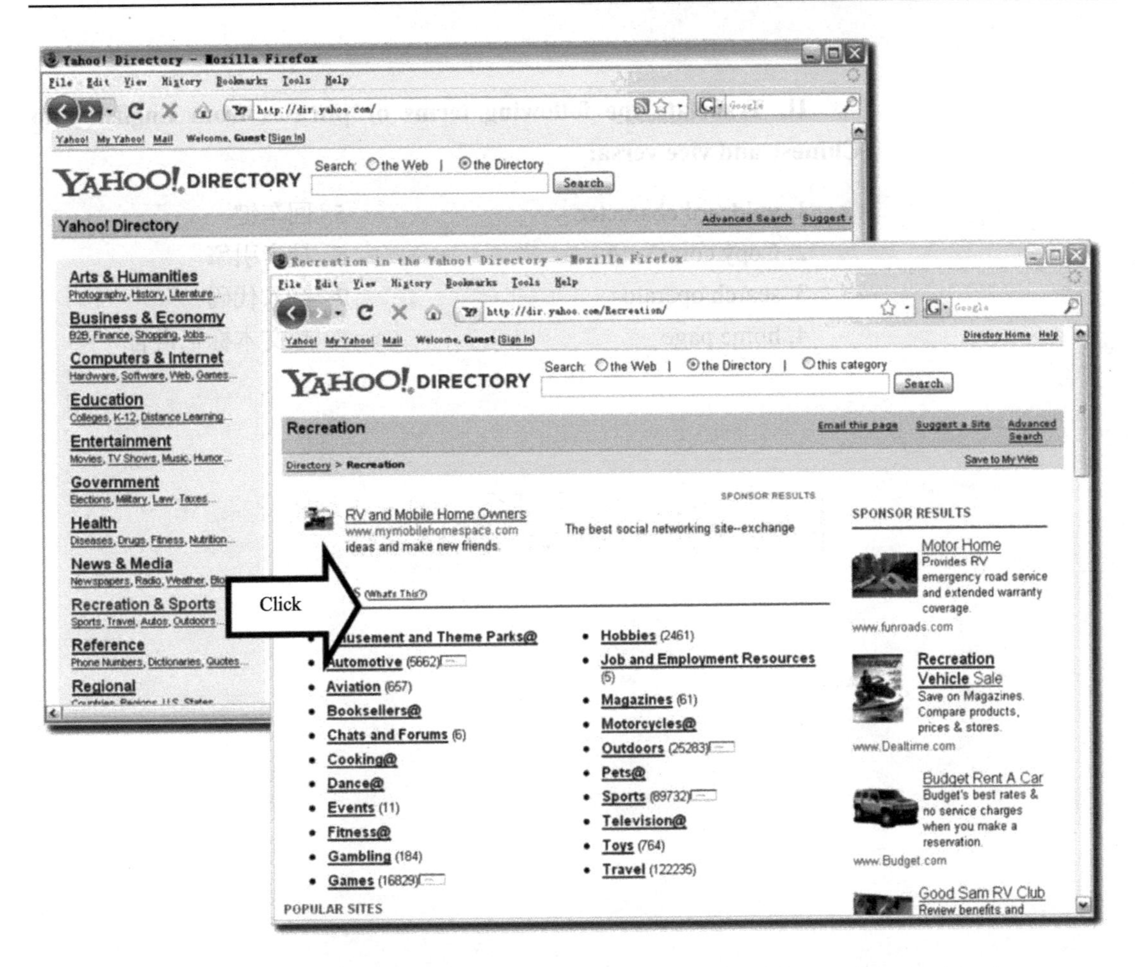


**Figure 8C-3: To use a topic directory, simply click a general topic. When a list of subtopics appears, click the one that's most relevant to the information you are trying to locate**

---

## Exercises

**I. Fill in the blanks with the information given in the text:**

1. Most Web pages contain links, sometimes called ________, to related documents and media files.

2. The main page of a Web site is sometimes called its ________ page.

3. HTML ________ are sets of instructions inserted into an HTML document to provide formatting and display information to a Web browser.

4. When you use a search engine, a query includes one or more ________ and can also include search ________.

**II. Translate the following terms or phrases from English into Chinese and vice versa:**

1. wildcard character
2. Copy command
3. search operator
4. home page
5. 回车键
6. 搜索引擎
7. 嵌入代码
8. 超文本标记语言

# Unit 9 Mobile and Cloud Computing

# （移动与云计算）

## Section A

## Cloud Computing

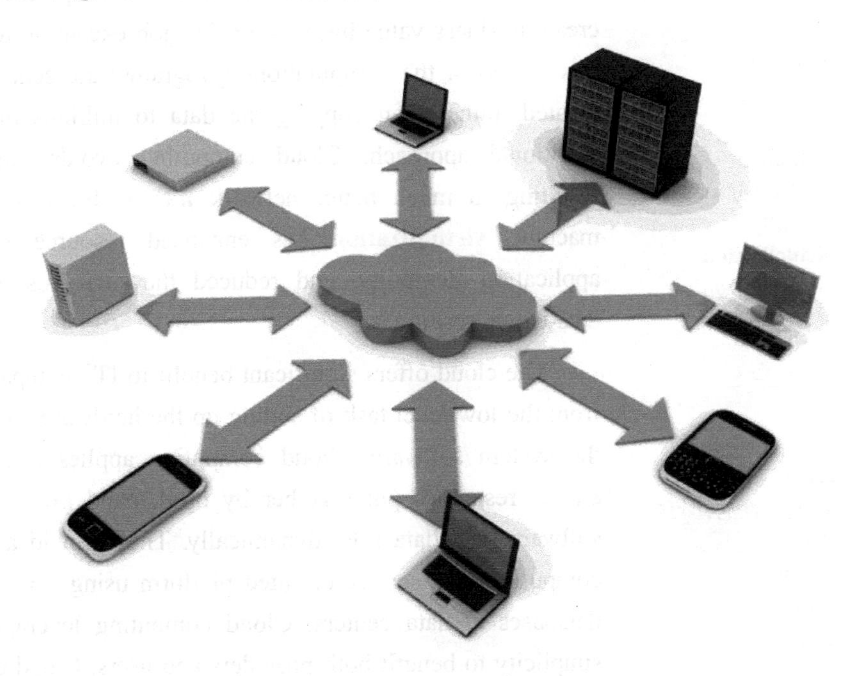

**cloud computing**
云计算
**virtualize**
/ ˈvəːtʃuəlaiz /
v. 虚拟化
**workload**
/ ˈwəːkləud /
n. 工作量；工作负荷
**batch** / bætʃ /
n. 批，批量，成批
**backend** / ˈbækend /
n. & a. 后端（的）
**grid** / grid /
n. 网格，格网
**leverage**
/ ˈliːvəridʒ; ˈle- /
v. 充分利用

### I. Introduction

**Cloud computing** has been defined differently by many users and designers. IBM, a major player in cloud computing, has defined it as follows: "*A cloud is a pool of **virtualized** computer resources. A cloud can host a variety of different **workloads**, including **batch**-style **backend** jobs and interactive and user-facing applications*[1]."

The concept of cloud computing has evolved from cluster, **grid**, and utility computing[2]. Cluster and grid computing **leverage** the use of many

---

1 *"A cloud is a pool of virtualized computer resources. A cloud can host a variety of different workloads, including batch-style backend jobs and interactive and user-facing applications."*：一个云是一个虚拟化计算机资源池。云可托管各种不同的工作负荷，包括批处理的后端作业和面向用户的交互式应用。

2 *The concept of cloud computing has evolved from cluster, grid, and utility computing.*：云计算这个概念从集群计算、网格计算和效用计算发展而来。

computers in parallel to solve problems of any size. Utility and Software as a Service (SaaS[1]) provide computing resources as a service with the notion of pay per use. Cloud computing leverages dynamic resources to deliver large numbers of services to end users. Cloud computing is a high-throughput computing (HTC[2]) paradigm **whereby** the infrastructure provides the services through a large data center or **server farms**. The cloud computing model enables users to share access to resources from anywhere at any time through their connected devices.

The cloud will free users to focus on user application development and create business value by **outsourcing** job execution to cloud providers. In this **scenario**, the computations (programs) are sent to where the data is located, rather than copying the data to millions of desktops as in the traditional approach. Cloud computing avoids large data movement, resulting in much better network **bandwidth utilization**. Furthermore, machine **virtualization** has enhanced resource utilization, increased application flexibility, and reduced the total cost of using virtualized data-center resources.

The cloud offers significant benefit to IT[3] companies by freeing them from the low-level task of setting up the hardware (servers) and managing the system software. Cloud computing applies a virtual platform with **elastic** resources put together by on-demand **provisioning** of hardware, software, and data sets, dynamically. The main idea is to move desktop computing to a service-oriented platform using **server clusters** and huge databases at data centers. Cloud computing leverages its low cost and simplicity to benefit both providers and users. Cloud computing intends to leverage multitasking to achieve higher throughput by serving many heterogeneous applications, large or small, simultaneously.

## II. Public, Private, and Hybrid Clouds

A *public cloud* is built over the Internet and can be accessed by any user who has paid for the service. Public clouds are owned by service providers and are accessible through a **subscription**. The **callout box** in top of Figure 9A-1 shows the architecture of a typical public cloud. Many public clouds are available, including Google App Engine[4] (GAE), Amazon

**whereby** / hwɛə'bai /
ad. 靠那个；借以
**server farm**
服务器农场
**outsource** / aut'sɔːs /
v.（将…）外包
**scenario** / si'nɑːriəu /
n. 方案；情况；脚本
**bandwidth** / 'bændwidθ /
n. 带宽
**utilization** / ˌjuːtilai'zeiʃən; -li'z- /
n. 利用
**virtualization** / vəːtʃuəlai'zeiʃən /
n. 虚拟化
**elastic** / i'læstik /
a. 有弹性的；可伸缩的
**provision** / prə'viʒən /
v. 供应，提供
**server cluster**
服务器集群
**hybrid** / 'haibrid /
a. 混合的
**subscription** / səb'skripʃən /
n. 订购；订阅
**callout** / 'kɔːlaut /
n. 标注
**callout box**
标注框

1 *SaaS*：软件即服务（*S*oftware *a*s *a* *S*ervice 的首字母缩略）。
2 *HTC*：高吞吐（量）计算（*h*igh-*t*hroughput *c*omputing 的缩略）。
3 *IT*：信息技术（*i*nformation *t*echnology 的首字母缩略）。
4 *Google App Engine*：谷歌应用引擎，一种平台即服务的云计算平台，可让用户在 Google 的基础架构上运行其网络应用程序。

**aforementioned**
/ əˈfɔːˌmenʃənd /
a. 前面提到的

**frontend**
/ ˈfrʌntˈend /
n. & a. 前端（的）

Web Services[1] (AWS), Microsoft Azure[2], IBM Blue Cloud[3], and Salesforce.com's Force.com[4]. The providers of the **aforementioned** clouds are commercial providers that offer a publicly accessible remote interface for creating and managing VM[5] instances within their proprietary infrastructure. A public cloud delivers a selected set of business processes. The application and infrastructure services are offered on a flexible price-per-use basis.

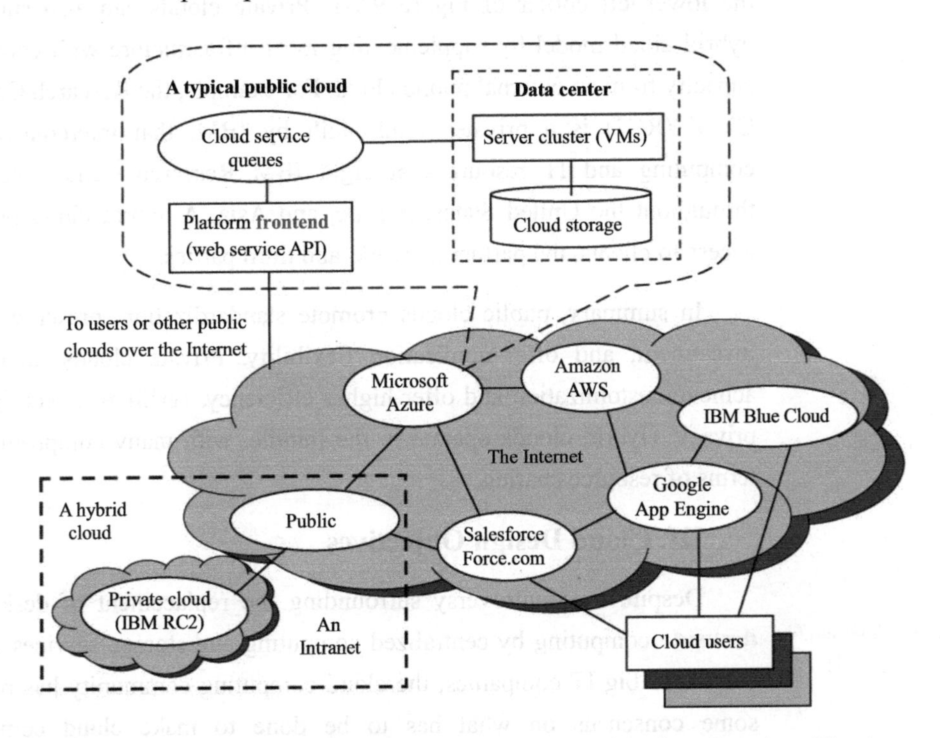


**Figure 9A-1: Public, private, and hybrid clouds illustrated by functional architecture and connectivity of representative clouds**

**agile** / ˈædʒaɪl /
a. 敏捷的；灵活的

A *private cloud* is built within the domain of an intranet owned by a single organization. Therefore, it is client owned and managed, and its access is limited to the owning clients and their partners. Its deployment was not meant to sell capacity over the Internet through publicly accessible interfaces. Private clouds give local users a flexible and **agile** private

1 *Amazon Web Services*：亚马逊网络服务，亚马逊公司提供的云计算服务。

2 *Microsoft Azure*：即 Windows Azure，微软提供的一个开放而灵活的云平台，用户通过该平台可在微软管理的数据中心的全球网络中快速生成、部署和管理应用程序。

3 *IBM Blue Cloud*：国际商用机器公司（IBM）的“蓝云”云计算解决方案。

4 *Salesforce.com's Force.com*：企业云计算公司 Salesforce.com 的 Force.com 网站，该网站拓展了该公司在企业云计算领域的领导地位。Salesforce 是创立于 1999 年的一家客户关系管理（CRM）软件服务提供商，总部位于美国加利福尼亚州旧金山。Salesforce 译作“软件营销部队”或“软营”。

5 *VM*：虚拟机（virtual machine 的首字母缩略）。

infrastructure to run service workloads within their administrative domains. A private cloud is supposed to deliver more efficient and convenient cloud services. It may impact the cloud standardization, while retaining greater **customization** and organizational control. Intranet-based private clouds are linked to public clouds to get additional resources.

**customization** /kʌstəmai'zeiʃən/
n. 定制，用户化

A *hybrid cloud* is built with both public and private clouds, as shown at the lower-left corner of Figure 9A-1. Private clouds can also support a hybrid cloud model by supplementing local infrastructure with computing capacity from an external public cloud. For example, the Research Compute Cloud (RC2) is a private cloud, built by IBM, that interconnects the computing and IT resources at eight IBM Research Centers scattered throughout the United States, Europe, and Asia. A hybrid cloud provides access to clients, the partner network, and third parties.

In summary, public clouds promote standardization, preserve capital investment, and offer application flexibility. Private clouds attempt to achieve customization and offer higher efficiency, **resiliency**, security, and privacy. Hybrid clouds operate in the middle, with many compromises in terms of resource sharing.

**resiliency** /ri'ziliənsi/
n. 弹性；复原力

## III. Cloud Design Objectives

Despite the controversy surrounding the replacement of desktop or **deskside** computing by centralized computing and storage services at data centers or big IT companies, the cloud computing community has reached some consensus on what has to be done to make cloud computing universally acceptable. The following list highlights six design objectives for cloud computing:

**deskside** /'desksaid/
a. 桌边（型）的

- **Shifting computing from desktops to data centers:** Computer processing, storage, and software delivery is shifted away from desktops and local servers and toward data centers over the Internet.
- **Service provisioning and cloud economics:** Providers supply cloud services by signing service-level agreements (SLAs[1]) with consumers and end users. The services must be efficient in terms of computing, storage, and power consumption. Pricing is based on a **pay-as-you-go** policy.
- **Scalability in performance:** The cloud platforms and software and infrastructure services must be able to **scale** in performance as the

**pay-as-you-go** /'peiəzju'gəu/
n. & a. 即付即用（的）

**scale** /skeil/
v. 攀登；（阶梯）逐步升高

[1] *SLA*：服务水平（或级别、等级）协议（*s*ervice-*l*evel *a*greement 的缩略）。

number of users increases.

- **Data privacy protection:** Can you trust data centers to handle your private data and records? This concern must be addressed to make clouds successful as trusted services.
- **High quality of cloud services:** The Quality of Service (QoS[1]) of cloud computing must be standardized to make clouds **interoperable** among multiple providers.
- **New standards and interfaces:** This refers to solving the data **lock-in** problem associated with data centers or cloud providers. Universally accepted APIs and access protocols are needed to provide high portability and flexibility of virtualized applications.

**interoperable** / ˌintərˈɔpərəbl /
a. 可互操作的，能互用的
**lock-in** / ˈlɔkin /
n. 锁定；同步

## IV. Enabling Technologies for Clouds

Clouds are enabled by the progress in hardware, software, and networking technologies summarized in Table 9A-1.

| Technology | Requirements and Benefits |
|---|---|
| Fast platform deployment | Fast, efficient, and flexible deployment of cloud resources to provide dynamic computing environment to users |
| Virtual clusters on demand | Virtualized cluster of VMs provisioned to satisfy user demand and virtual cluster reconfigured as workload changes |
| **Multitenant** techniques | SaaS for distributing software to a large number of users for their simultaneous use and resource sharing if so desired |
| Massive data processing | Internet search and web services which often require massive data processing, especially to support **personalized** services |
| Web-scale communication | Support for e-commerce, distance education, **telemedicine**, **social networking**, digital government, and digital entertainment applications |
| Distributed storage | Large-scale storage of personal records and public **archive** information which demands distributed storage over the clouds |
| **Licensing** and **billing** services | License management and billing services which greatly benefit all types of cloud services in utility computing |

Table 9A-1: Cloud-Enabling Technologies in Hardware, Software, and Networking

**multitenant** / ˌmʌltiˈtenənt /
a. 多租户的
**personalize** / ˈpəːsənəlaiz /
v. 使个性化
**telemedicine** / ˈteliˌmedisin /
n. 远程医学
**social networking**
社交网络
**archive** / ˈɑːkaiv /
n.［常作~s］档案；档案馆，档案室
**licensing** / ˈlaisənsiŋ /
n. 发给许可证；许可，特许
**billing** / ˈbiliŋ /
n. 开（账）单；记账
**disk array**
磁盘阵列

These technologies play instrumental roles in making cloud computing a reality. Most of these technologies are mature today to meet increasing demand. In the hardware area, the rapid progress in multi-core CPUs, memory chips, and **disk arrays** has made it possible to build faster data

[1] *QoS*：服务质量（*Q*uality *o*f *S*ervice 的首字母缩略）。

centers with huge amounts of storage space. Resource virtualization enables rapid cloud deployment and disaster recovery. Service-oriented architecture (SOA) also plays a vital role.

Progress in providing SaaS, Web 2.0 standards, and Internet performance have all contributed to the emergence of cloud services. Today's clouds are designed to serve a large number of **tenants** over massive volumes of data. The availability of large-scale, distributed storage systems is the foundation of today's data centers. Of course, cloud computing is greatly benefitted by the progress made in license management and automatic billing techniques.

**tenant** / 'tenənt /
n. 租户，承租人

### V. A Generic Cloud Architecture

Figure 9A-2 shows a security-aware cloud architecture[1]. The Internet cloud is **envisioned** as a massive cluster of servers. These servers are provisioned on demand to perform collective web services or distributed applications using data-center resources. The cloud platform is formed dynamically by provisioning or deprovisioning servers, software, and database resources. Servers in the cloud can be physical machines or VMs. User interfaces are applied to request services. The provisioning tool **carves** out the cloud system to deliver the requested service[2].

**envision** / in'viʒən /
v. 想象；设想

**carve** / kɑːv /
v. 雕刻，刻

In addition to building the server cluster, the cloud platform demands distributed storage and accompanying services. The cloud computing resources are built into the data centers, which are typically owned and operated by a third-party provider. Consumers do not need to know the underlying technologies. In a cloud, software becomes a service. The cloud demands a high degree of trust of massive amounts of data retrieved from large data centers. We need to build a framework to process large-scale data stored in the storage system. This demands a distributed file system over the database system. Other cloud resources are added into a cloud platform, including storage area networks (SANs[3]), database systems, **firewalls**, and security devices. Web service providers offer special APIs that enable developers to exploit Internet clouds. Monitoring and metering units[4] are used to track the usage and performance of provisioned resources.

**firewall** / 'faiəwɔːl /
n. 防火墙

---

1 *Figure 9A-2 shows a security-aware cloud architecture.*：图9A-2显示了一种具有安全意识的云体系结构。

2 *The provisioning tool carves out the cloud system to deliver the requested service.*：服务提供工具塑造出云系统，来提供所请求的服务。carve out 意为“（雕）刻出”“开辟出”。

3 *SAN*：存储区域网（络）(*s*torage *a*rea *n*etwork 的首字母缩略）。

4 *monitoring and metering units*：指“监控与计量设备”。

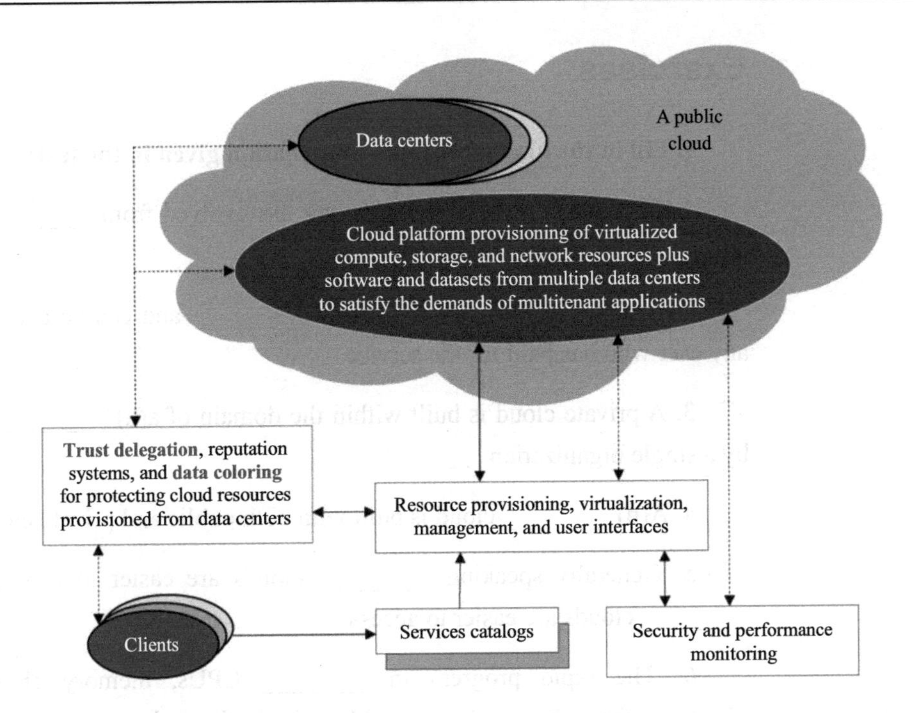


**Figure 9A-2: A security-aware cloud platform built with a virtual cluster of VMs, storage, and networking resources over the data-center servers operated by providers**

The software infrastructure of a cloud platform must handle all resource management and do most of the maintenance automatically. Software must detect the status of each node server joining and leaving, and perform relevant tasks accordingly. Cloud computing providers, such as Google and Microsoft, have built a large number of data centers all over the world. Each data center may have thousands of servers. The location of the data center is chosen to reduce power and cooling costs. Thus, the data centers are often built around **hydroelectric** power. The cloud physical platform builder is more concerned about the performance/price ratio and reliability issues than sheer speed performance.

**trust delegation** 信任委托
**data coloring** 数据着色

**hydroelectric** /ˌhaidrəuiˈlektrik/ a. 水力发电的；（电）由水力发的

In general, private clouds are easier to manage, and public clouds are easier to access. The trends in cloud development are that more and more clouds will be hybrid. This is because many cloud applications must go beyond the boundary of an intranet. One must learn how to create a private cloud and how to interact with public clouds in the open Internet. Security becomes a critical issue in **safeguarding** the operation of all cloud types.

**safeguard** /ˈseifgɑːd/ v. 保护，维护

## Exercises

**I. Fill in the blanks with the information given in the text:**

1. The concept of cloud computing has evolved from ________, grid, and ________ computing.

2. A public cloud is built over the ________ and can be accessed by any user who has paid for the service.

3. A private cloud is built within the domain of a(n) ________ owned by a single organization.

4. A(n) ________ cloud is built with both public and private clouds.

5. Generally speaking, ________ clouds are easier to manage, and ________ clouds are easier to access.

6. The rapid progress in ________ CPUs, memory chips, and ________ arrays has made it possible to build faster data centers with huge amounts of storage space.

7. One of the design objectives for cloud computing is shifting computing from ________ to data centers.

8. The Internet cloud is envisioned as a massive cluster of servers, which can be ________ machines or VMs.

**II. Translate the following terms or phrases from English into Chinese and vice versa:**

1. server farm
2. access protocol
3. storage area network
4. high-throughput computing
5. server cluster
6. public cloud
7. grid computing
8. security-aware cloud architecture
9. social networking
10. utility computing
11. 云计算提供商
12. 存储芯片
13. 基于内部网的私有云
14. 网络带宽
15. 混合云
16. 磁盘阵列
17. 软件即服务
18. 集群计算
19. 虚拟化计算机资源
20. 多核处理器

**III. Fill in each of the blanks with one of the words given in the following list, making changes if necessary:**

| | | | |
|---|---|---|---|
| *service* | *cloud* | *marketing* | *storage* |
| *automated* | *layer* | *resource* | *management* |
| *software* | *result* | *develop* | *platform* |
| *application* | *SaaS* | *foundation* | *serve* |

The architecture of a cloud is developed at three layers: infrastructure, platform, and application. The infrastructure ________ is deployed first to support IaaS (Infrastructure as a Service) services. This layer serves as the ________ for building the platform layer of the cloud for supporting PaaS (Platform as a Service) ________. In turn, the platform layer is a foundation for implementing the ________ layer for SaaS applications.

The infrastructure layer is built with virtualized compute, ________, and network resources. The abstraction of these hardware ________ is meant to provide the flexibility demanded by users. Internally, virtualization realizes ________ provisioning of resources and optimizes the infrastructure management process. The ________ layer is for general-purpose and repeated usage of the collection of ________ resources. This layer provides users with an environment to ________ their applications, to test operation flows, and to monitor execution ________ and performance. In a way, the virtualized cloud platform ________ as a "system *middleware* (中间件)" between the infrastructure and application layers of the ________. The application layer is formed with a collection of all needed software modules for ________ applications. Service applications in this layer include daily office ________ work. The application layer is also heavily used by enterprises in business ________ and sales, consumer relationship management, financial transactions, and supply chain management.

**IV. Translate the following passage from English into Chinese:**

Cloud computing applies a virtualized platform with elastic resources on demand by provisioning hardware, software, and data sets dynamically. The idea is to move desktop computing to a service-oriented platform using server clusters and huge databases at data centers. Cloud computing leverages its low cost and simplicity to benefit both users and providers. Machine virtualization has enabled such cost-effectiveness. Cloud computing intends to satisfy many user applications simultaneously. The cloud *ecosystem* (生态系统) must be designed to be secure, trustworthy, and dependable. Some computer users think of the cloud as a centralized resource pool. Others consider the cloud to be a server cluster which practices distributed computing over all the servers used.

Traditionally, a distributed computing system tends to be owned and operated by an *autonomous* (自主的) administrative domain (e.g., a research laboratory or company) for *on-premises* (机构内部的) computing needs. However, these traditional systems have encountered several performance *bottlenecks* (瓶颈): constant system maintenance, poor utilization, and increasing costs associated with hardware/software *upgrades* (升级). Cloud computing as an on-demand computing paradigm resolves or relieves us from these problems.

# Section B

# Fog and Edge Computing

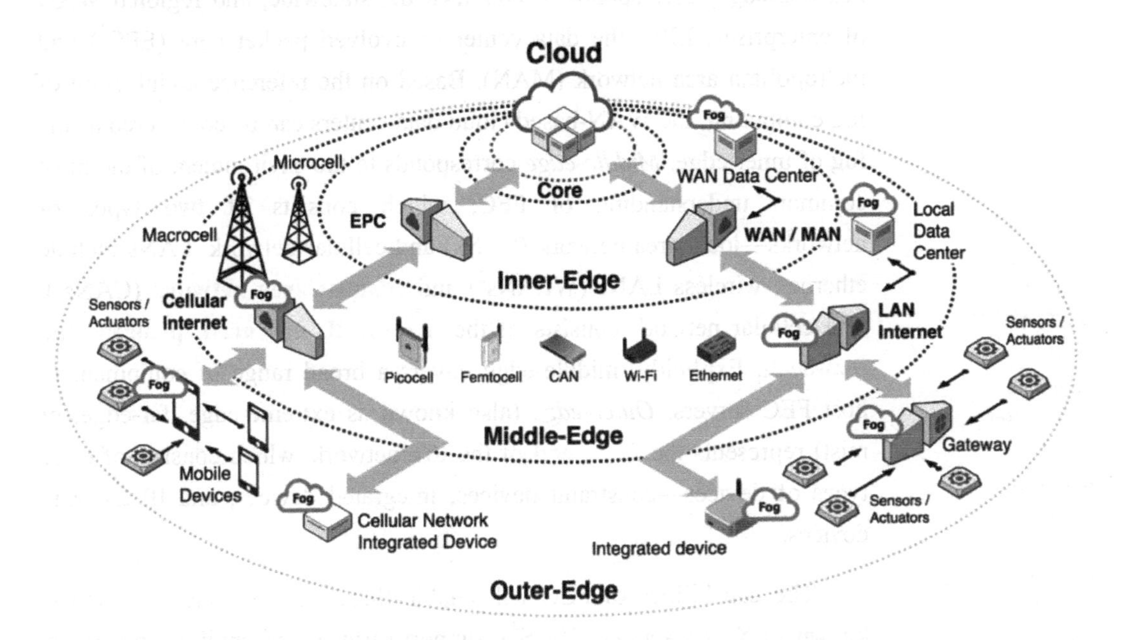


Figure 9B-1: Hierarchy of Fog and Edge Computing

## I. Introduction

**fog computing** 雾计算
**edge computing** 边缘计算
**latency** / ˈleitənsi / n. 潜伏；等待时间，时延，延迟

Commonly, an IoT [1] system follows the architecture of the Cloud-centric Internet of Things (CIoT[2]) in which the physical objects are represented in the form of Web resources that are managed by the servers in the global Internet. Although the CIoT model is a common approach to implement IoT systems, it is facing growing challenges, specifically in BLURS [3]—bandwidth, **latency**, uninterrupted, resource-constraint, and security. Fog and edge computing (FEC[4]) enhances CIoT by extending the cloud computing model to the edge networks of IoT where the network intermediate nodes such as routers, switches, hubs, and IoT devices are participating with the information-processing and decision-making toward

[1] *IoT*：物联网（*I*nternet *o*f *T*hings 的首字母缩略）。
[2] *CIoT*：以云为中心的物联网（*C*loud-centric *I*nternet *o*f *T*hings 的首字母缩略）。
[3] *BLURS*：带宽、时延、不间断、资源约束和安全性（*b*andwidth, *l*atency, *u*ninterrupted, *r*esource-constraint, and *s*ecurity 的首字母缩略）。
[4] *FEC*：雾与边缘计算（*f*og and *e*dge *c*omputing 的首字母缩略）。

**agility** / ə'dʒiliti /
n. 敏捷；机敏

improving security, cognition, **agility**, latency, and efficiency.

In general, from the perspective of central cloud in the core network, CIoT systems can deploy FEC servers at three edge layers–inner-edge, middle-edge, and outer-edge (see Figure 9B-1). *Inner-edge* (also known as near-the-edge) corresponds to countrywide, statewide, and regional WAN of enterprises, ISPs, the data center of evolved packet core (EPC[1]) and metropolitan area network (MAN). Based on the reference architecture of fog computing, the WAN-based cloud data centers can be considered as the fog of inner-edge. *Middle-edge* corresponds to the environment of the most common understanding of FEC, which consists of two types of networks—local area network (LANs) and cellular network. LANs include ethernet, wireless LANs (WLANs[2]) and **campus area network** (CANs[3]). The cellular network consists of the **macrocell**, **microcell**, **picocell**, and **femtocell**. Explicitly, middle-edge covers a broad range of equipment to host FEC servers. *Outer-edge* (also known as extreme-edge, far-edge, or mist) represents the **front-end** of the IoT network, which consists of three types of devices—constraint devices, integrated devices, and IP gateway devices.

**campus area network**
校园区域网络，校园网

**macrocell** / mækrəu'sel /
n. 宏蜂窝

**microcell** / maikrəu'sel /
n. 微蜂窝

**picocell** / pi:kəu'sel /
n. 微微蜂窝

**femtocell** / femtəu'sel /
n. 毫微微蜂窝

**front-end** / 'frʌnt'end /
n. 前端

**caching** / 'kæʃiŋ /
n. 缓存，高速缓存

The capabilities of FEC will enable three types of business models known as X as a service (XaaS[4]), support service, and application service. In summary, XaaS corresponds to the model that provides IaaS[5], PaaS[6], SaaS and S/CaaS[7] (storage or **caching** as a service), which are similar to existing cloud service models; support service corresponds to FEC software installation, configuration, and maintenance service that helps clients to set up their FEC on their own equipment; application service denotes the service providers that cater the complete solution that serves FEC mechanism to clients without them needing to configure their own FEC system[8].

---

1 *EPC*：演进分组核心（*e*volved *p*acket *c*ore 的首字母缩略）。
2 *WLAN*：无线局域网（*w*ireless *l*ocal *a*rea *n*etwork 的首字母缩略）。
3 *CAN*：校园区域网络，校园网（*c*ampus *a*rea *n*etwork 的首字母缩略）。
4 *XaaS*：X 即服务，一切皆服务（*X* *a*s *a* *s*ervice 的首字母缩略）。
5 *IaaS*：基础设施即服务（*i*nfrastructure *a*s *a* *s*ervice 的首字母缩略）。
6 *PaaS*：平台即服务（*p*latform *a*s *a* *s*ervice 的首字母缩略）。
7 *S/CaaS*：存储或缓存即服务（*s*torage *or* *c*aching *a*s *a* *s*ervice 的缩略）。
8 *application service denotes the service providers that cater the complete solution that serves FEC mechanism to clients without them needing to configure their own FEC system*：应用服务表示服务提供商为客户提供实现FEC机制的完整解决方案，而无须由客户配置自己的FEC系统。句中有两个that引导的定语从句，分别修饰providers和solution。此外，cater和serve在句中都表示“提供”之意。

## II. Advantages of FEC: SCALE

FEC offers five main advantages—security, cognition, agility, latency, and efficiency (SCALE[1]).

### 1. Security

FEC supports additional security to IoT devices to ensure safety and **trustworthiness** in transactions. For example, today's wireless sensors deployed in outdoor environments often require a remote wireless source code update in order to resolve the security-related issues. However, due to various dynamic environmental factors such as unstable signal strength, interruptions, constraint bandwidth etc., the distant central backend server may face challenges to perform the update swiftly and, hence, increases the chance of **cybersecurity** attack. On the other hand, if the FEC infrastructure is available, the backend can configure the best routing path among the entire network via various FEC nodes in order to rapidly perform the software security update to the wireless sensors.

**trustworthiness**
/ 'trʌst,wəːðinis /
n. 值得信任；可信；可靠

**cybersecurity**
/ ,saibəsi'kjuəriti /
n. 网络安全

### 2. Cognition

FEC enables the awareness of the objectives of clients toward supporting **autonomous** decision-making in terms of where and when to deploy computing, storage, and control functions[2]. Essentially, the awareness of FEC, which involves a number of mechanisms in terms of self-adaptation, self-organization, self-healing, self-expression, and so forth, shifts the role of IoT devices from passive to active smart devices that can continuously operate and react to customer requirements without relying on the decision from the distant Cloud.

**autonomous**
/ ɔː'tɔnəməs /
a.（独立）自主的；自治的

### 3. Agility

FEC enhances the agility of the large scope IoT system deployment. In contrast to the existing utility Cloud service business model, which relies on the large business holder to establish, deploy, and manage the fundamental infrastructure, FEC brings the opportunity to individual and small businesses to participate in providing FEC services using the common open

[1] *SCALE*：安全、认知、敏捷、低延迟和高效率（*s*ecurity, *c*ognition, *a*gility, *l*atency, and *e*fficiency 的首字母缩略）。

[2] *FEC enables the awareness of the objectives of clients toward supporting autonomous decision-making in terms of where and when to deploy computing, storage, and control functions.*：FEC 使客户能够意识到在何时何地部署计算、存储和控制功能，从而支持自主决策。

software interfaces or open Software Development Kits (SDKs[1]).

### 4. Latency

The common understanding of FEC is to provide rapid responses for the applications that require **ultra**-low latency. Specifically, in many **ubiquitous** applications and industrial automation, the system needs to collect and process the sensory data continuously in the form of the **data stream** in order to identify any event and to perform timely actions. Explicitly, by applying FEC, these systems are capable of supporting time-sensitive functions. Moreover, the softwarization feature of FEC, in which the behavior of physical devices can be fully configured by the distant central server using software abstraction, provides a highly flexible platform for rapid re-configuration of the IoT devices.

**ultra-** /ˈʌltrə/
pref. 表示"极端""超"

**ubiquitous** /juːˈbikwitəs/
a. 普遍存在的，无所不在的

**data stream**
数据流

### 5. Efficiency

FEC enhances the efficiency of CIoT in terms of improving performance and reducing the unnecessary costs. For example, by applying FEC, the ubiquitous **healthcare** or **eldercare** system can distribute a number of tasks to the Internet gateway devices of the healthcare sensors and utilize the gateway devices to perform the sensory data **analytics** tasks. Ideally, since the process happens near the data source, the system can generate the result much faster. Further, since the system utilizes gateway devices to perform most of the tasks, it highly reduces the unnecessary cost of outgoing communication bandwidth.

**healthcare** /ˈhelθkɛə/
n. 医疗保健

**eldercare** /ˈeldəkɛə/
n. 老年保健，老年护理

**analytics** /ˌænəˈlitiks/
n. 分析方法；分析学

## III. How FEC Achieves These Advantages: SCANC

These advantages are realized by the five basic mechanisms supported by FEC-enabled devices (FEC nodes), which can be termed as SCANC[2]—storage, compute, acceleration, networking, and control.

### 1. Storage

The mechanism of storage in FEC corresponds to the temporary data storing and caching at the FEC nodes in order to improve the performance of information or content delivery. For example, content service providers can perform **multimedia** content caching at the FEC nodes that are most close to their customers in order to improve the quality of experience.

**multimedia** /ˌmʌltiˈmiːdiə/
a.（使用）多媒体的

---

[1] *SDK*：软件开发工具包（*S*oftware *D*evelopment *K*it 的首字母缩略）。

[2] *SCANC*：存储、计算、加速、组网和控制（*s*torage, *c*ompute, *a*cceleration, *n*etworking, and *c*ontrol 的首字母缩略）。

Further, in connected vehicle scenarios, the connected vehicles can utilize the roadside FEC nodes to fetch and to share the information collected by the vehicles continuously.

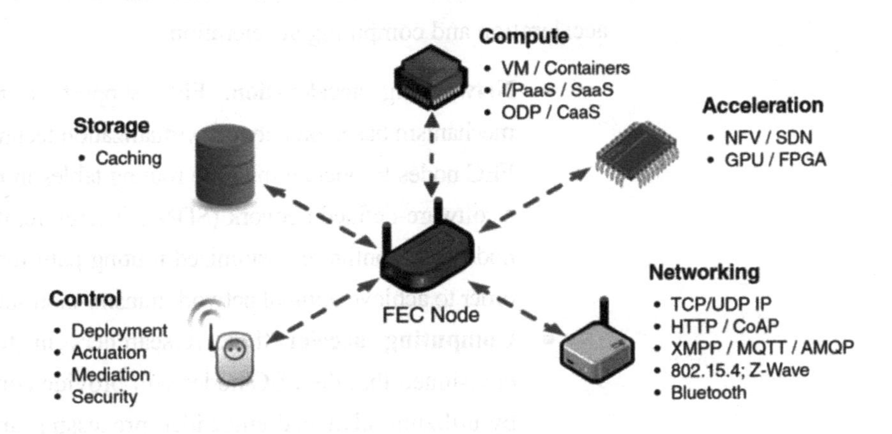


**Figure 9B-2: FEC nodes support five basic mechanisms—storage, compute, acceleration, networking, and control**

### 2. Compute

FEC nodes provide the computing mechanisms mainly in two models—infrastructure or platform as a service (I/PaaS[1]) and software as a service (SaaS). In general, FEC providers offer I/PaaS based on two approaches—**hypervisor virtual machines** (VMs) or containers engines (CEs[2]), which enable flexible platforms for FEC clients to deploy the **customized** software they need in a **sandbox** environment hosted in FEC nodes[3]. SaaS is also promising in FEC service provision. SaaS providers can offer two types of services—on-demand data processing (ODP[4]) and context as a service (CaaS[5]). Specifically, an ODP-based service has pre-installed methods that can process the data sent from the client in the request/response manner. Whereas, the CaaS-based service provides a customized data provision method in which the FEC nodes can collect and process the data to generate meaningful information for their clients.

**hypervisor**
/ ˈhaipəvaizə /
n.（系统）管理程序

**virtual machine**
虚拟机

**customized**
/ ˈkʌstəmaizd /
a. 定制的，用户化的

**sandbox**
/ ˈsændɔks /
n.（供儿童游戏的）沙池；沙箱，沙盒

1 *I/PaaS*：基础设施或平台即服务（*i*nfrastructure *or p*latform *as a s*ervice 的缩略）。
2 *CE*：容器引擎（*c*ontainers *e*ngine 的首字母缩略）。
3 *which enable flexible platforms for FEC clients to deploy the customized software they need in a sandbox environment hosted in FEC nodes*：这些方法可为 FEC 客户端提供灵活平台，用于在 FEC 节点中托管的沙箱环境下部署所需的定制软件。
4 *ODP*：按需数据处理（*o*n-demand *d*ata *p*rocessing 的首字母缩略）。
5 *CaaS*：上下文即服务（*c*ontext *as a s*ervice 的首字母缩略）。

### 3. Acceleration

FEC provides acceleration with a key concept—programmable. Fundamentally, FEC nodes support acceleration in two aspects—networking acceleration and computing acceleration.

- **Networking acceleration.** FEC supports a network acceleration mechanism based on network virtualization technology, which enables FEC nodes to operate multiple routing tables in parallel and to realize a software-defined network (SDN[1]). Therefore, the clients of the FEC nodes can configure customized routing path for their applications in order to achieve optimal network transmission speed.
- **Computing acceleration.** Researchers in fog computing have envisioned that the FEC nodes will provide computing acceleration by utilizing advanced embedded processing units such as graphics processing units (GPUs[2]) or field programmable gate arrays (FPGA[3]) units. Utilizing GPUs to enhance the process of complex algorithms has become a common approach in general cloud computing. FEC providers may also provide the equipment that contains middle- or high-performance independent GPUs. FPGA units allow users to redeploy program codes on them in order to improve or update the functions of the host devices.

### 4. Networking

Networking of FEC involves vertical and horizontal connectivities. Vertical networking interconnects things and cloud with the IP networks; whereas, horizontal networking can be heterogeneous in network signals and protocols, depending on the supported hardware specification of the FEC nodes.

- **Vertical networking.** FEC nodes enable the vertical network using IP network-based standard protocols. Specifically, the IoT devices can operate server-side functions that allow FEC nodes, which act as the **proxy** of cloud, to collect data from them and then forward the data to the cloud. FEC nodes can also operate as the message broker of publish-**subscribe**-based protocol that allows the IoT devices to publish data streams to the FEC nodes and enables the cloud backend to subscribe the data streams from the FEC nodes.

**proxy** /ˈprɔksi/
n. 代理（人）

**subscribe** /səbˈskraib/
v. 订阅；订购

[1] *SDN*：软件定义网络（*s*oftware-*d*efined *n*etwork 的首字母缩略）。
[2] *GPU*：图形处理单元，图形处理器（*g*raphics *p*rocessing *u*nit 的首字母缩略）。
[3] *FPGA*：现场可编程门阵列（*f*ield *p*rogrammable *g*ate *a*rray 的首字母缩略）。

- **Horizontal networking.** IoT systems often use heterogeneous cost-effective networking approaches. In particular, smart home, smart factories, and connected vehicles commonly utilize **Bluetooth**, ZigBee[1], and Z-Wave[2] on the IoT devices and connect them to an IP network gateway toward enabling the connectivity between the devices and the backend cloud. In general, the IP network gateway devices are the ideal entities to host FEC servers since they have the connectivity with the IoT devices in various signals. For example, the cloud can request that an FEC server hosted in a connected car communicate with the roadside IoT equipment using ZigBee in order to collect the environmental information needed for analyzing the real-time traffic situation.

**Bluetooth**
/ ˈbluːtuːθ /
n. 蓝牙（技术）

**5. Control**

The control mechanism supported by FEC consists of four basic types–deployment, **actuation**, **mediation**, and security:

**actuation**
/ ˌæktjuˈeiʃən, -tʃu- /
n. 驱动；启动
**mediation**
/ ˌmiːdiˈeiʃən /
n. 调解
**customizable**
/ ˈkʌstəmaizəbəl /
a. 可定制的

**Deployment control** allows clients to perform **customizable** software program deployment dynamically. Further, clients can configure FEC nodes to control which program the FEC node should execute and when it should execute it. Further, FEC providers can also provide a complete FEC network topology as a service that allows clients to move their program from one FEC node to another. Moreover, the clients may also control multiple FEC nodes to achieve the optimal performance for their applications.

**Actuation control** represents the mechanism supported by the hardware specification and the connectivities between the FEC nodes and the connected devices. Specifically, instead of performing direct interaction between the cloud and the devices, the cloud can delegate certain decisions to FEC nodes to directly control the behavior of IoT devices.

**Mediation control** corresponds to the capability of FEC in terms of interacting with external entities owned by different parties. In particular, the connected vehicles supported by different service providers can communicate with one another, though they may not have a common

1 *ZigBee*：译为“紫蜂”，是一种与蓝牙相类似的短距离、低速率、低功耗的无线网络技术，其先天性优势使其在物联网行业逐渐成为一种主流技术，在工业、农业、智能家居等领域得到大规模的应用。

2 *Z-Wave*：一种基于射频的、低成本、低功耗、高可靠、适于网络的短距离无线通信技术，系丹麦公司Zensys一手主导的无线组网规格。虽然Z-Wave联盟（Z-Wave Alliance）没有ZigBee联盟强大，但是其成员均是在智能家居领域拥有现成产品的厂商，并广泛分布在全球各地。

**interoperability**
/ ˈintər.ɔpərəˈbiliti /
n. 互操作性，互用性
**authentication**
/ ɔː.θentiˈkeiʃən /
n. 验证，鉴别
**authorization**
/ ˌɔːθəraiˈzeiʃən; -riˈz- /
n. 授权；委托

protocol initially. With the softwarization feature of FEC node, the vehicles can have on-demand software update toward enhancing their **interoperability**.

**Security control** is the basic requirement of FEC nodes that allows clients to control the **authentication**, **authorization**, identity, and protection of the virtualized runtime environment operated on the FEC nodes.

---

## Exercises

**I. Fill in the blanks with the information given in the text:**

1. From the perspective of central ________ in the core network, CIoT systems can deploy FEC servers at three edge layers, namely ________, middle-edge, and outer-edge.

2. The capabilities of FEC will enable three types of business models known as XaaS, ________ service, and application service.

3. SCALE stands for security, ________, agility, latency, and ________, which are the five main advantages offered by FEC.

4. The control mechanism supported by FEC consists of four basic types, which are ________, actuation, ________, and security.

**II. Translate the following terms or phrases from English into Chinese and vice versa:**

1. campus area network
2. fog and edge computing
3. X as a service
4. virtual machine
5. cybersecurity attack
6. 存储或缓存即服务
7. 按需数据处理
8. 软件定义网络
9. 图形处理单元
10. 软件开发工具包

# Section C
# Mobile Users

Mobile computers, such as laptop and handheld computers, are one of the fastest-growing segments of the computer industry. Their sales have already overtaken those of desktop computers. Why would anyone want one? People on the go[1] often want to use their mobile devices to read and send email, **tweet**, watch movies, download music, play games, or simply to **surf** the Web for information. They want to do all of the things they do at home and in the office. Naturally, they want to do them from anywhere on land, sea or in the air.

**tweet** / twiːt /
v. 上推特，发微博
**surf** / səːf /
v.（在…）冲浪

Connectivity to the Internet enables many of these mobile uses. Since having a wired connection is impossible in cars, boats, and airplanes, there is a lot of interest in wireless networks. Cellular networks operated by the telephone companies are one familiar kind of wireless network that **blankets** us with coverage for **mobile phones**. Wireless hotspots based on the 802.11 standard are another kind of wireless network for mobile computers. They have sprung up everywhere that people go, resulting in a **patchwork** of coverage at cafes, hotels, airports, schools, trains and planes. Anyone with a laptop computer and a wireless modem can just turn on their

**blanket** / ˈblæŋkit /
v. 用毯子（或毯状物）盖（或裹）；覆在…的上面
**mobile phone**
移动电话
**patchwork**
/ ˈpætʃwəːk /
n. 拼缀物；拼凑的东西，杂烩

[1] *on the go*：〈口〉忙个不停。这里指“在旅行中”。

computer and be connected to the Internet through the hotspot, as though the computer were plugged into a wired network.

Wireless networks are of great value to fleets of trucks, taxis, delivery vehicles, and repairpersons for keeping in contact with their home base. For example, in many cities, taxi drivers are independent businessmen, rather than being employees of a taxi company. In some of these cities, the taxis have a display the driver can see. When a customer calls up, a central **dispatcher** types in the **pickup** and destination points. This information is displayed on the drivers' displays and a **beep** sounds. The first driver to hit a button on the display gets the call.

**dispatcher** / dis'pætʃə /
n.（车辆）调度员
**pickup** / 'pikʌp /
n. 搭车；接人；提货
**beep** / bi:p /
n. 短促而尖厉的声音，嘟

Wireless networks are also important to the military. If you have to be able to fight a war anywhere on Earth at short notice, counting on using the local networking infrastructure is probably not a good idea. It is better to bring your own.

Although wireless networking and mobile computing are often related, they are not identical, as Figure 9C-1 shows. Here we see a distinction between fixed wireless and mobile wireless networks. Even notebook computers are sometimes wired. For example, if a traveler plugs a notebook computer into the wired network **jack** in a hotel room, he has mobility without a wireless network.

**jack** / dʒæk /
n. 插座；插口

| Wireless | Mobile | Typical Applications |
|---|---|---|
| No | No | Desktop computers in offices |
| No | Yes | A notebook computer used in a hotel room |
| Yes | No | Networks in unwired buildings |
| Yes | Yes | Store **inventory** with a handheld computer |

**Figure 9C-1: Combinations of Wireless Networks and Mobile Computing**

**inventory**
/ 'invəntəri; -tɔ:ri /
n. 存货（清单），库存

Conversely, some wireless computers are not mobile. In the home, and in offices or hotels that lack suitable **cabling**, it can be more convenient to connect desktop computers or media players wirelessly than to install wires. Installing a wireless network may require little more than[1] buying a small box with some electronics in it, unpacking it, and plugging it in. This solution may be far cheaper than having workmen put in cable **ducts** to wire the building.

**cabling** / 'keibliŋ /
n. 电缆

**duct** / dʌkt /
n. 管道；导管

Finally, there are also true mobile, wireless applications, such as

[1] *little more than*：只是…而已；只有，仅仅。

people walking around stores with handheld computers recording inventory. At many busy airports, car **rental** return clerks work in the **parking lot** with wireless mobile computers. They scan the **barcodes** or RFID[1] chips of returning cars, and their mobile device, which has a built-in printer, calls the main computer, gets the rental information, and prints out the bill on the spot.

**rental** / 'rentəl /
n. 租赁，出租
**parking lot**
〈美〉停车场
**barcode** / 'bɑ:kəud /
n. 条形码

Perhaps the key driver of mobile, wireless applications is the mobile phone. Text messaging or **texting** is tremendously popular. It lets a mobile phone user type a short message that is then delivered by the cellular network to another mobile **subscriber**. Few people would have predicted that having teenagers tediously typing short text messages on mobile phones would be an immense money maker for telephone companies. But texting (or **Short Message Service** as it is known outside the U.S.) is very **profitable** since it costs the **carrier** but a tiny fraction of one cent to **relay** a text message, a service for which they charge far more.

**texting** / 'tekstiŋ /
n. 发短信
**subscriber**
/ səb'skraibə /
n. 用户，订户
**Short Message Service**
短信（服务）
**profitable**
/ 'prɔfitəbl /
a. 有利可图的，有盈利的
**carrier** / 'kæriə /
n. 运输公司；运输工具；通信公司
**relay** / ri:'lei /
v. 接力传送；传递；转播

The **convergence** of telephones and the Internet has accelerated the growth of mobile applications. Smart phones, such as the popular iPhone, combine aspects of mobile phones and mobile computers. The cellular networks to which they connect can provide fast data services for using the Internet as well as handling phone calls. They can also connect to wireless hotspots too, and automatically switch between networks to choose the best option for the user.

**convergence**
/ kən'və:dʒəns /
n. 会聚；结合；收敛

Other **consumer electronics** devices can also use cellular and hotspot networks to stay connected to remote computers. **Electronic book readers** can download a newly purchased book or the next edition of a magazine or today's newspaper wherever they **roam**. **Electronic picture frames** can update their displays on cue[2] with fresh images.

**consumer electronics**
消费电子产品
**electronic book reader**
电子（图）书阅读器
**roam** / rəum /
v. 漫步；漫游
**electronic picture frame**
电子相框

Since mobile phones know their locations, often because they are equipped with GPS (Global Positioning System) receivers, some services are intentionally location dependent. Mobile maps and directions are an obvious candidate[3] as your GPS-enabled phone and car probably have a better idea of where you are than you do. So, too, are searches for a nearby bookstore or Chinese restaurant, or a local weather forecast. Other services

[1] *RFID*：射频识别（*r*adio-*f*requency *id*entification 的缩略）。
[2] *on cue*：恰好在这时候；恰好在得到提示的那一刻；接受了提示（或暗示）。在句中意思相当于“根据信号（或指令）”。
[3] *Mobile maps and directions are an obvious candidate*：移动地图与路线是显而易见的适选对象。

may record location, such as **annotating** photos and videos with the place at which they were made. This **annotation** is known as "geo-tagging."

An area in which mobile phones are increasingly used is m-commerce (mobile-commerce). They are used to make payments instead of cash and **credit cards**. When equipped with NFC[1] (Near Field Communication) technology, they can act as an RFID **smartcard** and interact with a nearby reader for payment. One huge thing that m-commerce has going for it[2] is that mobile phone users are accustomed to paying for everything (in contrast to Internet users, who expect everything to be free). If an Internet Web site charged a fee to allow its customers to pay by credit card, there would be an immense **howling** noise from the users. If, however, a mobile phone operator allowed its customers to pay for items in a store by waving the phone at the **cash register** and then **tacked** on a fee for this convenience, it would probably be accepted as normal. Time will tell[3].

No doubt, the uses of mobile and wireless computers will grow rapidly in the future as the size of computers shrinks, probably in ways no one can now foresee. Let us take a quick look at some possibilities. **Sensor networks** are made up of nodes that gather and wirelessly relay information they sense about the state of the physical world. The nodes may be part of familiar items such as cars or phones, or they may be small separate devices. For example, your car might gather data on its location, speed, vibration, and fuel efficiency from its **on-board diagnostic** system and **upload** this information to a database. Those data can help find **potholes**, plan trips around **congested** roads, and tell you if you are a "**gas guzzler**" compared to other drivers on the same stretch of road.

Sensor networks are revolutionizing science by providing a wealth of data on behavior that could not previously be observed. One example is tracking the migration of individual **zebras** by placing a small sensor on each animal. Researchers have packed a wireless computer into a cube 1 mm on edge[4]. With mobile computers this small[5], even small birds,

**annotate** /ˈænəuteit/
v. 给…做注解（或注释、评注）
**annotation** /ˌænəuˈteiʃən/
n. 注解，注释，评注
**credit card** 信用卡
**smartcard** /ˈsmɑːtkɑːd/
n. 智能卡
**howl** /haul/
v. 吼叫；怒吼
**cash register** 现金出纳机，收银机
**tack** /tæk/
v. 附加，追加（*on*）
**sensor network** 传感器网络，传感网，感知网
**on-board** /ˈɔnˈbɔːd, ˈɔːn-/
a. 车载的；机载的；舰载的
**diagnostic** /ˌdaiəgˈnɔstik/
a. 诊断的
**upload** /ˌʌpˈləud/
v. 上传，上载
**pothole** /ˈpɔthəul/
n. 路面凹坑，坑洞
**congest** /kənˈdʒest/
v. 拥挤；拥塞
**guzzler** /ˈgʌzlə/
n. 狂饮者；滥吃者；大量消耗者
**gas guzzler** 耗油量大的汽车，油老虎
**zebra** /ˈziːbrə, ˈze-/
n. 斑马

1 *NFC*：近场通信，近距离无线通信（Near Field Communication 的首字母缩略）。

2 *One huge thing that m-commerce has going for it*：移动电子商务拥有的一个巨大优势。that m-commerce has going for it 是一个定语从句，修饰 one huge thing，one huge thing 作 has 的宾语，it 指 m-commerce（相当于 m-commerce has one huge thing going for it）。

3 *Time will tell.*：时间会证明一切。

4 *Researchers have packed a wireless computer into a cube 1 mm on edge.*：研究人员已经将一个无线计算机装入一个边长为 1 毫米的立方体中。

5 *With mobile computers this small*：有了这么小的移动计算机。this 可作副词，与形容词或副词连用，表示“达到这样的程度”“这样”“这么”。

**rodents**, and insects can be tracked.

Even **mundane** uses, such as in **parking meters**, can be significant because they make use of data that were not previously available. Wireless parking meters can accept credit or **debit card** payments with instant verification over the wireless link. They can also report when they are in use over the wireless network. This would let drivers download a recent parking map to their car so they can find an available spot more easily. Of course, when a meter **expires**, it might also check for the presence of a car (by bouncing a signal off it) and report the **expiration** to parking **enforcement**[1].

Wearable computers are another promising application. Smart watches with radios have been part of our mental space since their appearance in the Dick Tracy[2] **comic strip** in 1946[3]; now you can buy them. Other such devices may be **implanted**, such as **pacemakers** and **insulin pumps**. Some of these can be controlled over a wireless network. This lets doctors test and reconfigure them more easily. It could also lead to some nasty problems if the devices are as insecure as the average PC and can be **hacked** easily.

---

**rodent** / 'rəudənt /
n. 啮齿目动物（如鼠、松鼠、河狸等）
**mundane**
/ mʌn'dein /
a. 世俗的；平凡的
**parking meter**
汽车停放计时器，汽车停放收费计
**debit** / 'debit /
n. 借方；借记，借入
**debit card**
借记卡，借方卡
**expire** / ik'spaiə /
v. 满期，届期
**expiration**
/ ˌekspi'reiʃən /
n. 满期，届期
**enforcement**
/ in'fɔːsmənt /
n. 实施；强制执行
**comic strip**（报刊上的）连环漫画
**implant**
/ im'plɑːnt; -'plænt /
v. 植入，埋置
**pacemaker**
/ 'peisˌmeikə /
n. 起搏器
**insulin**
/ 'insjulin; 'insə- /
n. 胰岛素（制剂）
**insulin pump**
胰岛素泵
**hack** / hæk /
v. 非法闯入（计算机网络），黑客攻击

## Exercises

**I. Fill in the blanks with the information given in the text:**

1. The sales of mobile computers have already overtaken those of ________ computers.

2. Perhaps the key driving force behind mobile, ________ applications is the mobile phone.

3. The ________ of sensor networks may be part of familiar items such as cars or phones, or they may be small separate devices.

4. Combining aspects of mobile ________ and mobile ________,

---

[1] *parking enforcement*：这里指停车执法者（或机构）。

[2] *Dick Tracy*：《迪克·特雷西》，一部美国长篇连载漫画，描绘虚构人物迪克·特雷西，一个智勇双全、枪法奇快神准的天才警探，与形形色色的歹徒斗智斗勇的传奇故事。由漫画家切斯特·古尔德（Chester Gould）创作，1931 年 10 月 4 日在《底特律镜报》（*Detroit Mirror*）上首次刊载。古尔德一直执笔创作《迪克·特雷西》到 1977 年。该漫画 1990 年被翻拍为电影，亦译为《至尊神探》。

[3] *Smart watches with radios have been part of our mental space since their appearance in the Dick Tracy comic strip in 1946*：带有收音机的智能手表自 1946 年出现在《迪克·特雷西》连环漫画中以来，一直是我们梦寐以求之物。

smart phones can use the Internet as well as make phone calls.

**II. Translate the following terms or phrases from English into Chinese and vice versa:**

1. notebook computer
2. wireless hotspot
3. Short Message Service
4. wearable computer
5. 移动电话
6. 条形码阅读器
7. 网站
8. 智能手机

# Unit 10 Computer Security

# （计算机安全）

## Section A

## Computer Security

### I. A Definition of Computer Security

The NIST[1] Computer Security Handbook defines *computer security* as follows:

The protection afforded to an automated information system in order to attain the applicable objectives of preserving the integrity, availability, and **confidentiality** of information system resources (including hardware, software, firmware,

**confidentiality**
/ ˈkɔnfiˌdenʃiˈæliti /
n. 机密性

---

1 *NIST*：（美国）国家标准与技术研究院（*N*ational *I*nstitute of *S*tandards and *T*echnology 的首字母缩略）。

information/data, and telecommunications).

This definition introduces three key objectives that are at the heart of computer security:

- **Confidentiality:** This term covers two related concepts:
  — **Data confidentiality:** Assures that private or **confidential** information is not made available or **disclosed** to unauthorized individuals.
  — **Privacy:** Assures that individuals control or influence what information related to them may be collected and stored and by whom and to whom that information may be disclosed[1].
- **Integrity:** This term covers two related concepts:
  — **Data integrity:** Assures that information and programs are changed only in a specified and authorized manner.
  — **System integrity:** Assures that a system performs its intended function in an **unimpaired** manner, free from deliberate or inadvertent unauthorized manipulation of the system.
- **Availability:** Assures that systems work promptly and service is not denied to authorized users.

These three concepts form what is often referred to as the CIA[2] **triad**. Although the use of the CIA triad to define security objectives is well established, some in the security field feel that additional concepts are needed to present a complete picture. Two of the most commonly mentioned are as follows:

- **Authenticity:** The property of being genuine and being able to be verified and trusted; confidence in the **validity** of a transmission, a message, or a message **originator**. This means verifying that users are who they say they are and that each input arriving at the system came from a trusted source.
- **Accountability:** The security goal that generates the requirement for actions of an entity to be traced uniquely to that entity. This supports **nonrepudiation**, **deterrence**, **fault isolation**, intrusion detection and prevention, and after-action recovery and legal action. Because truly secure systems aren't yet an achievable goal, we

**confidential** / ˌkɔnfiˈdenʃəl /
a. 秘密的，机密的
**disclose** / disˈkləuz /
v. 使显露；泄露，透露
**unimpaired** / ˌʌnimˈpɛəd /
a. 未受损的；未削弱的
**triad** / ˈtraiəd, -æd /
n. 三件一套（或组）；三位一体
**authenticity** / ˌɔːθenˈtisiti /
n. 可靠性，真实性
**validity** / vəˈliditi /
n. 有效（性）
**originator** / əˈridʒineitə /
n. 创始人；起源
**accountability** / əˌkauntəˈbiliti /
n. 负有责任
**nonrepudiation** / ˈnɔnriˌpjuːdiˈeiʃən /
n. 不拒绝；不否认；不可抵赖
**deterrence** / diˈterəns /
n. 威慑
**fault isolation**
故障隔离

[1] *Assures that individuals control or influence what information related to them may be collected and stored and by whom and to whom that information may be disclosed.*：确保个人能够控制或影响哪些与其有关的信息可被收集和存储，以及该信息可由谁透露和透露给谁。

[2] *CIA*：机密性、完整性与可用性（confidentiality, *i*ntegrity, and *a*vailability 的首字母缩略）。

must be able to trace a security **breach** to a responsible party. Systems must keep records of their activities to permit later **forensic** analysis to trace security breaches or to aid in transaction disputes.

**breach** / briːtʃ /
n. 破坏；违反
**forensic** / fəˈrensik /
a.（用于）法庭的；法医的

## II. Threats and Assets

The assets of a computer system can be categorized as hardware, software, data, and communication lines and networks. We will briefly describe these four categories and relate them to the concepts of integrity, confidentiality, and availability.

### 1. Hardware

A major threat to hardware is the threat to availability. Hardware is the most vulnerable to attack and the least **susceptible** to automated controls. Threats include accidental and deliberate damage to equipment as well as theft. The **proliferation** of personal computers and workstations and the widespread use of LANs increase the potential for losses in this area. Theft of CD-ROMs[1] and DVDs can lead to loss of confidentiality. Physical and administrative security measures are needed to deal with these threats.

**susceptible** / səˈseptəbl /
a. 易受影响的（*to*）
**proliferation** / prəuˌlifəˈreiʃən /
n. 激增；扩散

### 2. Software

A key threat to software is an attack on availability. Software, especially application software, is often easy to delete. Software can also be altered or damaged to render it useless. Careful software configuration management, which includes making backups of the most recent version of software, can maintain high availability. A more difficult problem to deal with is software modification that results in a program that still functions but behaves differently than before, which is a threat to integrity/authenticity. Computer viruses and related attacks fall into this category. A final problem is protection against software **piracy**. Although certain **countermeasures** are available, by and large[2] the problem of unauthorized copying of software has not been solved.

**piracy** / ˈpaiərəsi /
n. 侵犯版权，盗版，剽窃
**countermeasure** / ˈkauntəˌmeʒə /
n. 对策，对抗手段

### 3. Data

Security concerns with respect to data are broad, encompassing availability, **secrecy**, and integrity. In the case of availability, the concern is with the destruction of data files, which can occur either accidentally or

**secrecy** / ˈsiːkrəsi /
n. 秘密；保密

1 *CD-ROM*：只读光盘（存储器）（compact *d*isc *r*ead-*o*nly *m*emory 的首字母缩略）。
2 *by and large*：大体上，总的说来，一般地说。

maliciously.

The obvious concern with secrecy is the unauthorized reading of data files or databases, and this area has been the subject of perhaps more research and effort than any other area of computer security. A less obvious threat to secrecy involves the analysis of data and manifests itself in the use of so-called statistical databases, which provide summary or **aggregate** information. As the use of statistical databases grows, there is an increasing potential for **disclosure** of personal information. For example, if one table records the **aggregate** of the incomes of **respondents** A, B, C, and D and another records the aggregate of the incomes of A, B, C, D, and E, the difference between the two aggregates would be the income of E.

**aggregate**
/ 'ægrigət /
a. 聚集的；合计的
**disclosure**
/ dis'kləuʒə /
n. 泄露，透露
**aggregate**
/ 'ægrigət /
n. 总数，合计；聚集体
**respondent**
/ ri'spɔndənt /
n. 调查对象，(调查表的）回答者

Finally, data integrity is a major concern in most installations. Modifications to data files can have consequences ranging from minor to disastrous.

**4. Communication Lines and Networks**

Network security attacks can be classified as *passive attacks* and *active attacks*. A passive attack attempts to learn or make use of information from the system but does not affect system resources. An active attack attempts to alter system resources or affect their operation.

**Passive attacks** are in the nature of **eavesdropping** on, or monitoring of, transmissions. The goal of the attacker is to obtain information that is being transmitted. Two types of passive attacks are release of message contents and traffic analysis.

**eavesdrop**
/ 'i:vzdrɔp /
v. 偷听，窃听（***on***）

The **release of message contents** is easily understood. A telephone conversation, an electronic mail message, and a transferred file may contain sensitive or confidential information. We would like to prevent an opponent from learning the contents of these transmissions.

**Traffic analysis** is subtler. If we had encryption protection in place, an opponent might still be able to observe the pattern of messages. The opponent could determine the location and identity of communicating hosts and could observe the frequency and length of messages being exchanged. This information might be useful in guessing the nature of the communication that was taking place.

Passive attacks are very difficult to detect because they do not involve any alteration of the data. Thus, the emphasis in dealing with passive attacks is on prevention rather than detection.

**Active attacks** involve some modification of the data stream or the creation of a false stream and can be subdivided into four categories: replay, **masquerade**, modification of messages, and denial of service.

masquerade
/ ˌmæskəˈreid, ˌmɑːs- /
n. 假面舞会；伪装；假扮

**Replay** involves the passive capture of a data unit and its subsequent retransmission to produce an unauthorized effect.

A **masquerade** takes place when one entity pretends to be a different entity. A masquerade attack usually includes one of the other forms of active attack. For example, authentication sequences can be captured and replayed after a valid authentication sequence has taken place, thus enabling an authorized entity with few privileges to obtain extra privileges by **impersonating** an entity that has those privileges.

impersonate
/ imˈpəːsəneit /
v. 扮演；模仿；假冒

**Modification of messages** simply means that some portion of a **legitimate** message is altered, or that messages are delayed or reordered, to produce an unauthorized effect.

legitimate
/ liˈdʒitimit /
a. 合法的；正当的

The **denial of service** prevents or **inhibits** the normal use or management of communications facilities. This attack may have a specific target; for example, an entity may **suppress** all messages directed to a particular destination. Another form of service denial is the **disruption** of an entire network, either by disabling the network or by **overloading** it with messages so as to **degrade** performance.

inhibit / inˈhibit /
v. 抑制，约束

suppress / səˈpres /
v. 压制；抑制；阻止

disruption
/ disˈrʌpʃən /
n. 扰乱；中断

overload
/ ˌəuvəˈləud /
v. 使过载，使超载

degrade / diˈgreid /
v. 使降级，使退化

It is quite difficult to prevent active attacks absolutely, because to do so would require physical protection of all communications facilities and paths at all times. Instead, the goal is to detect them and to recover from any disruption or delays caused by them.

## III. Computer Security Strategy

A comprehensive security strategy involves three aspects:

- **Specification/policy:** What is the security scheme supposed to do?
- **Implementation/mechanisms:** How does it do it?
- **Correctness/assurance:** Does it really work?

### 1. Security Policy

In developing a security policy, a security manager needs to consider the following factors:

- The value of the assets being protected
- The **vulnerabilities** of the system
- Potential threats and the **likelihood** of attacks

Further, the manager must consider the following tradeoffs:

- **Ease of use versus security:** Virtually all security measures involve some penalty in the area of ease of use. For example, virus-checking software reduces available processing power and introduces the possibility of **system crashes** or malfunctions due to improper interaction between the security software and the operating system.
- **Cost of security versus cost of failure and recovery:** In addition to ease of use and performance costs, there are direct **monetary** costs in implementing and maintaining security measures. All of these costs must be balanced against the cost of security failure and recovery if certain security measures are lacking.

**vulnerability** /ˌvʌlnərəˈbiliti/
n. 易遭攻击的地方；脆弱（性）；漏洞
**likelihood** /ˈlaiklihud/
n. 可能，可能性
**versus** /ˈvəːsəs/
prep. 对；与…相对（或相比）
**system crash**
系统崩溃
**monetary** /ˈmʌnitəri/
a. 货币的；金钱的

**2. Security Implementation**

Security implementation involves four complementary courses of action:

- **Prevention:** There is a wide range of threats in which prevention is a reasonable goal. For example, if a secure encryption algorithm is used, and if measures are in place to prevent unauthorized access to **encryption keys**, then attacks on confidentiality of the transmitted data will be prevented.
- **Detection:** In a number of cases, absolute protection is not feasible, but it is practical to detect security attacks. For example, there are intrusion detection systems designed to detect the presence of unauthorized individuals **logged** onto a system.
- **Response:** If security mechanisms detect an ongoing attack, the system may be able to respond in such a way as to halt the attack and prevent further damage.
- **Recovery:** An example of recovery is the use of backup systems, so that if data integrity is compromised, a prior, correct copy of the data can be reloaded.

**encryption key**
（加密）密钥
**log** /lɔg/
v. 登录，注册，进入系统（*on, in*）

**3. Assurance and Evaluation**

Those who are "consumers" of computer security services and

mechanisms desire a belief that the security measures in place work as intended. That is, security consumers want to feel that the security infrastructure of their systems meet security requirements and enforce security policies. These considerations bring us to the concepts of assurance and evaluation.

**Assurance** is the degree of confidence one has that the security measures, both technical and operational, work as intended to protect the system and the information it processes. This encompasses both system design and system implementation. Thus, assurance deals with the questions, "Does the security system design meet its requirements?" and "Does the security system implementation meet its specifications?" Note that assurance is expressed as a degree of confidence, not in terms of a formal proof that a design or implementation is correct. With the present **state of the art**, it is very difficult if not impossible to move beyond a degree of confidence to absolute proof.

**state of the art**（学科、技术等当前的或某一时期的）发展水平，最新水平

**Evaluation** is the process of examining a computer product or system with respect to certain criteria. Evaluation involves testing and may also involve formal analytic or mathematical techniques. The central **thrust** of work in this area is the development of evaluation criteria that can be applied to any security system (encompassing security services and mechanisms) and that are broadly supported for making product comparisons[1].

**thrust** /θrʌst/ n. 要点，要旨；目标

# Exercises

**I. Fill in the blanks with the information given in the text:**

1. The CIA triad used to define computer security objectives consists of ________, integrity, and ________.

2. In computer security, integrity covers the two related concepts of ________ integrity and system integrity.

3. In addition to the CIA triad, two of the most commonly mentioned concepts concerning security objectives are ________ and accountability.

---

[1] *that are broadly supported for making product comparisons*：在进行产品比较方面得到广泛支持。evaluation criteria 后面跟有两个定语从句，这是其中之一。关系代词 that 在两个定语从句中均为主语，与谓语动词属于被动关系。

4. Security concerns with respect to data encompass availability, secrecy, and ________.

5. ________ network security attacks attempt to learn or make use of information from the system without affecting system resources, whereas ________ attacks attempt to alter system resources or affect their operation.

6. A key threat to software is an attack on ________.

7. In developing a security policy, a security manager must balance ________ of use against security, and cost of ________ against cost of failure and recovery.

8. In the context of computer security, ________ is the degree of confidence one has that the security measures work as intended.

**II. Translate the following terms or phrases from English into Chinese and vice versa:**

| | |
|---|---|
| 1. backup system | 11. 软件版本 |
| 2. encryption key | 12. 数据完整性 |
| 3. data confidentiality | 13. 系统崩溃 |
| 4. system vulnerability | 14. 病毒检查软件 |
| 5. unauthorized access | 15. 综合安全策略 |
| 6. intrusion detection system | 16. 软件配置管理 |
| 7. after-action recovery | 17. 故障隔离 |
| 8. software piracy | 18. 统计数据库 |
| 9. authorized user | 19. 保密的加密算法 |
| 10. data unit | 20. 数据流 |

**III. Fill in each of the blanks with one of the words given in the following list, making changes if necessary:**

| | | | |
|---|---|---|---|
| *database* | *control* | *allow* | *utility* |
| *user* | *native* | *resource* | *determine* |
| *authentication* | *system* | *operating* | *record* |
| *access* | *component* | *robust* | *function* |

An access control mechanism *mediates* (调解) between a user (or a process executing on behalf of a user) and system ________, such as applications, operating systems, firewalls, routers, files, and databases. The

________ must first *authenticate* (验证) a user seeking access. Typically the ________ function determines whether the user is permitted to ________ the system at all. Then the access control ________ determines if the specific requested access by this ________ is permitted. A security administrator maintains an authorization ________ that specifies what type of access to which resources is ________ for this user. The access control function consults this database to ________ whether to grant access. An *auditing* (审计) function monitors and keeps a ________ of user accesses to system resources.

In practice, a number of ________ may cooperatively share the access control function. All ________ systems have at least a *rudimentary* (基本的), and in many cases a quite ________, access control component. Add-on security packages can add to the ________ access control capabilities of the OS. Particular applications or ________, such as a database management system, also incorporate access ________ functions. External devices, such as firewalls, can also provide access control services.

**IV. Translate the following passage from English into Chinese:**

*Intruder* (入侵者) attacks range from the *benign* (良性的，温和的) to the serious. At the benign end of the scale, there are many people who simply wish to explore internets and see what is out there. At the serious end are individuals who are attempting to read privileged data, perform unauthorized modifications to data, or disrupt the system.

The objective of the intruder is to gain access to a system or to increase the range of privileges accessible on a system. Most initial attacks use system or software vulnerabilities that allow a user to execute code that opens a back door into the system. Intruders can get access to a system by exploiting attacks such as *buffer overflows* (缓存溢出) on a program that runs with certain privileges.

Alternatively, the intruder attempts to acquire information that should have been protected. In some cases, this information is in the form of a user password. With knowledge of some other user's password, an intruder can log in to a system and exercise all the privileges accorded to the legitimate user.

# Section B

## Antivirus Software

**antivirus software**
防病毒软件
**trojan** / 'trəudʒən /
n. 特洛伊木马（程序或病毒），木马
**worm** / wə:m /
n. 蠕虫（病毒）
**malware** / 'mælwεə /
n. 恶意软件
**infallible**
/ in'fæləbəl /
a. 不可能出错的；绝对可靠的

### I. Introduction

**Antivirus software** is a type of utility software that looks for and eliminates viruses, **trojans**, **worms**, and other **malware**. It is available for all types of computers and data storage devices, including smartphones, tablets, personal computers, USB flash drives, servers, PCs[1], and Macs[2].

Today's antivirus software is quite dependable but not **infallible**. A fast-spreading worm can reach your digital device before a virus definition update arrives, and **cloaking** software can hide some **viral exploits**.

Despite occasional misses, however, antivirus software and other security software modules are constantly weeding out[3] malware that would

[1] *PC*：这里指一种特定类型的个人计算机，即由最初的 IBM PC 机发展而来的个人计算机。
[2] *Mac*：麦金塔电脑，麦克机，苹果公司推出的个人计算机系列产品。Mac 系 Macintosh 的简称。
[3] *weed out*：清除；剔除；淘汰。

otherwise infect your device. It is essential to use security software, but it is also important to take additional precautions, such as making regular backups of your data and avoiding **untrustworthy** software distribution outlets.

## II. Detecting Malware

Modern antivirus software runs as a background process and attempts to identify malware that exists on a device or is entering a device as a download, email message, **attachment**, or Web page. The process of searching for malware is sometimes referred to as scanning or performing a virus scan. To identify malware, antivirus software can look for a **virus signature** or perform **heuristic** analyses.

A virus signature is a section of program code that contains a unique series of instructions known to be part of a malware exploit. Although they are called virus signatures, the unique code may identify a virus, worm, trojan, or other type of malware.

Virus signatures are discovered by security experts who examine the bit sequences contained in malware program code. When discovered, virus signatures are added to a collection of virus definitions, which form a database that is used by antivirus software as it works to scan files that may harbor malware.

Antivirus software can use techniques called heuristic analysis to detect malware by analyzing the characteristics and behavior of **suspicious** files. These techniques are especially useful for detecting new malware for which signatures have yet to be collected and added to the virus database.

One method of heuristic analysis allows the suspicious file to run in a guarded environment called a sandbox. If the file exhibits malicious behavior, it is treated like a virus and **quarantined** or deleted.

A second method of heuristic analysis involves inspecting the contents of a suspicious file for commands that carry out destructive or **surveillance** activities.

Heuristic analysis requires time and system resources to examine files that arrive as downloads and email attachments. The process can slightly affect performance while the analysis is in progress.

**cloak** / kləuk /
v. 掩盖；掩饰；伪装

**viral** / 'vaiərəl /
a. 病毒（性）的；病毒引起的

**exploit** / 'eksplɔit, ik's- /
n. 漏洞利用

**untrustworthy** / ˌʌn'trʌstˌwəːði /
a. 不可信赖的；靠不住的

**attachment** / ə'tætʃmənt /
n. 附件

**signature** / 'signətʃə /
n.（人或物的）识别标志；鲜明特征

**virus signature**
病毒特征码

**heuristic** / hju'ristik /
a. 启发（式）的；探索的

**suspicious** / sə'spiʃəs /
a. 可疑的

**quarantine** / 'kwɔrəntiːn /
v. & n. 隔离

**surveillance** / səː'veiləns, sə- /
n. 监视

**heuristics** / hju'ristiks /
n. 启发法；探索法
**mistakenly** / mi'steikənli /
ad. 错误地，误解地
**perplexing** / pə'pleksiŋ /
a. 使人困惑的；令人费解的
**flag** / flæg /
v. 用标志表明，做标记

**Heuristics** may produce false positives[1] that **mistakenly** identify a legitimate file as malware. For example, a legitimate disk utility that contains routines for enhancing disk drive performance by deleting redundant files might be mistaken for a virus and prevented from being installed. Such a situation could be **perplexing** for users who download software, only for it to disappear upon arrival. Users who understand how antivirus software works should be able to quickly conclude that the legitimate application was mistakenly **flagged** as malware.

A manual scan is initiated by a user for the purpose of scanning one or more files. Manual scans are useful if you suspect that a virus has slipped into a device despite security measures. For example, a previously unknown attack might arrive undetected, but after an antivirus update, it can be detected by running a manual scan.

Manually scanning all the files stored on a device can slow performance, so schedule the scan for a time when you are not usually using your device, but it is turned on.

You can also run a manual scan of a specific file. For example, suppose you download an application and you want to make sure it is virus-free before you install and run it. Depending on your antivirus software, you might be able to simply right-click the file name[2] to start the scan. Otherwise, open your antivirus software and select the manual scan option.

## III. Handling Malware

When antivirus software detects malware, it can try to remove the infection, put the file into quarantine, or simply delete the file.

Antivirus software can sometimes remove the malware code from infected files. This strategy is beneficial for files containing important documents that have become infected. Many of today's malware exploits are embedded in executable files and are difficult to remove. When malware cannot be removed, the file should not be used.

**encrypt** / in'kript /
v. 把…加密

In the context of antivirus software, a quarantined file contains code that is suspected of being part of a virus. For your protection, most antivirus software **encrypts** the file's contents and isolates it in a quarantine folder so

[1] *produce false positives*：产生假的病毒警报。
[2] *right-click the file name*：（用鼠标）右键单击文件名。

it can't be inadvertently opened or accessed by a hacker. Quarantined files cannot be run, but they can be moved out of quarantine if they are later found to have been falsely identified as malware.

Quarantined files should eventually be deleted. Most antivirus software allows users to specify how long an infected file should remain in quarantine before it is deleted. Most users rarely retrieve files from quarantine because it is risky to work with files that are suspected of harboring malicious code. There is no need, therefore, to delay deletion for more than a few days.

Most antivirus software displays an alert when malware is detected. Antivirus software automatically takes action to protect your device by attempting to repair the file, place it in quarantine, or delete it. You do not have to take any action. However, the alert message is an important piece of information that may indicate you are connected to a malicious site or receiving email from an unreliable source that would be best avoided in the future.

## IV. Configuring and Updating

Once you have installed antivirus software, the best and safest practice is to keep it running full time in the background so that it checks every email message as it arrives and scans all files that attempt to install themselves or run. For the most extensive protection from malware, you should look for and enable the following features of your antivirus software:

- Start scanning when the device boots.
- Scan all programs when they are launched, and scan **document files** when they are opened.
- Scan other types of files, such as graphics, if you engage in some risky computing behaviors and are not concerned with the extra time required to open files as they are scanned.
- Scan incoming email and attachments.
- Scan incoming **instant message** attachments.
- Scan **outgoing** email for worm activity such as **mass-mailing worms**.
- Scan **zipped** (compressed) files.
- Scan for **spyware** and PUAs[1] (potentially unwanted applications).

**document file**
文档文件
**instant message**
即时消息
**outgoing**
/ 'aut,gəuiŋ /
a. 往外去的，离去的
**mass-mailing worm**
群发邮件蠕虫
**zip** / zip /
v. 压缩（文件）
**spyware** / 'spaiwεə /
n. 间谍软件

[1] *PUA*：可能有害的应用程序（*p*otentially *u*nwanted *a*pplication 的首字母缩略）。

**storage volume**
存储卷

- Scan all files on the device's **storage volume** at least once a week.

The location for configuration settings depends on the antivirus software. Usually, there is a Settings menu or a Preferences option[1]. It is important to examine the settings after installing new antivirus software and after getting updates to make sure the desired level of protection is in place.

Also check for exclusions. If files, processes, and locations are excluded, they will not be scanned for malware. This feature is available to enhance performance from trusted sites, but it can be a doorway for malware exploits. For maximum protection, make sure that there are no exclusions listed in your antivirus software settings. Settings for Windows Defender[2] are accessed from the Settings **tab**[3].

**tab** / tæb /
n. 标记，标签；制表键；工作表选项卡

Antivirus software is an aspect of our digital lives that we tend to take for granted. We assume that it is installed and carrying out its work. However, antivirus software can be inadvertently disabled. Its configuration can be changed by malware that manages to **infiltrate** a device. It can expire at the end of a trial period or subscription. Ensuring that antivirus software is performing correctly may require **periodic intervention** from users.

**infiltrate**
/ ˈinfiltreit, inˈfil- /
v. 渗入，渗透
**periodic** / ˌpiəriˈɔdik /
a. 周期（性）的；定期的
**intervention**
/ ˌintəˈvenʃən /
n. 干涉，干预
**taskbar**
/ ˈtɑːskbɑː; ˈtæsk- /
n. 任务条，任务栏

Many antivirus products display an icon in the **taskbar** or notification area. The icon may offer a visual clue to indicate when the antivirus utility is active, scanning, or updating. Glancing at the icon can assure you that the software is running properly. Some targeted malware attacks may alter the icons, however, leading you to believe that the antivirus software is active when, in fact, it has been disabled by a malware attack. Opening antivirus software, such as Windows Defender, periodically to view its status is a good practice.

Two aspects of your antivirus software periodically need to be updated. First, the antivirus program itself might need a patch or an update to fix bugs or improve features. Second, the list of virus signatures must be updated to keep up with the latest malware developments.

Antivirus program updates and revised virus definitions are packaged into a file that can be manually or automatically downloaded. Most

---

1 *Usually, there is a Settings menu or a Preferences option.*：通常，有一个“设置”菜单或一个“首选”选项。
2 *Windows Defender*：Windows 守卫者，微软公司出品的杀毒软件。
3 *Settings for Windows Defender are accessed from the Settings tab.*：Windows Defender 的“设置”是从“设置”选项卡访问的。

antivirus products are preconfigured to regularly check for updates, download them, and install them without user intervention. If you would rather control the download and installation process yourself, you can configure your antivirus software to alert you when updates are ready. In any case, you should manually check for updates periodically just in case the auto-update function has become disabled by malware or if your subscription has expired.

**hoax** / həuks /
n. 恶作剧；骗局

### V. Virus Hoax

Some virus threats are very real, but you're also likely to get email messages about so-called viruses that don't really exist. A virus hoax usually arrives as an email message containing **dire** warnings about a **supposedly** new virus on the loose[1]. It typically provides a link to download some type of detection and protection software. It may include removal instructions that actually delete parts of the operating system. And, of course, you are encouraged to forward this "crucial" information to your friends.

**dire** / 'daiə /
a. 可怕的；极紧迫的；极端的
**supposedly** / sə'pəuzidli /
ad. 据推测；据称；大概，可能

When you receive an email message about a virus or any other type of malware, don't panic. It could be a hoax. You can check one of the many **hoaxbuster** or antivirus software Web sites to determine whether you've received a hoax or information about a real threat.

**hoaxbuster** / 'həuks,bʌstə /
n. 恶作剧揭穿者，辟谣者

These Web sites also provide security or virus alerts, which list the most recent legitimate malware threats. If the virus is a real threat, the Web site can provide information to help determine whether your device has been infected. You can also find instructions for **eradicating** the virus. If the virus threat is a hoax, by no means should you forward the email message to others.

**eradicate** / i'rædikeit /
v. 根除；消灭

---

## Exercises

**I. Fill in the blanks with the information given in the text:**

1. Antivirus software can detect viruses by looking for signatures or by ________ analysis.

2. When antivirus software detects malware, it can try to remove the

---

[1] *on the loose*：（动物等由于没有关好而）到处乱跑；行动不受限制，自由自在。

infection, put the file into ________, or simply delete the file.

3. Antivirus software produces what is referred to as a false ________ when a legitimate program is mistakenly identified as a virus.

4. A virus ________ usually arrives as an email alert that warns against an approaching virus attack.

**II. Translate the following terms or phrases from English into Chinese and vice versa:**

| | |
|---|---|
| 1. virus scan | 6. 启发式分析 |
| 2. infected file | 7. 群发邮件蠕虫 |
| 3. quarantined file | 8. 病毒特征码 |
| 4. virus hoax | 9. 防病毒软件 |
| 5. viral exploit | 10. 压缩文件 |

# Section C

## Types of Malicious Software

Malicious software can be divided into two categories: those that need a **host program**, and those that are independent. The former, referred to as **parasitic**, are essentially fragments of programs that cannot exist independently of some actual application program, utility, or system program. Viruses, **logic bombs**, and **backdoors** are examples. The latter are **self-contained** programs that can be scheduled and run by the operating system. Worms and **bot** programs are examples.

**host program**
主机程序，宿主程序，主程序
**parasitic**
/ ˌpærəˈsitik /
a. 寄生的
**logic bomb**
逻辑炸弹
**backdoor** / ˌbækˈdɔː /
n. 后门
**self-contained**
/ ˌselfkənˈteind /
a. 自给的，独立自足的
**bot** / bɔt /
n. 机器人；僵尸程序（= robot）
**trigger** / ˈtrigə /
n. 引爆器；触发器

We can also differentiate between those software threats that do not replicate and those that do. The former are programs or fragments of programs that are activated by a **trigger**. Examples are logic bombs, backdoors, and bot programs. The latter consist of either a program fragment or an independent program that, when executed, may produce one or more copies of itself to be activated later on the same system or some other system. Viruses and worms are examples.

In the following, we briefly survey some of the key categories of malicious software.

### I. Virus

A computer virus is a piece of software that can "infect" other programs by modifying them; the modification includes **injecting** the original program with a routine to make copies of the virus program, which can then go on to infect other programs. Computer viruses first appeared in the early 1980s, and the term itself is attributed to Fred Cohen[1] in 1983. Cohen is the author of a **groundbreaking** book on the subject.

**inject** / inˈdʒekt /
v. 注射；注入；引入

**groundbreaking**
/ ˈgraundˌbreikiŋ /
a. 开辟新天地的，开拓性的

Like its biological counterpart, a computer virus carries in its instructional code the recipe for making perfect copies of itself. The typical virus becomes embedded in a program on a computer. Then, whenever the infected computer comes into contact with an uninfected piece of software, a fresh copy of the virus passes into the new program. Thus, the infection can be spread from computer to computer by unsuspecting users who either

[1] *Fred Cohen*：弗雷德·科恩（1957—），美国计算机科学家，1987 年在其著名论文《计算机病毒》（*Computer Viruses*）中首先提出了关于"计算机病毒"的概念。

**swap** / swɔp /
v. 交换

**swap** disks or send programs to one another over a network. In a network environment, the ability to access applications and system services on other computers provides a perfect culture for the spread of a virus[1].

A virus can do anything that other programs do. The difference is that a virus attaches itself to another program and executes secretly when the host program is run. Once a virus is executing, it can perform any function, such as erasing files and programs.

**vector** / 'vektə /
n. 传病媒介；矢量
**payload** / 'peiləud /
n. 有效载荷，有效负载
**dormant** / 'dɔ:mənt /
a. 休眠的；暂停活动（或作用）的
**propagation** / ˌprɔpə'geiʃən /
n. 传播；繁殖；蔓延
**propagate** / 'prɔpəgeit /
v. 传播；繁殖；蔓延

A computer virus has three parts: infection mechanism (also referred to as infection **vector**), trigger, and **payload**. During its lifetime, a typical virus goes through the following four phases: **dormant** phase, **propagation** phase, triggering phase, and execution phase.

## II. Worm

A worm is a program that can replicate itself and send copies from computer to computer across network connections. Upon arrival, the worm may be activated to replicate and **propagate** again. In addition to propagation, the worm usually performs some unwanted function. An e-mail virus has some of the characteristics of a worm because it propagates itself from system to system. However, we can still classify it as a virus because it uses a document modified to contain viral macro content and requires human action. A worm actively seeks out more machines to infect and each machine that is infected serves as an automated **launching pad** for attacks on other machines.

**launching pad**
发射台；跳板；起点

The concept of a computer worm was introduced in John Brunner's[2] 1975 SF[3] novel *The Shockwave Rider*. The first known worm implementation was done in Xerox[4] Palo Alto Labs[5] in the early 1980s. It was nonmalicious search for idle systems to use to run a computationally intensive task.

Network worm programs use network connections to spread from system to system. Once active within a system, a network worm can behave as a computer virus or bacteria, or it could implant Trojan horse programs

1 *provides a perfect culture for the spread of a virus*：为传播病毒提供了理想的条件。

2 *John Brunner*：约翰·布伦纳（1934—1995），英国科幻小说作家，据认为他在1975年创作的小说《冲击波骑手》（*The Shockwave Rider*）中创造了worm（蠕虫病毒）一词并预言了计算机病毒的出现，该书首次描写了在信息社会中计算机作为正义和邪恶双方斗争的工具的故事，成为当年最佳畅销书之一。

3 *SF*：科学幻想小说，科幻小说（*s*cience *f*iction的首字母缩略）。

4 *Xerox*：施乐公司，一家数字与信息技术产品生产商，系复印技术的发明公司，创立于1906年，总部位于美国康涅狄格州诺沃克（Norwalk）。

5 *Palo Alto Labs*：（施乐公司）帕洛阿尔托实验室。

**disruptive**
/ dis'rʌptiv /
a. 扰乱性的，制造混乱的

or perform any number of **disruptive** or destructive actions. A network worm exhibits the same characteristics as a computer virus: a dormant phase, a propagation phase, a triggering phase, and an execution phase.

## III. Backdoor

**trapdoor** / 'træpdɔː /
n. 陷阱门，天窗

**maintenance hook**
维护陷阱

A backdoor, also known as a **trapdoor**, is a secret entry point into a program that allows someone who is aware of the backdoor to gain access without going through the usual security access procedures. Programmers have used backdoors legitimately for many years to debug and test programs; such a backdoor is called a **maintenance hook**. This usually is done when the programmer is developing an application that has an authentication procedure, or a long setup, requiring the user to enter many different values to run the application. To debug the program, the developer may wish to gain special privileges or to avoid all the necessary setup and authentication. The programmer may also want to ensure that there is a method of activating the program should something be wrong with the authentication procedure[1] that is being built into the application. The backdoor is code that recognizes some special sequence of input or is triggered by being run from a certain **user ID**[2] or by an unlikely sequence of events.

**user ID**
用户标识（符）

**unscrupulous**
/ ˌʌn'skruːpjuləs /
a. 肆无忌惮的；无耻的

Backdoors become threats when **unscrupulous** programmers use them to gain unauthorized access. It is difficult to implement operating system controls for backdoors. Security measures must focus on the program development and software update activities.

## IV. Logic Bomb

**predate** / priː'deit /
v. 在日期上早于（或先于）

One of the oldest types of program threat, **predating** viruses and worms, is the logic bomb. The logic bomb is code embedded in some legitimate program that is set to "explode" when certain conditions are met. Examples of conditions that can be used as triggers for a logic bomb are the presence or absence of certain files, a particular day of the week or date, or a particular user running the application. Once triggered, a bomb may alter or delete data or entire files, cause a machine halt, or do some other damage.

## V. Trojan Horse

A Trojan horse is a useful, or apparently useful, program or command

[1] *should something be wrong with the authentication procedure*：万一验证过程有点毛病。should 可用于倒装句，放在主语前面，表示语气较强的假设，意思相当于 if 引导的条件状语从句（if something is wrong with the authentication procedure）。

[2] *ID*：身份证明，身份识别，标识（*id*entification 的缩略）。

**invoke** / in'vəuk /
v. 调用；激活

procedure containing hidden code that, when **invoked**, performs some unwanted or harmful function.

Trojan horse programs can be used to accomplish functions indirectly that an unauthorized user could not accomplish directly. For example, to gain access to the files of another user on a shared system, a user could create a Trojan horse program that, when executed, changes the invoking user's file permissions so that the files are readable by any user. The author could then induce users to run the program by placing it in a common directory and naming it such that it appears to be a useful utility program or application. An example of a Trojan horse program that would be difficult to detect is a compiler that has been modified to insert additional code into certain programs as they are compiled, such as a system login program. The code creates a backdoor in the login program that permits the author to log on to the system using a special password. This Trojan horse can never be discovered by reading the source code of the login program.

**motivation**
/ ˌməuti'veiʃ ən /
n. 动机，诱因
**calculator program**
计算（器）程序

Another common **motivation** for the Trojan horse is data destruction. The program appears to be performing a useful function (e.g., a **calculator program**), but it may also be quietly deleting the user's files.

## VI. Mobile Code

**semantics**
/ si'mæntiks /
n. 语义（学）

Mobile code refers to programs (e.g., script, macro, or other portable instruction) that can be shipped unchanged to a heterogeneous collection of platforms and execute with identical **semantics**. The term also applies to situations involving a large homogeneous collection of platforms (e.g., Microsoft Windows).

Mobile code is transmitted from a remote system to a local system and then executed on the local system without the user's explicit instruction. Mobile code often acts as a mechanism for a virus, worm, or Trojan horse to be transmitted to the user's workstation. In other cases, mobile code takes advantage of vulnerabilities to perform its own exploits, such as unauthorized data access or root compromise[1]. Popular vehicles for mobile code include Java applets, ActiveX[2], JavaScript, and

1 *In other cases, mobile code takes advantage of vulnerabilities to perform its own exploits, such as unauthorized data access or root compromise.*：在其他情况下，移动代码利用漏洞为所欲为，如未经授权的数据访问或根权限入侵。exploit 作名词用时，常用作复数，表示“业绩”“功绩”“英勇（或令人激动、令人关注）的行为”等意思。

2 *ActiveX*：在广义上是指微软公司的整个 COM（组件对象模型）架构，但是现在通常用来称呼基于标准 COM 接口来实现对象连接与嵌入的 ActiveX 控件。

VB-Script[1]. The most common ways of using mobile code for malicious operations on local systems are cross-site scripting[2], interactive and dynamic Web sites, e-mail attachments, and downloads from untrusted sites or of untrusted software.

## VII. Bot

**zombie** / ˈzɔmbi /
n. 还魂尸，僵尸；僵（进程）
**drone** / drəun /
n. 游手好闲者，寄生虫
**botnet** / ˈbɔtnet /
n. 僵尸网络

A bot (robot), also known as a **zombie** or **drone**, is a program that secretly takes over another Internet-attached computer and then uses that computer to launch attacks that are difficult to trace to the bot's creator. The bot is typically planted on hundreds or thousands of computers belonging to unsuspecting third parties. The collection of bots often is capable of acting in a coordinated manner; such a collection is referred to as a **botnet**.

A botnet exhibits three characteristics: the bot functionality, a remote control facility, and a spreading mechanism to propagate the bots and construct the botnet.

## VIII. Rootkit

**rootkit** / ˈruːtkit /
n. 根工具包
**stealthy** / ˈstelθi /
a. 偷偷摸摸的，暗中进行的，秘密的

A rootkit is a set of programs installed on a system to maintain administrator (or root) access to that system. Root access provides access to all the functions and services of the operating system. The rootkit alters the host's standard functionality in a malicious and **stealthy** way. With root access, an attacker has complete control of the system and can add or change programs and files, monitor processes, send and receive network traffic, and get backdoor access on demand.

A rootkit can make many changes to a system to hide its existence, making it difficult for the user to determine that the rootkit is present and to identify what changes have been made. In essence, a rootkit hides by subverting the mechanisms that monitor and report on[3] the processes, files, and registries on a computer.

## IX. Multiple-Threat Malware

Viruses and other malware may operate in multiple ways. The terminology is far from uniform; this subsection gives a brief introduction to several related concepts that could be considered multiple-threat malware.

---

[1] *Popular vehicles for mobile code include Java applets, ActiveX, JavaScript, and VB-Script.*：传播移动代码的流行媒介包括 Java 小应用程序、ActiveX、Java 脚本语言和 Visual Basic 脚本语言。

[2] *cross-site scripting*：跨网站脚本（攻击），跨站脚本（攻击）。

[3] *report on*：报告，汇报。

**multipartite** /ˌmʌltiˈpɑːtait/
a. 分成多部分的
**eradication** /iˌrædiˈkeiʃən/
n. 根除；消灭
**contagion** /kənˈteidʒən/
n. (接)触(传)染
**severity** /siˈveriti/
n. 剧烈；严重；严厉

A **multipartite** virus infects in multiple ways. Typically, the multipartite virus is capable of infecting multiple types of files, so that virus **eradication** must deal with all of the possible sites of infection.

A blended attack uses multiple methods of infection or transmission, to maximize the speed of **contagion** and the **severity** of the attack. Some writers characterize a blended attack as a package that includes multiple types of malware. An example of a blended attack is the Nimda attack[1], erroneously referred to as simply a worm. Nimda uses four distribution methods: e-mail, Windows shares[2], Web servers, and Web clients. Thus, Nimda has worm, virus, and mobile code characteristics. Blended attacks may also spread through other services, such as instant messaging and peer-to-peer file sharing.

## Exercises

**I. Fill in the blanks with the information given in the text:**

1. A computer virus has three parts: ________ mechanism, trigger, and ________.

2. A backdoor is also known as a(n) ________.

3. A collection of bots capable of acting in a coordinated manner is referred to as a(n) ________.

4. A rootkit is a set of programs installed on a system to maintain ________ or ________ access to that system.

**II. Translate the following terms or phrases from English into Chinese and vice versa:**

| | |
|---|---|
| 1. maintenance hook | 5. 系统登录程序 |
| 2. multipartite virus | 6. 逻辑炸弹 |
| 3. authentication procedure | 7. 多威胁恶意软件 |
| 4. instant messaging | 8. 源代码 |

[1] *Nimda attack*：尼姆达病毒攻击。尼姆达病毒用 JavaScript 脚本语言编写，通过电子邮件、共享网络资源、IIS（互联网信息服务）服务器传播，同时也是一种感染本地文件的病毒。

[2] *Windows shares*：Windows 共享文件夹。

# Unit 11　Cyberculture

# （计算机文化）

## Section A

## Using E-Mail

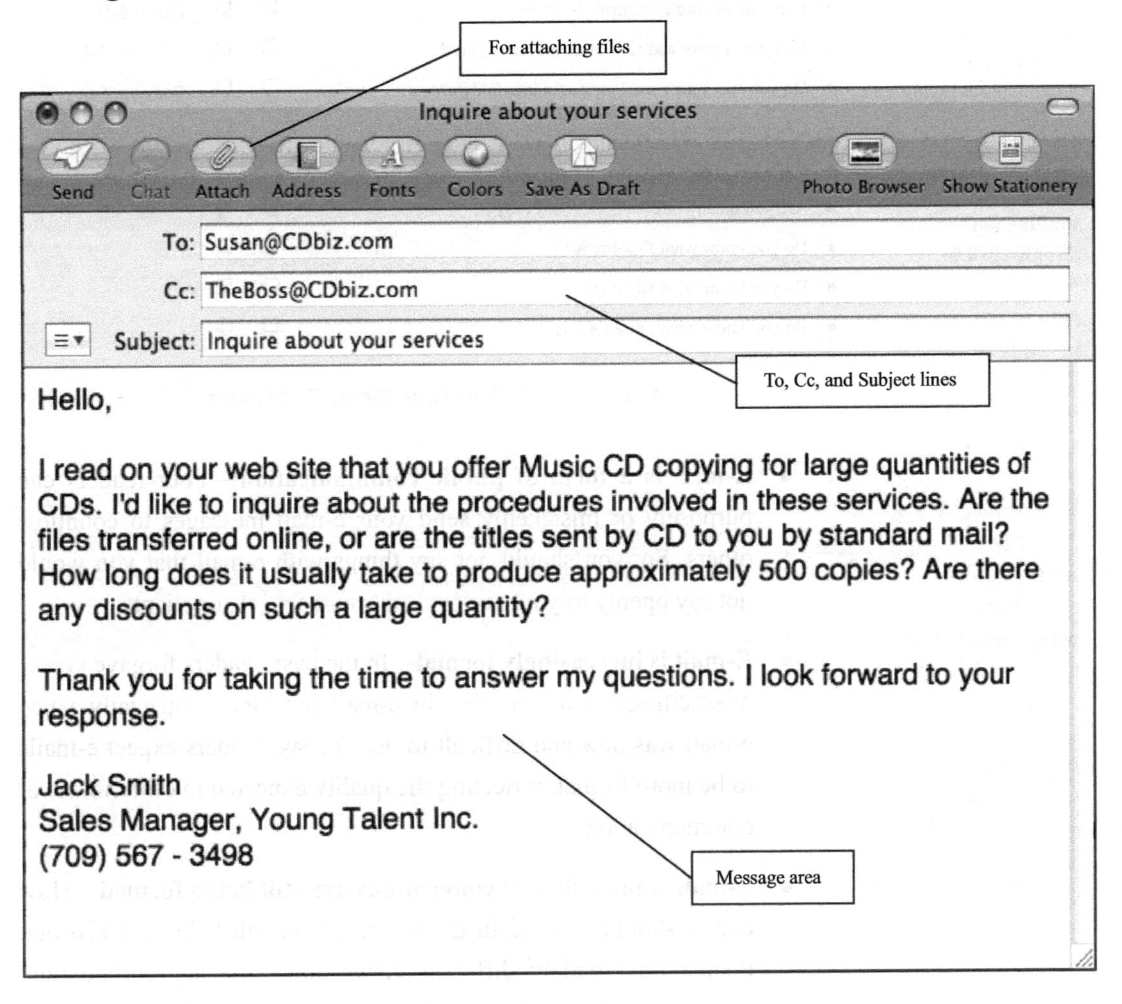


### I. Introduction

E-mail is an electronic system for sending and receiving messages and files over a computer network. Compared to the phone or paper-based documents, e-mail is still relatively new in the workplace. When you are using e-mail, here are a few guidelines to keep in mind:

**chatty** / 'tʃæti /
a. 聊天式的，轻松而亲切的；爱闲聊的

**frivolous** / 'frivələs /
a. 轻薄的；琐屑的

**achiever** / ə'tʃi:və /
n. 事业成功的人

**signature file**
签名文件

**mailing list**
邮件发送清单，邮件列表

**listserv** / 'listsə:v /
n. 邮件发送清单（或邮件列表）管理程序

**flaming** / 'fleimiŋ /
n. 争论（特指在邮件讨论组或网络论坛中争论）

**spam** / spæm /
n. 垃圾邮件

**netiquette** / 'netiket /
n. 网络礼节，网规（*net*work et*iquette* 的缩合）

**supervisor** / 'sju:pəvaizə /
n. 监督人；管理人；指导者

**coworker** / kəu'wə:kə /
n. 同事

**typo** / 'taipəu /
n. 打字（或排印）错误

**blooper** / 'blu:pə /
n. 过失，失礼

- **E-mail is increasingly used for professional purposes**—Not long ago, e-mail was considered a secondary form of communication. It was spontaneous and **chatty**, used mostly for quick comments or nonessential information. Today, e-mail is a principal form of communication in most workplaces, so people expect e-mail messages to be professional and **nonfrivolous**.

| | Yes | No | Score |
|---|---|---|---|
| • Can you receive and send messages? | ☐ | ☐ | Yes Marks |
| • Can you forward documents to others? | ☐ | ☐ | Expert 8-10 |
| • Can you receive and send attached documents? | ☐ | ☐ | Achiever 7-8 |
| • Do you sort your e-mail messages into folders? | ☐ | ☐ | Beginner 5-6 |
| • Can you send one e-mail message to multiple people? | ☐ | ☐ | Get to Work 0-5 |
| • Have you created and used a **signature file**? | ☐ | ☐ | |
| • Are you on any **mailing lists (listservs)**[1]? | ☐ | ☐ | |
| • Do you know what **flaming** is? | ☐ | ☐ | |
| • Do you know what **spam** is? | ☐ | ☐ | |
| • Do you know what **netiquette** is? | ☐ | ☐ | |

**Figure 11A-1: Test Your E-Mail Readiness**

- **E-mail is a form of public communication**—Your readers can purposely or mistakenly send your e-mail messages to countless others. So, you should not say things with e-mail that you would not say openly to your **supervisors**, **coworkers**, or clients.

- **E-mail is increasingly formal**—In the past, readers forgave **typos**, misspellings, and **bloopers** in e-mail messages, especially when e-mail was new and difficult to use. Today, readers expect e-mails to be more formal, reflecting the quality expected in other forms of communication.

- **E-mail standards and conventions are still being formed**—How e-mail should be used in the workplace is still being worked out. People hold widely different views about the appropriate (and inappropriate) use of e-mail. So, you need to pay close attention to how e-mail is used in your company and your readers' companies. Many companies are developing policies explaining how e-mail should be used. If your company has a policy on e-mail usage, you

[1] *mailing lists (listservs)*：在文中指邮件讨论组（新闻组）。

should read it and follow it.

You should also keep in mind that legal constraints shape how e-mail is used in the workplace. E-mail, like any other written document, is protected by copyright law. So, you need to be careful not to use e-mails in any way that might **violate** copyright law. For example, if you receive an e-mail from a client, you cannot immediately post it to your company's website without that client's permission.

**violate** / 'vaiəleit /
v. 违反，违犯

Also, lawyers and courts treat e-mail as written communication, equivalent to a **memo** or letter. For example, much of the **antitrust** case against Microsoft in the late 1990s was built on recovered e-mail messages in which Bill Gates[1] and other executives chatted informally about aggressively competing with other companies.

**memo**
/ 'meməu, 'mi:- /
n. 〈口〉备忘录（= memorandum）

**antitrust**
/ ˌænti'trʌst /
a. 反托拉斯的，反垄断的

Legally, any e-mail you send via the employer's computer network belongs to the employer. So, your employers are within their rights to read your e-mail without your knowledge or permission. Also, deleted e-mails can be retrieved from the company's servers, and they can be used in a legal case.

Increasingly, **harassment** and **discrimination** cases **hinge** on evidence found in e-mails. Careless e-mails about personal relationships or appearances can be saved and used against the sender in a court case. **Indiscreet** comments about gender, race, or **sexual orientation** can also have unexpected consequences. Your "harmless" dirty jokes sent to your coworkers might end up being used by a lawyer to prove that you are creating a "hostile workplace environment."

**harassment**
/ 'hærəsmənt, hə'ræs- /
n. 骚扰；烦扰

**discrimination**
/ disˌkrimi'neiʃən /
n. 区别；歧视

**hinge** / hindʒ /
v. 依…而定，以…为转移（*on, upon*）

**indiscreet**
/ ˌindi'skri:t /
a. 不慎重的，轻率的

**orientation**
/ ˌɔ:rien'teiʃən /
n. 方向；定向

**sexual orientation**
（对同性或异性的）性取向，性倾向

## II. Basic Features of E-Mail

An e-mail is formatted similarly to a memo. Typical e-mail messages will have a *header* and *body*. They also have additional features like *attachments* and *signatures*.

### 1. Header

The header has lines for *To* and *Subject*. Usually, there are also lines like *cc*[2], *bcc*[3], and *Attachments*, which allow you to expand the capabilities

[1] *Bill Gates*：比尔·盖茨（1955—），从哈佛大学辍学后，与好友共同创立微软公司，曾任微软董事长、首席执行官和首席软件设计师。

[2] *cc*：抄送（*c*arbon *c*opy 的首字母缩略）。

[3] *bcc*：密送（*b*lind *c*arbon *c*opy 的首字母缩略）。

of the message.

- **To line**—Here is where you type the e-mail address of the person to whom you are sending the e-mail. You can put multiple addresses on this line, allowing you to send your message to many people.

- **cc and bcc lines**—They are used to copy the message to people who are not the primary readers, like your supervisors or others who might be interested in your conversation. The cc line shows your message's **recipient** that others are receiving copies of the message too. The bcc line ("blind cc") allows you to copy your messages to others without anyone else knowing.

**recipient** / ri'sipiənt /
n. 接受者；收件人

- **Subject line**—It signals the topic of the e-mail. Usually a small phrase is used. If the message is a response to a prior message, e-mail programs usually automatically insert a "Re:[1]" into the subject line. If the message is being forwarded, a "Fwd:[2]" is inserted in the subject line.

- **Attachments line**—It signals whether there are any additional files, pictures, or programs attached to the e-mail message. You can attach whole documents created on your **word processor**, **spreadsheet program**, or **presentation software**. An attached document retains its original formatting and can be downloaded right to the reader's computer.

**word processor**
文字处理软件
**spreadsheet**
/ 'spredʃi:t /
n. 电子表格，电子数据表
**spreadsheet program**
电子数据表程序
**presentation software**
演示软件

**2. Message Area**

After the header, the *message area* is where you can type your comments to your readers. It should have a clear introduction, body, and conclusion.

- The *introduction* should minimally (1) define the subject, (2) state your purpose, and (3) state your main point. Also, if you want the reader to do something, you should mention it up front[3], not at the end of the e-mail.

- The *body* should provide the information needed to prove or

[1] *Re*：response 的缩略，表示"回复"。
[2] *Fwd*：forward 的缩略，表示"转发"。
[3] *up front*：〈口〉在最前面（或最突出）的位置；直率地；预先。

support your e-mail's purpose.

- The *conclusion* should restate the main point and look to[1] the future. Most readers of e-mail never reach the conclusion, so you should tell them any action items early in the message and then *restate* them in the conclusion.

The message area might also include these other kinds of text:

- **Reply text**—When you reply to a message, most e-mail programs allow you to copy parts of the original message into your message. These parts are often identified with > arrows running down the left margin[2].
- **Links**—You can also add in direct links to websites. Most programs will automatically recognize a **webpage** address like "http://www.predatorconservation.org" and make it a live link in the e-mail's message area.
- **Attachments**—If you attach a file to your e-mail message, you should tell the readers in the message area that a file is attached. Otherwise, they may not notice it.

<table>
<tr><th>Symbols</th><th>Meaning</th></tr>
<tr><td>:-)</td><td>Happy face</td></tr>
<tr><td>;-)</td><td>Winking happy face</td></tr>
<tr><td>:-o</td><td>Surprised face</td></tr>
<tr><td>:-|</td><td>Grim face</td></tr>
<tr><td>:-\</td><td>Smirking face</td></tr>
<tr><td>:-(</td><td>Unhappy face</td></tr>
<tr><td>;-(</td><td>Winking unhappy face</td></tr>
<tr><td>:-x</td><td>Silent face</td></tr>
<tr><td>:'-(</td><td>Crying</td></tr>
</table>

**Figure 11A-2: Some Commonly Used Emoticons**

- **Emoticons**—Another common feature in the message area is the use of *emoticons* (Figure 11A-2). Used **sparingly**, they can help you signal emotions that are hard to convey in written text. When

**webpage** /ˈwebpeidʒ/
n. 网页

**wink** /wiŋk/
v. 眨眼；眨眼示意，使眼色

**grim** /grim/
a. 严厉的；阴森的

**smirk** /sməːk/
v. 假笑；得意地笑

**emoticon** /iˈməutikɔn/
n. 情感符（*emotion icon* 的缩合）

**sparing** /ˈspɛəriŋ/
a. 节约的；有节制的

[1] *look to*：盼望，展望。

[2] *These parts are often identified with > arrows running down the left margin.*：这些部分常常用沿着左面的页边空白向下排列的箭头“>”标出。

overused, emoticons can become annoying to some readers. In most workplace situations, emoticons should not be used. They are playful and informal; therefore, they are really only appropriate in e-mails between close colleagues or friends.

**3. Signature**

E-mail programs usually let you create a *signature file* that automatically puts a *signature* at the end of your messages. Signature files can be both simple and complex. They allow you to personalize your message and add in additional contact information. By creating a signature file, you can avoid typing your name, title, phone number, etc. at the end of each message you write.

**4. Attachments**

One of the advantages of e-mail over other forms of communication is the ability to send and receive attachments. Attachments are files, pictures, or programs that readers can download to their own computer.

- **Sending attachments**—If you would like to add an attachment to your e-mail message, click on the button that says "Attach Document" or "Attachment" in your e-mail software program. Most programs will then open a **textbox** that allows you to find and select the file you want to attach.

**textbox** / 'tekstbɔks /
n. 文本（或正文、文字）框

- **Receiving attachments**—If someone sends you an attachment, your e-mail program will use an icon to signal that a file is attached to the e-mail message. Click on that icon. Most e-mail programs will then allow you to save the document to your hard disk. From there, you can open the file.

## III. E-Mail Netiquette

E-mail has its own **etiquette**, or "netiquette." Here are a dozen netiquette guidelines that you can follow in the workplace:

**etiquette** / 'etiket /
n. 礼节；（行业中的）道德规范；规矩

**concise** / kən'sais /
a. 简明的，简要的

- **Be concise**—Keep the length of your messages under a screen and a half. If you have more than a screen and a half of content, pick up the phone, write a memo, or send the information as an attachment.
- **Provide only need-to-know information**—Decide who needs to know what you have to say, and send them only information they need—nothing more.

- **Treat the security of the message about the same as a message on a postcard**—Recognize that anyone can pass your message along or use it against you. If the message is confidential or proprietary, e-mail is not an appropriate way to send it.

- **Don't say anything over e-mail that you would not say in a meeting or to legal authorities**—Hey, accidents happen. You might send something embarrassing to the entire organization. Or, someone might do it for you.

- **Never immediately respond to a message that made you angry or upset**— Give yourself time to cool off. You should never write e-mail when you are angry, because these little **critters** have a way of returning to bite you.

- **Avoid using too much humor, especially irony or sarcasm**—In an e-mail, your **witty quips** rarely come off exactly as you intended[1]. Remember that your e-mails can be easily misinterpreted or taken out of context. The American sense of humor, meanwhile, is often disturbing to international audiences; so if your message is going overseas, keep the humor to a minimum.

- **Be extra careful about excerpting or forwarding the e-mail of others**—If there is the remotest chance that someone else's message could be misunderstood, **paraphrase** it in your own message instead of forwarding it or copying parts of it.

- **Don't be unpleasant over e-mail if you would not be so face to face**—E-mail provides a false sense of security, much like driving in a car. But in reality, those are real people on that **information superhighway**[2], and they have feelings. **Plus** they can save your **abusive** or **harassing** messages and use them as evidence against you.

- **Never send anything that could be proprietary**—There have

**critter** / 'kritə /
n.〈美口〉生物；动物
**irony** / 'aiərəni /
n. 反语；讽刺文体
**sarcasm** / 'sɑːkæzəm /
n. 讽刺，挖苦
**witty** / 'witi /
a. 诙谐的；说话风趣的
**quip** / kwip /
n. 妙语，俏皮话
**excerpt** / ek'səːpt/
v. 摘录；引用
**paraphrase** / 'pærəfreiz /
v. 将…释义（或意译）
**information superhighway**
信息高速公路
**plus** / plʌs /
ad.〈口〉外加地；另外
**abusive** / ə'bjuːsiv /
a. 谩骂的；毁谤的
**harass** / 'hærəs /
v. 骚扰；烦扰

[1] *In an e-mail, your witty quips rarely come off exactly as you intended.*：在电子邮件中，你的诙谐俏皮话难得达到你想要的确切效果。come off 表示“以某种方式结束”“获得某种结果”“结果是”等。

[2] *those are real people on that information superhighway*：信息高速公路上的那些人都是真人。句中的 those 与其定语（on that information superhighway）被谓语（系表结构）分隔开。如果把 on that information superhighway 作为 people 的定语，那么 those 是指哪些人则不好解释，因为上下文没有提到。这实际上是一个分隔结构。分隔结构指语法关系密切的两个句子成分被其他句子成分分隔开的现象。分隔结构的产生多是为了保持句子平衡、避免头重脚轻，或是为了语义严密、结构紧凑等。

**classified** / 'klæsifaid /
a. 归入密级的，保密的
**courier** / 'kuriə /
n. 信使
**flame** / fleim /
v.（向…）发送争论（或争辩）邮件
**spam** / spæm /
v.（向…）发送垃圾邮件
**chain-letter** / 'tʃein,letə /
v. 向…发送连锁信（或连锁邮件）
**gullible** / 'gʌlibəl /
a. 易受骗的，易上当的
**sloppy** / 'slɔpi /
a.〈口〉马虎的；凌乱的

been some classic cases where **classified** information or trade secrets have been sent to competitors or the media over e-mail. If the information is proprietary, use traditional routes like the regular mail or a **courier**.

- **Don't flame, spam, or chain-letter[1] people at work**—If you must, do these things from your personal computer at home. At work, assume these activities will ultimately be used against you. You might be fired.

- **Think twice before you send "urgent" messages that are making their way across the net**—There are lots of hoaxes about computer viruses "erasing your hard drive." These hoaxes gain new life when some **gullible** persons start sending them to all their friends. Some viruses are real, but most are not. Wait a day or two to see if the threat is real before you warn all your friends and colleagues.

- **Be forgiving of the grammatical mistakes of others—but don't make them yourself**—Some people treat e-mail informally, so they rarely revise (though they should). Grammar mistakes and misspellings happen—so you can forgive the sender. On the other hand, these mistakes and typos make the sender look stupid. Don't send out **sloppy** e-mail messages.

---

# Exercises

**I. Fill in the blanks with the information given in the text:**

1. As used in the text, e-mail can refer both to a single electronic message and to the entire electronic ________ for sending and receiving messages and files over a computer network.

2. E-mail is protected by ________ law. If you receive an e-mail from a client, you cannot immediately post it to your company's website without his or her permission.

3. Legally, e-mail messages sent by employees via the employer's

---

[1] *chain-letter*：源自名词 chain letter，即"连锁信"（分别寄给数人的信，并要求他们复制此信若干份，再分别寄出）。

computer network belong to the employer, so the employer is within his ________ to read them without the employees' knowledge or permission.

4. When you are using e-mail, you should remember that ________ e-mail messages can be retrieved from your company's servers and used in a legal case.

5. An attached document retains its original ________ when it is received by the reader and can be downloaded right to his or her computer.

6. A(n) ________ file created with e-mail programs can automatically put a signature at the end of each e-mail message.

7. E-mail messages should be concise. Usually, their length should be kept under one and a half ________.

8. Because ________ are playful and informal, they are only appropriate in e-mails between close colleagues or friends and should not be used in most workplace situations.

**II. Translate the following terms or phrases from English into Chinese and vice versa:**

| | |
|---|---|
| 1. mailing list | 11. 常用情感符 |
| 2. proprietary software | 12. 已删除电子邮件 |
| 3. cc line | 13. 电子系统 |
| 4. bcc line | 14. 附件行 |
| 5. forwarded e-mail messages | 15. 版权法 |
| 6. e-mail convention | 16. 电子邮件网规 |
| 7. click on an icon | 17. 信息高速公路 |
| 8. confidential document | 18. 签名文件 |
| 9. classified information | 19. 电子数据表程序 |
| 10. recovered e-mail message | 20. 文字处理软件 |

**III. Fill in each of the blanks with one of the words given in the following list, making changes if necessary:**

| | | | |
|---|---|---|---|
| *access* | *in-person* | *respond* | *download* |
| *create* | *see* | *cost* | *window* |
| *e-mail* | *touch* | *message* | *audio* |
| *camera* | *website* | *program* | *computer* |

Instant messaging has always been a good way to stay in ________ with friends. As companies look to cut ________ and improve efficiency, instant messaging is a good way to replace ________ meetings, while allowing people to *collaborate* (合作) through their ________. Instant messaging is a valuable new tool that is becoming as common as ________ in the workplace.

There are companies which give you free ________ to instant messaging. To use their systems, you will need to ________ their software to your computer from their ________. Some instant messaging programs include the ability to send live ________ or video. With a microphone and/or digital ________ mounted on your computer, you can let other people ________ and hear you talking.

Most instant messaging ________ ask you to sign in with a password. Then, you can ________ a list of other people with whom you want to *converse* (交谈). To begin a conversation, send a ________ to the others to see if they are at their computers. Their computers will open a ________ or make a sound to tell them you want to talk to them. If they ________, you can start writing back and forth.

**IV. Translate the following passage from English into Chinese:**

The pace of change brought about by new technologies has had a significant effect on the way people live, work, and play worldwide. New and emerging technologies challenge the traditional process of teaching and learning, and the way education is managed. It is frequently claimed that the most *potent* (强有力的) agent of change on present-day society will prove to be information technology (IT). IT, while an important area of study in its own right, is having a major impact across all curriculum areas. Easy worldwide communication provides instant access to a vast array of data. Rapid communication, plus increased access to IT in the home, at work, and in educational establishments, could mean that learning becomes a truly lifelong activity—an activity in which the pace of technological change forces constant evaluation of the learning process itself.

# Section B

# Ethical Guidelines for Computer Professionals

## I. Special Aspects of Professional Ethics

**ethical** / 'eθikəl /
a. 道德的；伦理的
**ethics** / 'eθiks /
n.［用作单］伦理学；［用作单或复］道德准则

**expertise** / ˌekspəː'tiːz /
n. 专门知识（或技能），专长
**profound** / prə'faund /
a. 深邃的；深刻的
**incompetence** / in'kɔmpitəns /
n. 无能力；不胜任

Professional **ethics** have several characteristics different from general ethics since the role of a professional is special in several ways. First, the professional is an expert in a field that many customers know little about. Most of the people affected by the devices, systems, and services of professionals do not understand how they work and cannot easily judge their quality and safety. This creates responsibilities for the professional as customers must trust the professional's knowledge, **expertise**, and honesty. Second, the products of many professionals **profoundly** affect large numbers of people. A computer professional's work can affect the life, health, finances, freedom, and future of a client or members of the public. A professional can cause great harm through dishonesty, carelessness, or **incompetence**. Often, the victims have little ability to protect themselves; they are not the direct customers of the professional and have no direct control or decision-making role in choosing the product or making decisions about its quality and safety. Thus, computer professionals have special responsibilities, not only to their customers, but also to the general public, to the users of their products, regardless of whether they have a direct relationship with the users. These responsibilities include thinking

**diminish** / di'miniʃ /
v. 减少；降低

about potential risks to privacy, system security, safety, reliability, and ease of use, and then acting to **diminish** risks that are too high.

In some cases, people act in clearly unethical or irresponsible ways; however, in many cases, there is no ill intent. Software can be enormously complex, and the process of developing it involves communications between many people with diverse roles and skills. Because of the complexity, risks, and impact of computer systems, a professional has an ethical responsibility not simply to avoid intentional evil, but to exercise a high degree of care and follow good professional practices to reduce the likelihood of errors and other problems. That includes a responsibility to maintain an expected level of **competence** and be up to date on current knowledge, technology, and standards of the profession. Professional responsibility includes knowing or learning enough about the application field to do a good job. Responsibility for a noncomputer professional who manages or uses a sophisticated computer system includes knowing or learning enough about the system to understand potential problems.

**competence**
/ 'kɔmpitəns /
n. 能力；胜任

Although people often associate courage with heroic acts, we have many opportunities to display courage in day-to-day life by making good decisions that might be difficult or unpopular. Courage in a professional setting could mean admitting to a customer that your program is faulty, declining a job for which you are not qualified, or speaking out when you see someone else doing something wrong. In some situations, it could mean quitting your job.

## II. Professional Codes of Ethics

Many professional organizations have codes of professional conduct. These codes provide a general statement of ethical values and remind people in the profession that ethical behavior is an essential part of their job and that they have specific professional responsibilities. Professional codes provide valuable guidance for new or young members of the profession who want to behave ethically but do not know what is expected of them.

There are several organizations for the range of professions included in the general term "computer professional." The main ones are the ACM[1] and the IEEE Computer Society (IEEE CS[2]). They developed the Software Engineering Code of Ethics and Professional Practice (adopted jointly by

1 *ACM*：（国际）计算机协会（*A*ssociation for *C*omputing *M*achinery 的首字母缩略）。
2 *IEEE CS*：IEEE 计算机学会（*IEEE* Computer *S*ociety 的缩略）。

the ACM and IEEE CS) and the ACM Code of Ethics and Professional Conduct. The codes emphasize the basic ethical values of honesty and fairness. They cover many aspects of professional behavior, including the responsibility to respect confidentiality, maintain professional competence, be aware of relevant laws, and honor contracts and agreements[1]. In addition, the codes put special emphasis on areas that are particularly (but not uniquely) vulnerable from computer systems. They stress the responsibility to respect and protect privacy, to avoid harm to others, and to respect property rights. The SE Code[2] covers many specific points about software development, and is available in several languages. Numerous organizations have adopted it as their internal professional standard.

Managers have special responsibility because they oversee projects and set the ethical standards for employees. Principle 5 of the SE Code includes many specific guidelines for managers. Another important code of ethics for project managers is the Project Management Institute's (PMI[3]) Code of Ethics and Professional Conduct. This code provides **mandatory** standards for all project managers and **aspirational** standards that managers should strive to **uphold**.

**mandatory** / ˈmændətəri; -ˌtɔːri /
a. 命令的；强制的
**aspirational** / ˌæspəˈreiʃənəl /
a. 有志向的；有抱负的
**uphold** / ʌpˈhəuld /
v. 举起；支持；维护

## III. Guidelines and Professional Responsibilities

Here, we highlight only a few of the many principles for producing good systems. Most concern software developers, programmers, and consultants while some are for professionals in other areas who make decisions about computer systems. Many more specific guidelines appear in the SE Code and in the ACM Code[4].

**foul-up** / ˈfaulʌp /
n. 混乱，一团糟

*Understand what success means.* After the utter **foul-up** on opening day at Kuala Lumpur's[5] airport, blamed on clerks typing incorrect commands, an airport official said, "There's nothing wrong with the system." His statement is false, and the attitude behind the statement contributes to the development of systems that will fail. The official defined the role of the airport system narrowly: to do certain data manipulation correctly, assuming all input is correct. Its true role was to get passengers, crews, planes, luggage, and cargo to the correct gates on schedule—a goal

[1] *honor contracts and agreements*：执行合同与协议。honor 有“实践”“执行”之词义。
[2] *the SE Code*：指 the Software Engineering Code of Ethics and Professional Practice。
[3] *PMI*：项目管理协会（*P*roject *M*anagement *I*nstitute 的首字母缩略）。
[4] *the ACM Code*：指 the ACM Code of Ethics and Professional Conduct。
[5] *Kuala Lumpur*：吉隆坡（马来西亚首都），读作/ ˌkwɑːləˈlumpuə /。

**institutional**
/ ˌinstiˈtjuːʃənəl /
a.(公共)机构的

**premature**
/ ˌpreməˈtjuə; ˌpriː- /
a. 提早的；早产的

at which it did not succeed. Developers and **institutional** users of computer systems must view the system's role and their responsibility in a wide enough context.

*Include users in the design and testing stages to provide safe and useful systems.* In one case, a system for a newborn nursery at a hospital rounded each baby's weight to the nearest pound[1]. For **premature** babies, the difference of a few ounces is crucial information. The responsibility of developers to talk to users is not limited to systems that affect safety and health. Systems designed to manage stories for a news website, to manage inventory in a toy store, or to organize photos and video on a website could cause frustration, waste a client's money, and end up on the trash heap if designed without sufficient consideration of the needs of actual users. Numerous studies have found that user input and communication throughout the design and development of a system are critical to the system's success.

*Do a thorough, careful job when planning and scheduling a project and when writing bids or contracts.* This includes, among many other things, allocating sufficient time and budget for testing the software or system and its security. Inadequate planning leads to pressure to cut corners[2] later.

*Design for real users.* In so many cases, systems crashed because someone typed input incorrectly. Real people make typos, get confused, or are new at their jobs. It is the responsibility of the system designers and programmers to provide clear user interfaces and include appropriate checking of input. It is impossible for software to detect all incorrect input, but there are techniques for catching many kinds of errors and for reducing the damage that errors cause.

*Require a convincing case for safety.* One of the most difficult ethical problems that arise in safety-critical applications is deciding how much risk is acceptable. For the ethical decision maker, the policy should be to suspend or delay use of the system in the absence of a convincing case for safety, rather than to proceed in the absence of a convincing case for disaster.

---

1 *a system for a newborn nursery at a hospital rounded each baby's weight to the nearest pound*：一家医院的新生儿病房的系统把每个婴儿的体重四舍五入到最接近的整磅数。

2 *cut corners*：走捷径；（为节省时间、金钱等）用简便方法办事。

cobble /ˈkɒbəl/
v. 修补（鞋子）；匆忙草率地拼凑

*Require a convincing case for security.* Systems that have security patched or **cobbled** on later are seldom as secure as those where developers design security in from the start[1]. Many insecure devices, once deployed, cannot be recalled or upgraded, and thus remain vulnerable. Designers of every device or application that connects to the Internet should expect that someone with malicious intent will discover it and attempt to expose its data or take over its operations. As with safety, the policy should be to suspend or delay use of the system in the absence of a convincing case for security.

*Do not assume existing software is safe or correct.* If you use software from another application, verify its suitability for the current project. If the software was designed for an application where the degree of harm from a failure was small, the quality and testing standards might not have been as high as necessary in the new application. The software might have confusing user interfaces that were tolerable (though not admirable) in the original application but that could have serious negative consequences in the new application. A complete safety evaluation is important even for software from an earlier version of the same application if a failure would have serious consequences.

*Be open and honest about capabilities, safety, and limitations of software.* The line between emphasizing your best qualities and being dishonest is not always clear, but it should be clear that hiding known, serious flaws and lying to customers are on the wrong side of the line. Honesty includes taking responsibility for damaging or injuring others. If a business finds that its product caused injury, it should not hide that fact or attempt to put the blame on others. Honesty about system limitations is especially important for expert systems (also called decision systems). Developers must explain the limitations and uncertainties to users. Users must not **shirk** responsibility for understanding them and using the systems properly.

shirk /ʃɜːk/
v. 逃避（工作、责任等）

*Pay attention to defaults.* Default settings might not seem important, but they are critical. Many users do not know the options they can control or how best to configure the options. Very importantly, most users do not understand issues of security and do not take the time to change settings. As

[1] *Systems that have security patched or cobbled on later are seldom as secure as those where developers design security in from the start.*：事后打上安全补丁的系统很少像开发人员一开始就在其中进行了安全性设计的系统那样安全。

a result, system designers must give serious thought to default settings. Sometimes, protection (of privacy or from hackers, for example) is the ethical priority. Other times, ease of use and compatibility with user expectations is a priority. Balancing these priorities can lead to difficult conflicts.

*Develop communications skills.* There are many situations in which a computer professional must explain technical issues to customers and coworkers. Learning how to organize information, distinguishing what is important to communicate and what is not, engaging the listener actively in the conversation to maintain interest, and so on, will help make one's presentations more effective and help to ensure that the client or coworker is truly informed.

---

## Exercises

**I. Fill in the blanks with the information given in the text:**

1. Computer professionals should not only avoid ________ evil, but also exercise a high degree of care and follow good professional ________ to reduce the likelihood of errors and other problems.

2. An ethical decision maker should suspend or delay use of a system in the absence of a convincing case for ________, rather than to proceed in the absence of a convincing case for ________.

3. Honesty about system ________ is especially important for ________ systems (also called decision systems).

4. Most users do not understand issues of security and do not take the time to change settings, so system designers must give serious thought to ________ settings.

**II. Translate the following terms or phrases from English into Chinese and vice versa:**

| | |
|---|---|
| 1. safety-critical system | 6. 专家系统 |
| 2. professional competence | 7. 默认设置 |
| 3. code of professional conduct | 8. 产权 |
| 4. code of ethics | 9. 项目经理 |
| 5. data manipulation | 10. 计算机专业人员 |

# Section C

# Social Issues of Computer Networks

Computer networks, like the **printing press** 500 years ago, allow ordinary citizens to distribute and view content in ways that were not previously possible. But along with the good comes the bad, as this new-found freedom brings with it many unsolved social, political, and ethical issues. Let us just briefly mention a few of them; a thorough study would require a full book, at least.

**Social networks, message boards**, content sharing sites, and a host of other applications allow people to share their views with like-minded individuals. As long as the subjects are restricted to technical topics or hobbies like gardening, not too many problems will arise.

The trouble comes with topics that people actually care about, like politics, religion, or sex. Views that are publicly posted may be deeply offensive to some people. Worse yet, they may not be politically correct. Furthermore, opinions need not be limited to text; high-resolution color photographs and video clips are easily shared over computer networks. Some people take a **live-and-let-live** view, but others feel that posting certain material (e.g., **verbal** attacks on particular countries or religions, **pornography**, etc.) is simply unacceptable and that such content must be **censored**. Different countries have different and conflicting laws in this area. Thus, the debate **rages**.

In the past, people have **sued** network operators, claiming that they are responsible for the contents of what they carry, just as newspapers and magazines are. The inevitable response is that a network is like a telephone company or the post office and cannot be expected to **police** what its users say.

It should now come only as a slight surprise to learn that some network operators block content for their own reasons. Some users of peer-to-peer applications had their network service cut off because the network operators did not find it profitable to carry the large amounts of traffic sent by those applications. Those same operators would probably like to treat different companies differently. If you are a big company and pay well then you get

**printing press**
印刷机

**social network**
社交网络

**message board**
留言板，消息板

**live-and-let-live**
a. 自己活也让别人活的；互相宽容的；互不相扰的

**verbal** / ˈvɜːbəl /
a. 用言辞的；文字上的；口头的

**pornography**
/ pɔːˈnɒɡrəfi /
n. [总称] 色情（或淫秽）作品

**censor** / ˈsensə /
v. 审查，检查

**rage** / reidʒ /
v. 发怒；激烈进行

**sue** / sjuː; suː /
v. 控告，起诉

**police** / pəˈliːs /
v. 维持…的治安；管理；监督

good service, but if you are a **small-time** player, you get poor service. Opponents of this practice argue that peer-to-peer and other content should be treated in the same way because they are all just bits to the network. This argument for communications that are not differentiated by their content or source or who is providing the content is known as network **neutrality**. It is probably safe to say that this debate will go on for a while.

Many other parties are involved in the **tussle** over content. For instance, **pirated** music and movies fueled the massive growth of peer-to-peer networks, which did not please the copyright holders, who have threatened (and sometimes taken) legal action. There are now automated systems that search peer-to-peer networks and fire off[1] warnings to network operators and users who are suspected of **infringing** copyright. In the United States, these warnings are known as DMCA[2] takedown notices after the Digital **Millennium** Copyright Act[3]. This search is an **arms race** because it is hard to reliably catch copyright **infringement**. Even your printer might be mistaken for a **culprit**.

Computer networks make it very easy to communicate. They also make it easy for the people who run the network to **snoop** on the traffic. This sets up[4] conflicts over issues such as employee rights versus employer rights. Many people read and write email at work. Many employers have claimed the right to read and possibly censor employee messages, including messages sent from a home computer outside working hours. Not all employees agree with this, especially the latter part.

Another conflict is centered around government versus citizen's rights. The FBI[5] has installed systems at many Internet service providers to snoop on all incoming and outgoing email for **nuggets** of interest[6]. One early system was originally called **Carnivore**, but bad publicity caused it to be renamed to the more innocent-sounding DCS1000[7]. The goal of such systems is to spy on millions of people in the hope of perhaps finding information about illegal activities. Unfortunately for the spies, the Fourth

**small-time** / ˈsmɔːlˈtaim /
a. 〈口〉次要的，无关紧要的

**neutrality** / njuːˈtræliti /
n. 中立

**tussle** / ˈtʌsəl /
n. 争执，争辩

**pirate** / ˈpaiərət /
v. 剽窃；盗用；非法翻印

**infringe** / inˈfrindʒ /
v. 违犯，侵犯（权利等）

**millennium** / miˈleniəm /
n. 一千年；千禧年

**arms race**
军备竞赛

**infringement** / inˈfrindʒmənt /
n.（对他人权利等的）侵犯，侵害

**culprit** / ˈkʌlprit /
n. 罪犯；被控犯罪的人

**snoop** / snuːp /
v. 窥探；打探

**nugget** / ˈnʌgit /
n. 小块；天然金块；有价值的东西

**carnivore** / ˈkɑːnivɔː /
n. 食肉动物

[1] *fire off*：恶狠狠地发出；迅速发出。

[2] *DMCA*：《数字千年版权法》《千禧年数字版权法》（*D*igital *M*illennium *C*opyright *A*ct 的首字母缩略）。

[3] *In the United States, these warnings are known as DMCA takedown notices after the Digital Millennium Copyright Act.*：在美国，这些警告按照《数字千年版权法》称为 DMCA 取下通知。

[4] *set up*：引起，产生。

[5] *FBI*：（美国）联邦调查局（*F*ederal *B*ureau of *I*nvestigation 的首字母缩略）。

[6] *to snoop on all incoming and outgoing email for nuggets of interest*：窥探所有的来往电子邮件，寻找感兴趣的信息。

[7] *DCS*：数字收集系统（*D*igital *C*ollection *S*ystem 的首字母缩略）。

**Amendment** to the U.S. Constitution prohibits government searches without a search **warrant**, but the government often ignores it.

Of course, the government does not have a **monopoly** on threatening people's privacy. The private sector does its bit too by profiling users[1]. For example, small files called **cookies** that Web browsers store on users' computers allow companies to track users' activities in cyberspace and may also allow credit card numbers, social security numbers, and other confidential information to leak all over the Internet. Companies that provide Web-based services may maintain large amounts of personal information about their users that allows them to study user activities directly. For example, Google can read your email and show you advertisements based on your interests if you use its email service, Gmail[2].

A new twist with mobile devices is location privacy[3]. As part of the process of providing service to your mobile device the network operators learn where you are at different times of day. This allows them to track your movements. They may know which nightclub you **frequent** and which medical center you visit.

Computer networks also offer the potential to increase privacy by sending **anonymous** messages. In some situations, this capability may be desirable. Beyond preventing companies from learning your habits, it provides, for example, a way for students, soldiers, employees, and citizens to blow the whistle on[4] illegal behavior on the part of professors, officers, superiors, and politicians without fear of **reprisals**. On the other hand, in the United States and most other democracies, the law specifically permits an accused person the right to confront and challenge his accuser in court so anonymous **accusations** cannot be used as evidence.

The Internet makes it possible to find information quickly, but a great deal of it is ill considered, misleading, or **downright** wrong. That medical advice you **plucked** from the Internet about the pain in your chest may have come from a Nobel Prize winner or from a high-school dropout.

Other information is frequently unwanted. Electronic **junk** mail (spam) has become a part of life because **spammers** have collected millions of

**amendment** /əˈmendmənt/ n. 修正案，修正条款

**warrant** /ˈwɔrənt; ˈwɔː-/ n. 授权（令）

**monopoly** /məˈnɔpəli/ n. 垄断；独有

**cookie** /ˈkuki/ n. 小甜饼(指一种临时保存网络用户信息的结构)

**frequent** /friˈkwent/ v. 常到，常去，时常出入于

**anonymous** /əˈnɔniməs/ a. 匿名的

**reprisal** /riˈpraizəl/ n. 报复

**accusation** /ˌækjuːˈzeiʃən/ n. 指控，控告

**downright** /ˈdaunrait/ ad. 彻底地，完全地

**pluck** /plʌk/ v. 拔；摘，采

**junk** /dʒʌŋk/ n. 废旧物品；破烂，垃圾

**spammer** /ˈspæmə/ n. 垃圾邮件发送者

[1] *The private sector does its bit too by profiling users.*：私营部门通过用户资料挖掘（或用户分析）在这方面也有份。do one's bit 意为“做自己应做的一份工作”“尽本分”。

[2] *Gmail*：Google 提供的电子邮件服务。

[3] *A new twist with mobile devices is location privacy.*：移动设备的一个新花样是位置隐私。

[4] *to blow the whistle (on)*：告发，揭发。

email addresses and would-be marketers can cheaply send computer-generated messages to them. Fortunately, filtering software is able to read and discard the spam generated by other computers, with lesser or greater degrees of success.

Still other content is intended for criminal behavior. Web pages and email messages containing active content (basically, programs or macros that execute on the receiver's machine) can contain viruses that take over your computer. They might be used to steal your bank account passwords, or to have your computer send spam as part of a botnet or pool of compromised machines.

**phishing** /ˈfiʃiŋ/
n. 网络钓鱼
**masquerade** /ˌmæskəˈreid, ˌmɑːs-/
v. 假冒；假扮
**identity theft**
身份（信息）盗取

**Phishing** messages **masquerade** as originating from a trustworthy party, for example, your bank, to try to trick you into revealing sensitive information, for example, credit card numbers. **Identity theft** is becoming a serious problem as thieves collect enough information about a victim to obtain credit cards and other documents in the victim's name.

It can be difficult to prevent computers from impersonating people on the Internet. This problem has led to the development of CAPTCHAs[1], in which a computer asks a person to solve a short recognition task, for example, typing in the letters shown in a **distorted** image, to show that he is human. This process is a variation on the famous Turing test[2] in which a person asks questions over a network to judge whether the entity responding is human.

**distorted** /disˈtɔːtid/
a. 扭曲的，变形的
**authenticate** /ɔːˈθentikeit/
v. 验证，鉴别
**mischief** /ˈmistʃif/
n. 恶作剧；淘气；损害
**buggy** /ˈbʌgi/
a. 有（程序）错误的，有故障的
**refund** /ˈriːfʌnd/
n. 退款；退还
**defective** /diˈfektiv/
a. 有缺点的，有缺陷的，有毛病的

A lot of these problems could be solved if the computer industry took computer security seriously. If all messages were encrypted and **authenticated**, it would be harder to commit **mischief**. Such technology is well established. The problem is that hardware and software vendors know that putting in security features costs money and their customers are not demanding such features. In addition, a substantial number of the problems are caused by **buggy** software, which occurs because vendors keep adding more and more features to their programs, which inevitably means more code and thus more bugs. A tax on new features might help, but that might be a tough sell in some quarters[3]. A **refund** for **defective** software might be

---

[1] *CAPTCHA*：全自动区分计算机和人类的图灵测试，验证码（*C*ompletely *A*utomated *P*ublic *T*uring Test to Tell *C*omputers and *H*umans *A*part 的缩略），读作 /ˈkæptʃə/。

[2] *Turing test*：图灵测试。英国数学家和逻辑学家艾伦·图灵（1912—1954）于 1950 年提出的一种测试方法，用来检测一台计算机是否具有类似人类思维的能力。

[3] *A tax on new features might help, but that might be a tough sell in some quarters.*：对新特征征税可能发挥作用，但这样做可能让有些人士很难接受。

nice, except it would bankrupt the entire software industry in the first year.

Computer networks raise new legal problems when they interact with old laws. Electronic gambling provides an example. Computers have been simulating things for decades, so why not simulate **slot machines**, **roulette wheels**, **blackjack dealers**, and more gambling equipment? Well, because it is illegal in a lot of places. The trouble is, gambling is legal in a lot of other places (England, for example) and **casino** owners there have grasped the potential for Internet gambling. What happens if the gambler, the casino, and the server are all in different countries, with conflicting laws? Good question.

**slot** / slɔt /
n. 狭长孔，狭缝
**slot machine**
吃角子老虎（一种投硬币的赌具）
**roulette** / ruːˈlet /
n. 轮盘赌
**roulette wheel**
（轮盘赌台上的）轮盘
**blackjack**
/ ˈblækdʒæk /
n. 21 点（一种牌戏）
**dealer** / ˈdiːlə /
n. 发牌者；商人
**casino** / kəˈsiːnəu /
n.（有表演、舞池等的）卡西诺赌场

---

## Exercises

**I. Fill in the blanks with the information given in the text:**

1. The argument for communications not to be differentiated by their content or source or who is providing the content is known as network ________.

2. Social ________, message boards, ________ sharing sites, and a host of other applications allow people to share their views with like-minded individuals.

3. CAPTCHAs are used by a computer to ask a person to solve a short ________ task to show that he is human.

4. ________messages masquerade as originating from a trustworthy party to try to trick you into revealing sensitive information.

**II. Translate the following terms or phrases from English into Chinese and vice versa:**

| | |
|---|---|
| 1. message board | 5. 过滤软件 |
| 2. software vendor | 6. 版权侵犯 |
| 3. anonymous message | 7. 网络中立性 |
| 4. video clip | 8. 网络运营商 |

# Unit 12 Smart World

# (智能世界)

## Section A

## Artificial Intelligence

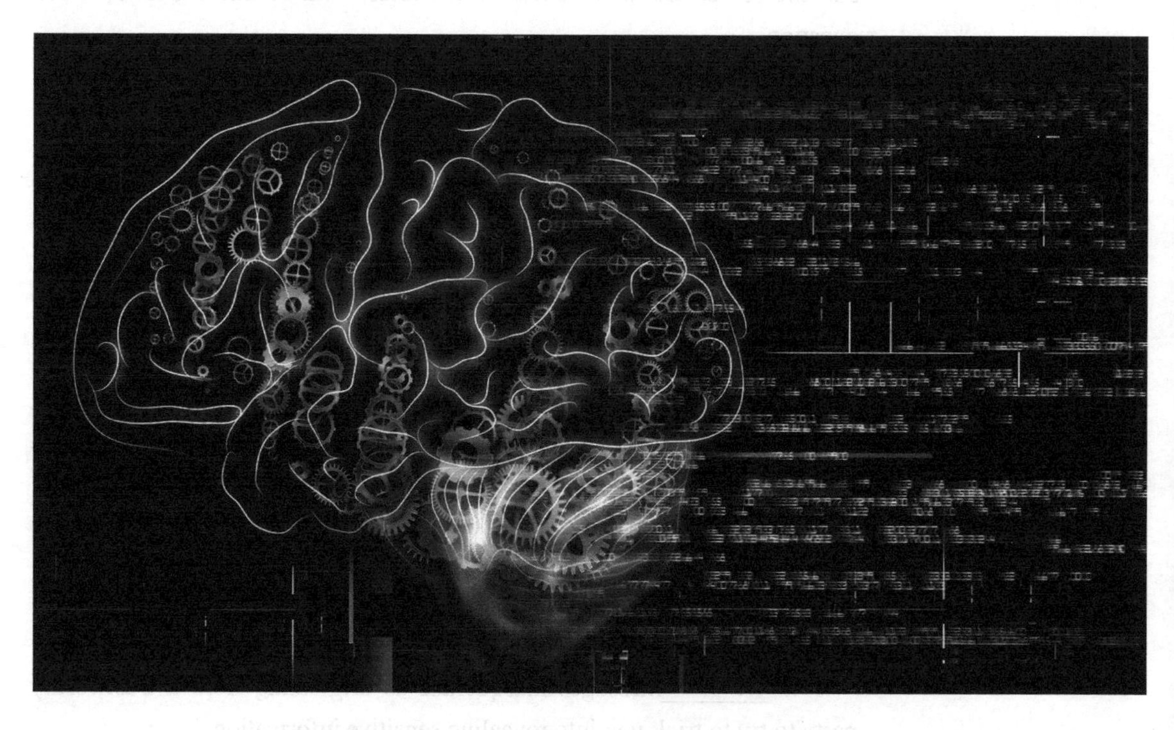

Artificial intelligence is the field of computer science that seeks to build autonomous machines—machines that can carry out complex tasks without human intervention. This goal requires that machines be able to perceive and reason. Such capabilities fall within the category of common sense activities that, although natural for the human mind, have historically proven difficult for machines. The result is that work in the field continues to be **challenging**.

**challenging**
/ 'tʃælindʒiŋ /
a. 挑战性的

**intelligent agent**
智能代理，
智能主体

**neurology**
/ ˌnjuə'rɔlədʒi /
n. 神经病学；
神经学

### I. Intelligent Agents

The field of artificial intelligence is quite large and merges with other subjects such as psychology, **neurology**, mathematics, linguistics, and electrical and mechanical engineering. To focus our thoughts, then, we

**stimulus** / 'stimjuləs /（[复] **-li** / -lai / 或 **-luses**）
n. 刺激（物），激励（物）
**actuator** / 'æktjueitə /
n. 执行器，执行机构；致动器
**range sensor**
距离传感器
**gripper** / 'gripə /
n. 夹子；抓爪（器）
**synthesizer** / 'sinθisaizə /
n. 合成器
**speech synthesizer**
语音合成器

begin by considering the concept of an agent and the types of intelligent behavior that an agent might exhibit. Indeed, much of the research in artificial intelligence can be categorized in terms of an agent's behavior.

An agent is a "device" that responds to **stimuli** from its environment. It is natural to envision an agent as an individual machine such as a robot, although an agent may take other forms such as an autonomous airplane, a character in an interactive video game, or a process communicating with other processes over the Internet (perhaps as a client, a server, or a peer). Most agents have sensors by which they receive data from their environments and **actuators** by which they can affect their environments. Examples of sensors include microphones, cameras, **range sensors**, and air or soil sampling devices. Examples of actuators include wheels, legs, wings, **grippers**, and **speech synthesizers**.

Much of the research in artificial intelligence can be characterized in the context of building agents that behave intelligently, meaning that the actions of the agent's actuators must be rational responses to the data received through its sensors. In turn, we can classify this research by considering different levels of these responses.

**reflex** / 'ri:fleks /
n. 反射(作用);(对刺激的）本能反应
**reflex action**
反射作用；本能反应；本能动作
**empower** / im'pauə /
v. 授权给；使能够

The simplest response is a **reflex action**, which is merely a predetermined response to the input data. Higher levels of response are required to obtain more "intelligent" behavior. For example, we might **empower** an agent with knowledge of its environment and require that the agent adjust its actions accordingly. The process of throwing a baseball is largely a reflex action but determining how and where to throw the ball requires knowledge of the current environment. How such real-world knowledge can be stored, updated, accessed, and ultimately applied in the decision-making process continues to be a challenging problem in artificial intelligence.

**maneuver**
/ mə'nu:və /
v.〈美〉调动；机动；移动
**passageway**
/ 'pæsidʒwei /
n. 走廊；通道

Another level of response is required if we want the agent to seek a goal such as winning a game of chess or **maneuvering** through a crowded **passageway**. Such goal-directed behavior requires that the agent's response, or sequence of responses, be the result of deliberately forming a plan of action or selecting the best action among the current options.

In some cases an agent's responses improve over time as the agent learns. This could take the form of developing procedural knowledge (learning "how") or storing declarative knowledge (learning "what").

**trial-and-error** / 'traiələnd'erə /
a. 试错法的；反复试验的

**checkers** / 'tʃekəz /
n.〈美〉西洋跳棋

Learning procedural knowledge usually involves a **trial-and-error** process by which an agent learns appropriate actions by being punished for poor actions and rewarded for good ones. Following this approach, agents have been developed that, over time, improve their abilities in competitive games such as **checkers** and chess. Learning declarative knowledge usually takes the form of expanding or altering the "facts" in an agent's store of knowledge. For example, a baseball player must repeatedly adjust his or her database of knowledge from which rational responses to future events are determined.

**gyroscope** / 'dʒaiərəskəup /
n. 陀螺仪，回转仪

To produce rational responses to stimuli, an agent must "understand" the stimuli received by its sensors. That is, an agent must be able to extract information from the data produced by its sensors, or in other words, an agent must be able to perceive. In some cases this is a straightforward process. Signals obtained from a **gyroscope** are easily encoded in forms compatible with calculations for determining responses[1]. But in other cases extracting information from input data is difficult. Examples include understanding speech and images. Likewise, agents must be able to formulate their responses in terms compatible with their actuators[2]. This might be a straightforward process or it might require an agent to formulate responses as complete spoken sentences—meaning that the agent must generate speech. In turn, such topics as image processing and analysis, natural language understanding, and speech generation are important areas of research.

**toehold** / 'təuhəuld /
n.（攀登悬崖等时脚趾大小的）立足点，支点

The agent attributes that we have identified here represent past as well as current areas of research. Of course, they are not totally independent of each other. We would like to develop agents that possess all of them, producing agents that understand the data received from their environments and develop new response patterns through a learning process whose goal is to maximize the agent's abilities. However, by isolating various types of rational behavior and pursuing them independently, researchers gain a **toehold** that can later be combined with progress in other areas to produce more intelligent agents.

## II. Research Methodologies

To appreciate the field of artificial intelligence, it is helpful to

1 *encoded in forms compatible with calculations for determining responses*：以与确定反应的计算兼容的形式进行编码。

2 *formulate their responses in terms compatible with their actuators*：以与其执行器兼容的词语确切表述其反应。

understand that it is being pursued along two paths. One is the engineering track in which researchers are trying to develop systems that exhibit intelligent behavior. The other is a theoretical track in which researchers are trying to develop a computational understanding of animal—especially human—intelligence. This **dichotomy** is clarified by considering the manner in which the two tracks are pursued. The engineering approach leads to a performance-oriented methodology because the underlying goal is to produce a product that meets certain performance goals. The theoretical approach leads to a **simulation**-oriented methodology because the underlying goal is to expand our understanding of intelligence and thus the emphasis is on the underlying process rather than the exterior performance.

**dichotomy** /daiˈkɔtəmi/
n. 一分成二；二分法

**simulation** /ˌsimjuˈleiʃən/
n. 模拟，仿真

As an example, consider the fields of natural language processing and linguistics. These fields are closely related and benefit from research in each other, yet the underlying goals are different. **Linguists** are interested in learning how humans process language and thus tend toward more theoretical **pursuits**. Researchers in the field of natural language processing are interested in developing machines that can manipulate natural language and therefore lean in the engineering direction. Thus, linguists operate in simulation-oriented mode—building systems whose goals are to test theories. In contrast, researchers in natural language processing operate in performance-oriented mode—building systems to perform tasks. Systems produced in this latter mode (such as document translators and systems by which machines respond to verbal commands) rely heavily on knowledge gained by linguists but often apply "shortcuts" that happen to work in the restricted environment of the particular system.

**linguist** /ˈliŋgwist/
n. 语言学家

**pursuit** /pəˈsjuːt/
n. 追求；追赶

As an elementary example, consider the task of developing a shell for an operating system that receives instructions from the outside world through verbal English commands. In this case, the shell (an agent) does not need to worry about the entire English language. More precisely, the shell does not need to distinguish between the various meanings of the word *copy*. (Is it a noun or a verb? Should it carry the **connotation** of **plagiarism**?) Instead, the shell needs merely to distinguish the word *copy* from other commands such as *rename* and *delete*. Thus the shell could perform its task just by matching its inputs to predetermined audio patterns. The performance of such a system may be satisfactory to an engineer, but the way it is obtained would not be **aesthetically** pleasing to a **theoretician**.

**connotation** /ˌkɔnəuˈteiʃən/
n. 内涵（意义），含义

**plagiarism** /ˈpleidʒjərizəm/
n. 剽窃，抄袭

**aesthetical** /iːsˈθetikəl/
a. 美学的；审美的

**theoretician** /ˌθiəriˈtiʃən/
n. 理论家

**Turing test**
图灵测试
**benchmark**
/ ˈbentʃmɑːk /
n. 基准
**folklore** / ˈfəuklɔː /
n. 民间传说；
民俗（学）
**interrogator**
/ inˈterəugeitə /
n. 讯问者；审问者

## III. The Turing Test

In the past the Turing test (proposed by Alan Turing in 1950) has served as a **benchmark** in measuring progress in the field of artificial intelligence. Today the significance of the Turing test has faded although it remains an important part of the artificial intelligence **folklore**. Turing's proposal was to allow a human, whom we call the **interrogator**, to communicate with a test subject by means of a typewriter system without being told whether the test subject was a human or a machine. In this environment, a machine would be declared to behave intelligently if the interrogator was not able to distinguish it from a human. Turing predicted that by the year 2000 machines would have a 30 percent chance of passing a five-minute Turing test—a conjecture that turned out to be surprisingly accurate.

**eerie** / ˈiəri; ˈiri /
a.（因怪诞或阴森而）引起恐惧的；怪异的

**restructure**
/ riːˈstrʌktʃə /
v. 重建；改组；调整
**psychotherapy**
/ ˌsaikəuˈθerəpi /
n. 心理疗法，心理治疗
**dismay** / disˈmei /
n. 失望；气馁；惊愕
**thesis** / ˈθiːsis /
（[复] **-ses** / -siːz /）
n. 命题；论点；（学位）论文
**therapeutic**
/ ˌθerəˈpjuːtik /
a. 治疗的；
有疗效的

One reason that the Turing test is no longer considered to be a meaningful measure of intelligence is that an **eerie** appearance of intelligence can be produced with relative ease. A well-known example arose as a result of the program DOCTOR (a version of the more general system called ELIZA) developed by Joseph Weizenbaum[1] in the mid-1960s. This interactive program was designed to project the image of a Rogerian analyst[2] conducting a psychological interview; the computer played the role of the analyst while the user played the patient. Internally, all that DOCTOR did was **restructure** the statements made by the patient according to some well-defined rules and direct them back to the patient. For example, in response to the statement "I am tired today," DOCTOR might have replied with "Why do you think you're tired today?" If DOCTOR was unable to recognize the sentence structure, it merely responded with something like "Go on" or "That's very interesting."

Weizenbaum's purpose in developing DOCTOR dealt with the study of natural language communication. The subject of **psychotherapy** merely provided an environment in which the program could "communicate." To Weizenbaum's **dismay**, however, several psychologists proposed using the program for actual psychotherapy. (The Rogerian **thesis** is that the patient, not the analyst, should lead the discussion during the **therapeutic** session,

---

[1] *Joseph Weizenbaum*：约瑟夫·魏岑鲍姆（1923—2008），出生于德国，麻省理工学院计算机科学荣誉退休教授。

[2] *Rogerian analyst*：罗杰斯式分析师。来源于卡尔·罗杰斯（1902—1987），美国心理学家，首创以患者为中心的心理疗法。

and thus, they argued, a computer could possibly conduct a discussion as well as a **therapist** could.) Moreover, DOCTOR projected the image of comprehension so strongly that many who "communicated" with it became **subservient** to the machine's question-and-answer dialogue. In a sense, DOCTOR passed the Turing test. The result was that ethical, as well as technical, issues were raised, and Weizenbaum became an advocate for maintaining human dignity in a world of advancing technology.

**therapist** / 'θerəpist /
n.（特定治疗法的）治疗专家
**subservient** / səb'sə:viənt /
a. 屈从的；恭顺的；顺从的

More recent examples of Turing test "successes" include Internet viruses that carry on "intelligent" dialogs with a human victim in order to trick the human into dropping his or her malware guard. Moreover, phenomena similar to Turing tests occur in the context of computer games such as chess-playing programs. Although these programs select moves merely by applying **brute**-force[1] techniques, humans competing against the computer often experience the **sensation** that the machine possesses creativity and even a personality. Similar sensations occur in robotics where machines have been built with physical attributes that project intelligent characteristics. Examples include toy robot dogs that project **adorable** personalities merely by **tilting** their heads or lifting their ears in response to a sound.

**brute** / bru:t /
a. 粗暴的；野蛮的
**sensation** / sen'seiʃən /
n. 感觉；轰动；激动
**adorable** / ə'dɔ:rəbəl /
a. 值得崇拜的；可爱的
**tilt** / tilt /
v.（使）倾斜，（使）倾侧

## Exercises

**I. Fill in the blanks with the information given in the text:**

1. Artificial intelligence seeks to build ________ machines to carry out complex tasks without human intervention.

2. Artificial intelligence is closely related with psychology, ________, mathematics, linguistics, and electrical and ________ engineering.

3. Most agents have ________ by which they receive data from their environments and ________ by which they can affect their environments.

4. An agent's learning could take the form of developing ________ knowledge (learning "how") or storing ________ knowledge (learning "what").

5. Artificial intelligence is being pursued along two paths, namely the

[1] *brute force*：暴力，蛮力。在计算机领域，指BF算法、暴力算法、蛮力法。

________ track and the ________ track.

6. The ________ methodology and the ________ methodology are adopted respectively for the two tracks pursued in artificial intelligence.

7. Turing predicted that by the year 2000 machines would have a 30 percent chance of passing a five-minute ________ test.

8. DOCTOR is a(n) ________ program developed in the mid-1960s to project the image of a Rogerian ________ conducting a psychological interview.

**II. Translate the following terms or phrases from English into Chinese and vice versa:**

| | |
|---|---|
| 1. autonomous machine | 11. 图灵测试 |
| 2. intelligent agent | 12. 语音合成器 |
| 3. performance-oriented methodology | 13. 测试对象 |
| 4. declarative knowledge | 14. 文档翻译程序 |
| 5. reflex action | 15. 交互程序 |
| 6. soil sampling device | 16. 距离传感器 |
| 7. physical attribute | 17. 面向模拟的方法论 |
| 8. goal-directed behavior | 18. 图像处理与分析 |
| 9. common sense | 19. 过程性知识 |
| 10. audio pattern | 20. 试错过程 |

**III. Fill in each of the blanks with one of the words given in the following list, making changes if necessary:**

| | | | |
|---|---|---|---|
| *machine* | *directly* | *observe* | *conjecture* |
| *strong* | *debate* | *combine* | *right* |
| *possible* | *same* | *different* | *component* |
| *weak* | *intelligence* | *consciousness* | *attribute* |

The conjecture that machines can be programmed to exhibit intelligent behavior is known as ________ AI and is accepted, to varying degrees, by a wide audience today. However, the ________ that machines can be programmed to possess intelligence and, in fact, ________, which is known as strong AI, is widely ________. Opponents of strong AI argue that a machine is inherently ________ from a human and thus can never feel love, tell ________ from wrong, and think about itself in the ________ way that a human does. However, *proponents* (支持者) of ________ AI argue that the human mind is constructed from small ________ that individually are not human and are not conscious but, when ________, are. Why, they argue, would the same phenomenon not be ________ with machines? The problem in resolving the strong AI debate is that such ________ as intelligence and consciousness are internal characteristics that cannot be identified ________. As Alan Turing pointed out, we credit other humans with ________ because they behave intelligently—even though we cannot ________ their internal mental states. Are we, then, prepared to grant the same *latitude* (自由度) to a ________ if it exhibits the external characteristics of consciousness?

**IV. Translate the following passage from English into Chinese:**

Early work in artificial intelligence approached the subject in the context of explicitly writing programs to simulate intelligence. However, many argue today that human intelligence is not based on the execution of complex programs but instead on simple stimulus-response functions that have evolved over generations. This theory of "intelligence" is known as behavior-based intelligence because "intelligent" stimulus-response functions appear to be the result of behaviors that caused certain individuals to survive and reproduce while others did not.

Behavior-based intelligence seems to answer several questions in the artificial intelligence community such as why machines based on the von Neumann architecture easily outperform humans in computational skills but struggle to exhibit common sense. Thus behavior-based intelligence promises to be a major influence in artificial intelligence research. Behavior-based techniques have been applied in the field of artificial neural networks to teach neurons to behave in desired ways, in the field of genetic algorithms to provide an alternative to the more traditional programming process, and in robotics to improve the performance of machines through reactive strategies.

# Section B

# Augmented Reality and Its Applications

## I. Introduction

**augment** /ɔ:g'ment/
v. 增强；增加；扩大
**augmented reality**
增强现实
**indistinguishable**
/ˌindis'tiŋgwiʃəbəl/
a. 难以区别的，难以分辨的
**detach** /di'tætʃ/
v. 拆卸；使分开；使分离
**proposition**
/ˌprɔpə'ziʃən/
n. 提议，建议

Virtual reality (VR[1]) is becoming increasingly popular, as computer graphics have progressed to a point where the images are often **indistinguishable** from the real world. However, the computer-generated images presented in games, movies, and other media are **detached** from our physical surroundings. This is both a virtue—everything becomes possible—and a limitation. The limitation comes from the main interest we have in our daily life, which is not directed toward some virtual world, but rather toward the real world surrounding us.

In many ways, enhancing mobile computing so that the association with the real world happens automatically seems an attractive **proposition**.

[1] *VR*：虚拟现实（*v*irtual *r*eality 的首字母缩略）。

Augmented reality (AR[1]) holds the promise of creating direct, automatic, and actionable links between the physical world and electronic information. It provides a simple and immediate user interface to an electronically enhanced physical world. AR can **overlay** computer-generated information on views of the real world, **amplifying** human perception and cognition in remarkable new ways.

**overlay** /ˌəuvəˈlei/
v. 置（一物）于他物之上；在…上覆盖
**amplify** /ˈæmplifai/
v. 放大；增强；扩大

## II. Definition and Scope

Whereas VR places a user inside a completely computer-generated environment, AR aims to present information that is directly registered to the physical environment. AR goes beyond mobile computing in that it bridges the gap between virtual world and real world, both spatially and **cognitively**. With AR, the digital information appears to become part of the real world, at least in the user's perception.

**cognitive** /ˈkɔgnitiv/
a. 认识的，认知的

Achieving this connection is a grand goal—one that draws upon knowledge from many areas of computer science, yet can lead to **misconceptions** about what AR really is. For example, many people associate the visual combination of virtual and real elements with the special effects in movies such as ***Jurassic Park***[2] and ***Avatar***[3]. While the computer graphics techniques used in movies may be applicable to AR as well, movies lack one crucial aspect of AR—**interactivity**. To avoid such confusion, we need to set a scope for the topic discussed. In other words, we need to answer a key question: What is AR?

**misconception** /ˌmiskənˈsepʃən/
n. 误解，错误想法
**Jurassic** /dʒuˈræsik/
n. & a. 侏罗纪（的）
**avatar** /ˈævətɑː/
n. 化身（印度教和佛教中化作人形或兽形的神）
**interactivity** /ˌintərækˈtiviti/
n. 交互性

The most widely accepted definition of AR was proposed by Azuma[4] in his 1997 survey paper. According to Azuma, AR must have the following three characteristics:

- Combines real and virtual
- Interactive in **real time**
- Registered in 3D[5]

**real time**
实时

That is, for a working definition, AR relies on three key components: (1) the combination of virtual and real information, with the real world as the primary place of action and (2) interactive, real-time updates of (3)

---

[1] *AR*：增强现实（*a*ugmented *r*eality 的首字母缩略）。
[2] *Jurassic Park*：《侏罗纪公园》，美国科幻电影。
[3] *Avatar*：《阿凡达》，美国科幻电影。
[4] *Azuma*：Ronald T. Azuma，罗纳德·T. 阿祖马，增强现实领域开拓者之一，现就职于英特尔研究院。其 1997 年的 survey paper（综述论文）指的是 *A Survey of Augmented Reality*。
[5] *3D*：三维，立体（*three d*imensions 的缩略），读作 /ˈθriːˈdiː/。

**haptic** / 'hæptik /
a. 触觉的
**olfactory** / ɔl'fæktəri /
a. 嗅觉的
**gustatory** / 'gʌstətəri /
a. 味觉的
**alignment** / ə'lainmənt /
n. 对准，对齐
**mandate** / 'mændeit /
n. 授权；命令；指令
**augmentation** / ˌɔːgmen'teiʃən /
n. 增强；增加；扩大
**situated** / 'sitjueitid, -tʃu- /
a. 位于…的；处于…境地的
**situated visualization**
情境可视化，位置可视化
**showcase** / 'ʃəukeis /
v. 陈列；展示

virtual information registered in 3D with the physical environment. This definition does not require a specific output device, such as a head-mounted display (HMD[1]), nor does it limit AR to visual media. Audio, **haptic**, and even **olfactory** or **gustatory** AR are included in its scope, even though they may be difficult to realize. Note that the definition does require real-time control and spatial registration, meaning precise real-time **alignment** of corresponding virtual and real information. This **mandate** implies that the user of an AR display can at least exercise some sort of interactive viewpoint control, and the computer-generated **augmentations** in the display will remain registered to the referenced objects in the environment[2].

While opinions on what qualifies as real-time performance may vary depending on the individual and on the task or application, interactivity implies that the human-computer interface operates in a tightly coupled feedback loop[3]. The user continuously navigates the AR scene and controls the AR experience. The system, in turn, picks up the user's input by tracking the user's viewpoint or pose. It registers the pose in the real world with the virtual content, and then presents to the user a **situated visualization** (a visualization that is registered to objects in the real world).

We can see that a complete AR system requires at least three components: a tracking component, a registration component, and a visualization component. A fourth component—a spatial model (i.e., a database)—stores information about the real world and about the virtual world (Figure 12B-1). The real-world model is required to serve as a reference for the tracking component, which must determine the user's location in the real world. The virtual-world model consists of the content used for the augmentation. Both parts of the spatial model must be registered in the same coordinate system.

## III. Application Examples

In this section, we continue our exploration of AR by examining a few examples, which **showcase** both AR technology and applications of that technology. Besides the following examples, many other specific

---

1 *HMD*：头戴式显示器（*h*ead-*m*ounted *d*isplay 的缩略）。

2 *the computer-generated augmentations in the display will remain registered to the referenced objects in the environment*：显示器中计算机生成的增强内容将持续地注册到环境中的参考对象。

3 *the human-computer interface operates in a tightly coupled feedback loop*：人机界面在紧密耦合的反馈回路中操作。

application possibilities exist, such as AR for industry and construction, surgery, navigation, and games.

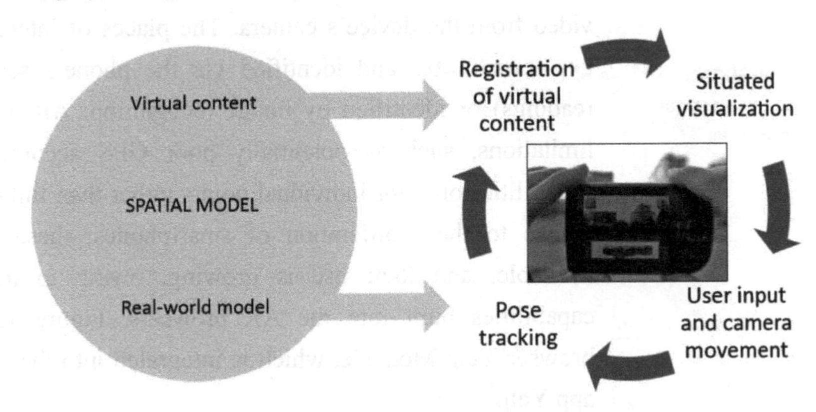


**Figure 12B-1: AR uses a feedback loop between human user and computer system. The user observes the AR display and controls the viewpoint. The system tracks the user's viewpoint, registers the pose in the real world with the virtual content, and presents situated visualizations**

### 1. Maintenance and Training

Understanding how things work, and learning how to assemble, disassemble, or repair them, is an important challenge in many professions. Maintenance engineers often devote a large amount of time to studying manuals and documentation, since it is often impossible to **memorize** all procedures in detail. AR, however, can present instructions directly **superimposed** in the field of view of the worker. This can provide more effective training, but, more importantly, allows personnel with less training to correctly perform the work.

If human support is sought, AR can provide a shared visual space for live mobile remote collaboration on physical tasks. With this approach, a remote expert can explore the scene independently of the local user's current camera position and can communicate via spatial annotations that are immediately visible to the local user in the AR view. This can be achieved with real-time visual tracking and reconstruction, eliminating the need for preparation or **instrumentation** of the environment. AR **telepresence** combines the benefits of live **video conferencing** and remote scene exploration into a natural **collaborative** interface.

**memorize**
/ 'memәraiz /
v. 记住；熟记

**superimpose**
/ ˌsju:pərim'pəuz /
v. 添加；附加；叠加

**instrumentation**
/ ˌinstrumen'teiʃ әn, ˌinstrә- /
n. 仪表化；装设仪器；（一套）仪器

**telepresence**
/ 'teliˌprezәns /
n. 远程呈现

**conferencing**
/ 'kɔnfərənsiŋ /
n. 召开会议；会议技术

**video conferencing**
（召开）视频会议；视频会议技术

**collaborative**
/ kә'læbәreitiv /
a. 合作的，协作的

### 2. Personal Information Display

A large variety of AR browser apps are already available on

smartphones. These apps are intended to deliver information related to places of interest in the user's environment, superimposed over the live video from the device's camera. The places of interest are either given in **geo-coordinates** and identified via the phone's sensors (GPS, compass readings) or identified by image recognition. AR browsers have obvious limitations, such as potentially poor GPS accuracy and augmentation capabilities only for individual points rather than full objects. Nevertheless, thanks to the proliferation of smartphones, these apps are universally available, and their use is growing, owing to the social networking capabilities built into the AR browsers. Figure 12B-2 shows the AR browser Yelp Monocle, which is integrated into the social business review app Yelp[1].

**geo-coordinate**
/ ˌdʒiːəukəuˈɔːdinət /
n. 地理坐标

**Figure 12B-2: AR browsers such as Yelp Monocle superimpose points of interest on a live video feed**

Another **compelling** use case for AR browsing is simultaneous translation of foreign languages. This utility is now widely available in the Google Translate app. The user just has to select the target language and point the device camera toward the printed text; the translation then appears

**compelling**
/ kəmˈpeliŋ /
a. 令人信服的；有强烈吸引力的

[1] *Figure 12B-2 shows the AR browser Yelp Monocle, which is integrated into the social business review app Yelp.*：图 12B-2 展示了集成到社交商务评论应用 Yelp 中的 AR 浏览器 Yelp Monocle。Yelp 是美国著名商户点评网站，创立于 2004 年，用户可在网站中给商户打分，提交评论，交流购物体验等。

**footage** / ˈfutidʒ /
n.（影片的）连续镜头，片段
**yardage** / ˈjɑːdidʒ /
n. 码数；以码计量的长度
**screencast** / ˈskriːnkɑːst /
n. 截屏视频，屏幕录像，录屏
**overlay** / ˈəuvəlei /
n. 覆盖物
**hockey** / ˈhɔki /
n. 曲棍球；冰上曲棍球，冰球
**ice hockey**
冰上曲棍球，冰球
**televise** / ˈtelivaiz /
v. 用电视播放
**incarnation** / ˌinkɑːˈneiʃən /
n.（某种品质、概念、思想等的）典型体现，化身
**albeit** / ɔːlˈbiːit /
conj. 尽管，即使

superimposed over the image.

**3. Television**

Many people likely first encountered AR as annotations to live camera **footage** brought to their homes via broadcast TV. The first and most prominent example of this concept is the virtual 1st & 10 line in American football, indicating the **yardage** needed for a first down[1], which is superimposed directly on the TV **screencast** of a game. The same concept of annotating TV footage with virtual **overlays** has successfully been applied to many other sports, including baseball, **ice hockey**, car racing, and sailing. Figure 12B-3 shows a **televised** soccer game with augmentations. The audience in this **incarnation** of AR has no ability to vary the viewpoint individually. Given that the live action on the playing field is captured by tracked cameras, interactive viewpoint changes are still possible, **albeit** not under the end-viewer's control.

**Figure 12B-3: Augmented TV Broadcast of a Soccer Game**

**informative** / inˈfɔːmətiv /
a. 提供有用信息的；增长知识的

Several competing companies provide augmentation solutions for various broadcast events, creating convincing and **informative** live annotations. The annotation possibilities have long since moved beyond just sports information or simple line graphics, and now include sophisticated

[1] *the virtual 1st & 10 line in American football, indicating the yardage needed for a first down*：美式橄榄球比赛中的虚拟 1 攻 10 码线，用来指示第一次进攻所需的码数。在美式橄榄球中，down 指“10 码进攻”，即进攻方为向前推进 10 码的连续四次进攻中的一次，每次进攻机会也称为一“档”进攻。first down 指“首档”，也称“首攻”；1st & 10 意为“1 攻 10 码”，即“这是第一档进攻，目标是向前推进 10 码”。连续四次进攻未能累计推进 10 码，便要把控球权在第四档进攻结束的位置交给对手。

**rendering** /ˈrendəriŋ/ n. 艺术处理；渲染
**branding** /ˈbrændiŋ/ n. 品牌推广，品牌宣传
**logo** /ˈləugəu/ n. 标识，标志，徽标
**moderator** /ˈmɔdəreitə/ n.（电视或广播节目的）主持人
**prop** /prɔp/ n.（戏剧、电影等中用的）道具
**preview** /ˈpriːvjuː/ n. & v. 预映；试演；预看
**compositing** /ˈkɔmpəzitiŋ, kəmˈpɔz-/ n. 混合；合成
**kiosk** /ˈkiːɔsk, kiˈɔsk/ n. 亭子；信息亭
**vantage** /ˈvɑːntidʒ; ˈvæn-/ n. 优势；有利地位
**vantage point** 有利位置；观点，看法
**flyer** /ˈflaiə/ n.（广告）传单
**prophet** /ˈprɔfit/ n. 预言者，预言家
**animated** /ˈænimeitid/ a. 动画（片）的；活生生的

3D graphics **renderings** of **branding logos** or product advertisements[1].

Using similar technology, it is possible—and, in fact, common in today's TV broadcasts—to present a **moderator** and other TV personalities in virtual studio settings. In this application, the moderator is filmed by tracked cameras in front of a green screen and inserted into a virtual rendering of the studio. The system even allows for interactive manipulation of virtual **props**.

Similar technologies are being used in the film industry, such as for providing a movie director and actors with live **previews** of what a film scene might look like after special effects or other **compositing** has been applied to the camera footage of a live set environment. This application of AR is sometimes referred to as Pre-Viz[2].

### 4. Advertising and Commerce

The ability of AR to instantaneously present arbitrary 3D views of a product to a potential buyer is already being welcomed in advertising and commerce. This technology can lead to truly interactive experiences for the customer. For example, customers in Lego[3] stores can hold a toy box up to an AR **kiosk**, which then displays a 3D image of the assembled Lego model. Customers can turn the box to view the model from any **vantage point**.

An obvious target for AR is the augmentation of printed material, such as **flyers** or magazines. Readers of the *Harry Potter*[4] novels know how pictures in the *Daily* ***Prophet*** newspaper come alive. This idea can be realized with AR by superimposing digital movies and animations on top of specific portions of a printed template. When the magazine is viewed on a computer or smartphone, the static pictures are replaced by **animated** sequences or movies.

AR can also be helpful for a sales person who is trying to demonstrate the virtues of a product. Especially for complex devices, it may be difficult to convey the internal operation with words alone. Letting a potential customer observe the animated interior allows for much more compelling

---

1 *sophisticated 3D graphics renderings of branding logos or product advertisements*：对品牌标志或产品广告的复杂三维图形渲染。

2 *Pre-Viz*：可视化预览，视觉预览，*pre-vi*sualization 的缩略。

3 *Lego*：乐高，系商标名称，也是乐高集团（LEGO Group）的名称。乐高集团创立于 1932 年，系世界著名玩具制造商，总部位于丹麦比隆（Billund），乐高积木是其最重要的产品。Lego 读作/ˈlegəu/。

4 *Harry Potter*：《哈利·波特》，英国作家 J.K.罗琳（J. K. Rowling）所著的魔幻文学系列小说。哈利·波特是其中的主角。

**show room**
（商品样品的）陈列室，展览室
**dressing room**
试衣室，试衣间
**garment** / 'gɑːmənt /
n.（一件）衣服
**render** / 'rendə /
v. 对…做艺术处理（或进行渲染）

presentations at trade shows and in **show rooms** alike.

Pictofit[1] is a virtual **dressing room** application that lets users preview garments from online fashion stores on their own body (Figure 12B-4). The garments are automatically adjusted to match the wearer's size. In addition, body measurements are estimated and made available to assist in the entry of purchase data.

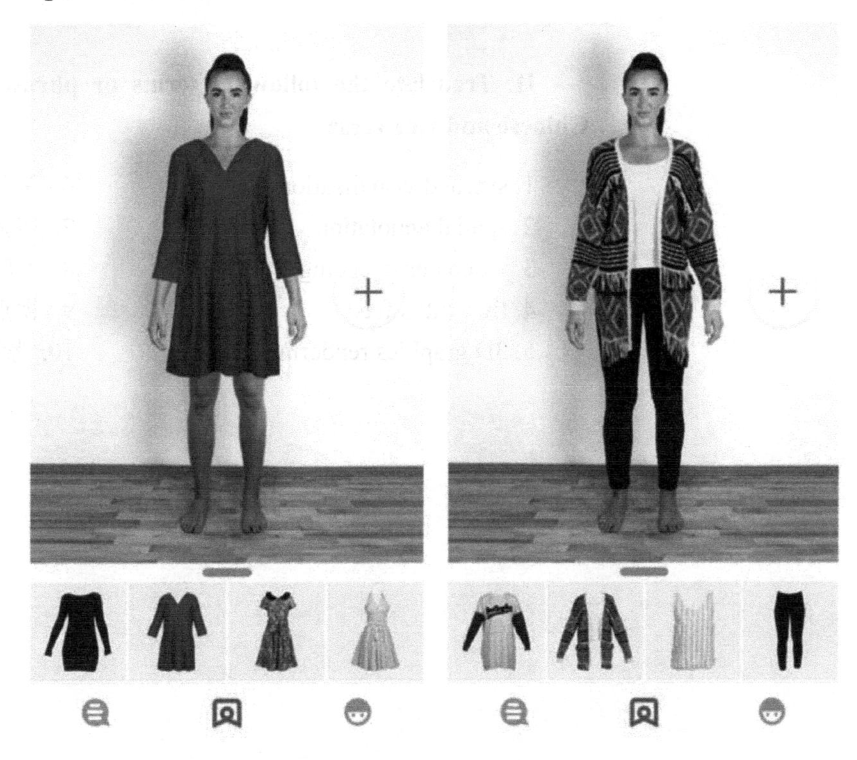

**Figure 12B-4: Pictofit can extract garment images from online shopping sites and render them to match an image of the customer**

## Exercises

**I. Fill in the blanks with the information given in the text:**

1. According to Azuma, AR must have three characteristics: combining real and ________, interactive in real time, and ________ in 3D.

2. A complete AR system requires at least three components: a(n) ________ component, a registration component, and a(n) ________

---

[1] *Pictofit*：一个虚拟试衣间应用程序，由总部位于奥地利格拉茨市（Graz）的 Reactive Reality 公司开发。

component.

3. AR telepresence combines the benefits of live video ________ and remote scene exploration into a natural collaborative ________.

4. Pictofit is a virtual dressing room application which allows users to preview garments from ________ fashion stores on their own body.

**II. Translate the following terms or phrases from English into Chinese and vice versa:**

1. situated visualization
2. spatial annotation
3. video conferencing
4. live video feed
5. 3D graphics rendering
6. 头戴式显示器
7. 增强现实
8. 实时
9. 视点控制
10. 紧密耦合的反馈回路

# Section C

## The Internet of Things

**limelight** / 'laimlait /
n. 公众注意的中心
**radio frequency**
射频；无线电频率
**radio-frequency identification**
射频识别
**ubiquitous computing**
普适计算

In 2005, the concept of the Internet of Things (IoT) entered the **limelight**. The IoT should be designed to connect the world's objects in a sensory manner. The approach is to tag things through **radio-frequency identification** (RFID), feel things through sensors and wireless networks, and think things by building embedded systems that interact with human activities.

### I. Ubiquitous Computing

**pervasive**
/ pə'veisiv /
a. 遍布的；普遍的

**minuscule**
/ 'minəskju:l, mi'nʌs- /
a. 非常小的，微小的

Ubiquitous computing is a post-desktop model of human-computer interaction in which information processing is integrated into everyday objects and activities. For daily activities, people may engage in using many **pervasive** devices simultaneously. They may not even be aware of the existence of the interactive devices. Although the idea is simple, its application is difficult. If all objects in the world were equipped with **minuscule** identifying devices, daily life on our planet could undergo a major transformation.

The IoT cannot be realized without systems design and engineering,

**command line**
命令行

**smart card**
智能卡

**cyber-** / ˈsaibə /
comb. form 表示“计算机（网络）的”

**cyber-physical**
/ ˌsaibəˈfizikəl /
a. 信息物理的

**e-label** / ˈiːleibəl /
n. 电子标签

**synergistic**
/ ˌsinəˈdʒistik /
a. 增效的；协同作用的；协作的

**modulate**
/ ˈmɔdjuleit /
v. 调制

**demodulate**
/ diːˈmɔdjuleit /
v. 解调

**antenna** / ænˈtenə /
n. 天线

and user interfaces. Contemporary human-computer interaction models, whether **command-line**, menu-driven, or GUI-based, are inappropriate and inadequate to meet ubiquitous computing demands. The natural IoT paradigm appropriate to a ubiquitous computing world has yet to emerge. Contemporary devices that lend support to ubiquitous computing include smartphones, tablet computers, sensor networks, RFID tags, **smart cards**, GPS devices, and others.

In ubiquitous computing, the IoT provides a network of sensor- or radio-connected devices that can be uniquely identified and located in the **cyber-physical** space. This IoT is mostly wirelessly connected as a self-configuring network of radio-frequency tags, low-cost sensors, or **e-labels**. The term “IoT” combines RFID technology with today’s IPv6[1]-based Internet technology. All things (objects) have IP addresses, which can be uniquely identified. The IP-identifiable objects are readable, recognizable, locatable, addressable, and/or controllable via the Internet, aided by RFID, Wi-Fi, ZigBee, mobile networks, and GPS[2].

## II. Enabling and Synergistic Technologies

Many technologies can be applied to build the IoT infrastructure and specific IoT systems for special application domains. Supportive technologies are divided into two categories. Enabling technologies build up the foundations of the IoT. Among the enabling technologies, tracking (RFID), sensor networks, and GPS are critical.

RFID is applied with electronic labels or RFID tags on any objects being monitored or tracked. The tags may be applied to any objects, such as merchandise, tools, smartphones, computers, animals, or people. The purpose is to identify and track the objects using radio waves or sensing signals. Some tags can be read from tens or hundreds of meters away via a wireless reader. Most RFID tags contain at least two major parts. One is an integrated circuit for storing and processing information, **modulating** and **demodulating** a radio-frequency (RF[3]) signal, and other special functions. The other part is an **antenna** for receiving and transmitting the radio signals.

---

1 *IPv6*：IP 协议第 6 版（*I*nternet *P*rotocol *v*ersion *6* 的缩略）。

2 *The IP-identifiable objects are readable, recognizable, locatable, addressable, and/or controllable via the Internet, aided by RFID, Wi-Fi, ZigBee, mobile networks, and GPS.*：在射频识别、Wi-Fi、ZigBee、移动网络及 GPS 的帮助下，可用 IP 地址标识的对象能够通过因特网来读取、识别、定位、寻址和（或）控制。

3 *RF*：射频；无线电频率（*r*adio *f*requency 的首字母缩略）。

Today's sensor networks are mostly wireless, and are known as wireless sensor networks (WSNs[1]). A typical WSN consists of spatially distributed autonomous sensors to **cooperatively** monitor physical or environmental conditions, such as temperature, sound, vibration, pressure, motion, or **pollutants**. The development of wireless sensor networks was motivated by military applications such as **battlefield** surveillance. WSN technology is now used in many industrial and civilian application areas, including process monitoring and control, machine health monitoring, environment and **habitat** monitoring, health care and home automation, and intelligent traffic control.

The GPS was developed in 1973 by the U.S. Air Force. Similar developments have also occurred in the European Union, Russia, and China. Since 1994, a degraded GPS has been made available for civilian applications in providing reliable positioning, navigation, and timing services. For anyone with a GPS receiver, the system will provide accurate location and time information for an unlimited number of users in all weather conditions, day and night, anywhere in the world.

| Enabling Technologies | Synergistic Technologies |
| --- | --- |
| Machine-to-machine interfaces | **Geotagging/geocaching** |
| Protocols of electronic communication | **Biometrics** |
| Microcontrollers | **Machine vision** |
| Wireless communication | Robotics |
| RFID | Augmented reality |
| **Energy harvesting** technologies | Telepresence and adjustable **autonomy** |
| Sensors and sensor networks | Life recorders and personal black boxes |
| Actuators | **Tangible** user interfaces |
| Positioning or location technology (GPS) | Clean technologies |
| Software engineering | Mirror worlds |

**Table 12C-1: Enabling and Synergistic Technologies for the IoT**

Synergistic technologies play supporting roles. For example, biometrics could be widely applied to personalize the interactions among humans, machines, and objects. Artificial intelligence, computer vision, robotics, and telepresence can make our lives more automated in the future.

As an emerging technology, the IoT will become more mature and more sophisticated. Figure 12C-1 shows the major technology advances and key applications that may benefit from the IoT. For example, **supply chains** are now better supported than before. Vertical market applications may

**cooperative** /kəu'ɔpərətiv/
a. 合作的，协作的

**pollutant** /pə'lu:tənt/
n. 污染性物质

**battlefield** /'bætlfi:ld/
n. 战场

**habitat** /'hæbitæt/
n.（动植物的）自然生存环境，栖息地

**geotagging** /ˌdʒi:əu'tægiŋ/
n. 地理标记，地理标签

**geocaching** /ˌdʒi:əu'kæʃiŋ/
n. 地理藏宝

**biometrics** /ˌbaiəu'metriks/
n. 生物统计学

**machine vision**
机器视觉

**energy harvesting**
能量收集

**autonomy** /ɔ:'tɔnəmi/
n. 自治（权），自主性

**tangible** /'tændʒəbl/
a. 可触摸的；有形的；明确的

**supply chain**
供应链

[1] *WSN*：无线传感器网络（*w*ireless *s*ensor *n*etwork 的首字母缩略）。

**societal** / sə'saiətəl /
a. 社会的
**software agent**
软件代理
**fusion** / 'fjuːʒən /
n. 熔合；融合
**miniaturization**
/ ˌminitʃərai'zeiʃən; ˌminiətʃərə- /
n. 小型化，微型化
**spectrum**
/ 'spektrəm /
n. 频谱；系列；范围
**teleoperation**
/ ˌteliɔpə'reiʃən /
n. 远程操作，遥操作
**geolocation**
/ ˌdʒiːəuləu'keiʃən /
n. 地理位置，地理定位
**diffusion**
/ di'fjuːʒən /
n. 扩散
**expedite**
/ 'ekspidait /
v. 迅速执行；加速
**inventory**
/ 'invəntəri; -tɔːri /
v. 为…开列存货清单；编制…的目录；盘存

represent the next wave of advances. Ubiquitous positioning is expected to become a reality as we move toward 2020. Beyond that, a physical IoT may be in place in a global scale. These advances will significantly upgrade human abilities, **societal** outcomes, national productivity, and quality of life.

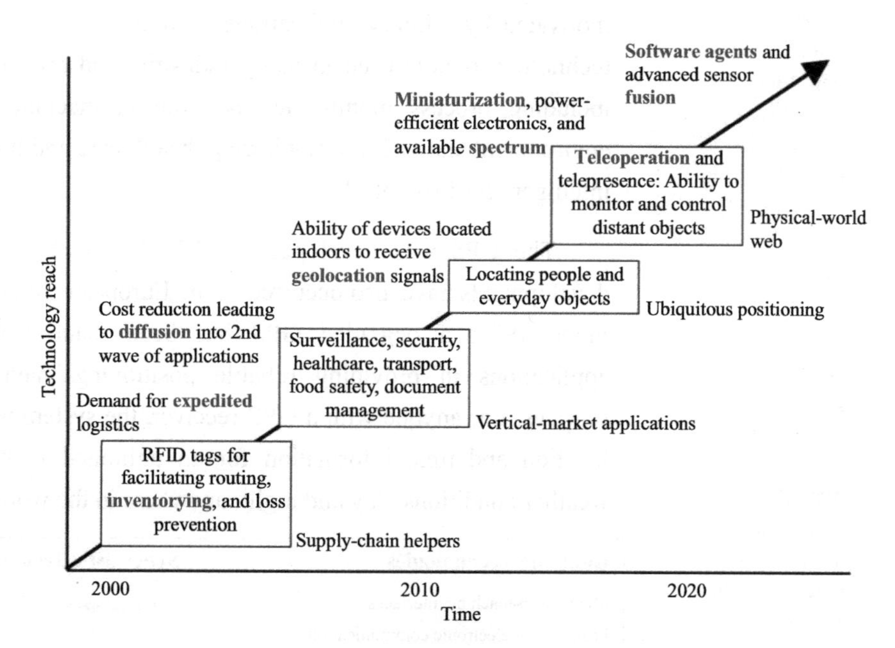


**Figure 12C-1: Technology Road Map of the IoT**

## III. Architecture of the Internet of Things

The IoT system is likely to have an event-driven architecture. In Figure 12C-2, IoT development is shown with a three-layer architecture. The top layer is formed by driven applications[1]. The application space of the IoT is huge. The bottom layers represent various types of sensing devices: namely RFID tags, ZigBee or other types of sensors, and road-mapping GPS **navigators**[2]. The sensing devices are locally or wide-area-connected in the form of RFID networks, sensor networks, and GPSs. Signals or information collected at these sensing devices are linked to the applications through the cloud computing platforms at the middle layer.

**navigator**
/ 'nævigeitə /
n. 导航仪，导航系统；航海者
**deterministic**
/ diˌtəːmi'nistik /
a. 决定论的；确定性的

The signal processing clouds are built over the mobile networks, the Internet backbone, and various information networks at the middle layer. In the IoT, the meaning of a sensing event does not follow a **deterministic** or

[1] *The top layer is formed by driven applications.*：顶层由受驱动的应用程序构成。
[2] *road-mapping GPS navigators*：标示路线图的 GPS 导航仪。

syntactic model[1]. In fact, the service-oriented architecture (SOA) model is adoptable here. A large number of sensors and filters are used to collect the raw data. Various compute and storage clouds and grids are used to process the data and transform it into information and knowledge formats. The sensed information is used to put together a decision-making system for intelligence applications. The middle layer is also considered as a **Semantic Web** or **Grid**. Some actors (services, components, avatars) are self-referenced[2].

**semantic** / si'mæntik /
a. 语义（学）的
**Semantic Web**
语义网
**Semantic Grid**
语义网格

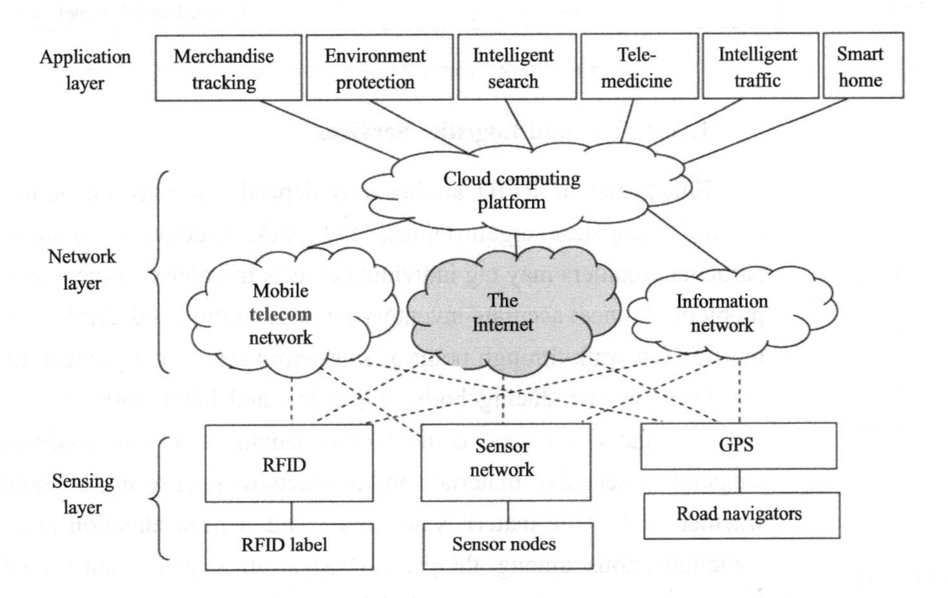


**Figure 12C-2: The architecture of an IoT consisting of sensing devices that are connected to various applications via mobile networks, the Internet, and processing clouds**

## IV. Applications of the Internet of Things

Table 12C-2 summarizes IoT applications in three major civilian application domains. Obviously, the IoT has a lot of military applications, which is beyond the scope of this section. In general, use of the IoT aims to promote industrial productivity and raise economic growth. The IoT plays important roles in environment protection, including pollution control, weather forecasting, and disaster avoidance and recovery. In terms of societal impacts, the IoT can make our lives more convenient and comfortable. Government services, law enforcement, and home and health improvements are the major **beneficiaries**. In the remaining space of this

**telecom** / 'telikɔm /
n. 电信（= telecommunication）

**beneficiary** / ˌbeni'fiʃiəri /
n. 受益人，受惠者

[1] *In the IoT, the meaning of a sensing event does not follow a deterministic or syntactic model.*：在物联网中，一个感知事件的意义不是遵循确定性或句法模型的。

[2] *Some actors (services, components, avatars) are self-referenced.*：有些行为体(服务、组件、化身)是自引用的。

**indicative**
/ in'dikətiv /
a. 标示的；指示的
**logistics**
/ ləu'dʒistiks /
n. 后勤（学），物流
**retailing** / 'ri:teiliŋ /
n. 零售业
**retailer** / 'ri:teilə /
n. 零售商；零售店
**unattended**
/ ˌʌnə'tendid /
a. 没人看管的；没人照料的
**checkout** / 'tʃekaut /
n. 结账（离去）；付款台
**shoplifting**
/ 'ʃɔpˌliftiŋ /
n.（混在顾客群中进行的）商店货物扒窃
**cold chain** 冷链
**audit** / 'ɔ:dit /
v. 审计，查账
**assurance** / ə'ʃuərəns /
n. 保证；把握，信心
**perishable**
/ 'periʃəbəl /
a. 易腐的，易烂的
**intact** / in'tækt /
a. 完整无缺的；未受损伤的
**entail** / in'teil /
v. 使成为必要，需要
**refrigeration**
/ riˌfridʒə'reiʃən /
n. 冷藏；制冷
**data logging**
数据（事件）记录；数据登录
**participant**
/ pɑ:'tisipənt /
n. 参加者，参与者

section, we will briefly discuss some of the application domains.

| Domain | Brief Descriptions | Indicative Examples |
|---|---|---|
| Growth in industry and economy | Activities involving financial or commercial transactions between companies and organizations | Manufacturing, **logistics**, service sector, banking, financial governmental authorities, intermediaries, etc. |
| Environment and natural resources | Activities regarding the protection, monitoring, and development of all natural resources | Agriculture and breeding, recycling, environmental management services, energy management, etc. |
| Society and daily life | Activities/initiatives regarding the development and inclusion of societies, cities, and people | Governmental services toward citizens and society structures, e-inclusion (aging, disabled people), etc.[1] |

Table 12C-2: Some Application Domains of the IoT

### 1. Retailing and Logistics Services

Emergence of RFID applications depends strongly on adoption by **retailers**, logistics organizations, and package-delivery companies. In particular, retailers may tag individual objects in order to solve a number of problems at once: accurate inventorying, loss control, and ability to support **unattended** walk-through point of sale terminals (which promise to speed **checkout** while reducing both **shoplifting** and labor costs). **Cold-chain auditing** and **assurance**[2] could require tagging food and medicine with temperature-sensitive materials and/or electronics; ensuring or monitoring whether **perishable** materials are **intact** and/or need attention may **entail** communications among things, **refrigeration** systems, automated **data logging** systems, and human technicians.

### 2. Supply Chain Management

Supply chain management is a process used by companies to ensure that their supply chain is efficient and cost-effective. Supply chain management can be aided by an IoT system. The idea is to manage a whole network of related businesses or partners involved in product manufacturing, delivery, and services as required by end customers. At any given time, market forces could demand changes from suppliers, logistics providers, locations and customers, and any number of specialized **participants** in a supply chain. This variability has significant effects on the supply chain infrastructure, ranging from the foundation layers of establishing the electronic communication between the trading partners to the more complex

[1] *Governmental services toward citizens and society structures, e-inclusion (aging, disabled people), etc.*：政府对公民的服务与社会结构、电子包容（衰老、残疾人士）等。

[2] *Cold-chain auditing and assurance*：冷链审计与保证。

configuration of the processes and the arrangement of workflows that are essential to a fast production process.

**power grid**
电网
**thermostat**
/ ˈθəːməustæt /
n. 温度自动调节器，恒温器
**residential**
/ ˌreziˈdenʃəl /
a. 居住的；住宅的
**mall** / mɔːl, mæl /
n.（非露天的）购物区，购物街
**shopping mall**
（同一建筑物里有许多商店的）购物中心
**high-rise** / ˈhaiˈraiz /
a.（建筑）高层的
**office tower**
办公大楼
**suppression**
/ səˈpreʃən /
n. 压制；抑制；阻止

**3. Smart Power Grid and Smart Buildings**

A critical IoT application is promotion of a smart power grid. Various power companies across the United States have or are in the process of upgrading their power management and distribution systems. Various sensors at individual homes (smart **thermostats**) can collect information that is sent via a network to main stations (perhaps even local "hubs") that can apply complex power management and send control signals back to the grid to save energy. The smart grid is made possible by applying sensing, measurement, and control devices to electricity production, transmission, distribution, and consumption.

The IoT has been suggested in construction of smart buildings in **residential**, commercial, industrial, and government settings. A smart building can be a **shopping mall** or a home, a hospital or a **high-rise office tower**. Smart buildings need monitoring and regulation of heating, air conditioning, lighting, and environmental changes. They can oversee building security, fire **suppression**, and elevator operations. Smart building technologies focus on bringing more detailed monitoring and sensing "awareness" to buildings.

---

## Exercises

**I. Fill in the blanks with the information given in the text:**

1. Supportive technologies for the IoT are divided into two categories: ________ and synergistic technologies.

2. Contemporary devices that lend support to ubiquitous computing include smartphones, ________ computers, sensor networks, RFID tags, ________ cards, and GPS devices.

3. The architecture of an IoT consists of ________ devices connected to various applications via mobile networks, the Internet, and processing ________.

4. A typical wireless sensor network consists of spatially distributed ________ sensors to cooperatively monitor physical or environmental conditions.

**II. Translate the following terms or phrases from English into Chinese and vice versa:**

| | |
|---|---|
| 1. computer vision | 5. 移动网络 |
| 2. ubiquitous computing | 6. 供应链 |
| 3. command line | 7. 物联网 |
| 4. smart power grid | 8. 传感器融合 |

# 练习参考答案

## Unit 1: Computer and Computer Science

### Unit 1/Section A

**I. Fill in the blanks with the information given in the text:**

1. Charles Babbage; Augusta Ada Byron
2. input; output
3. VLSI (very large-scale integrated)
4. workstations; mainframes
5. vacuum; transistors
6. instructions; software
7. digit; eight; byte
8. optical; quantum

**II. Translate the following terms or phrases from English into Chinese and vice versa:**

1. artificial intelligence 人工智能
2. paper-tape reader 纸带阅读器
3. optical computer 光计算机
4. neural computer 神经（元）计算机
5. instruction set 指令集
6. quantum computer 量子计算机
7. difference engine 差分机
8. versatile logical element 通用逻辑元件
9. silicon substrate 硅衬底
10. vacuum tube 真空管
11. 数据的存储与处理 the storage and handling of data
12. 超大规模集成电路 VLSI circuit (very large-scale integrated circuit)
13. 中央处理器 central processing unit
14. 个人计算机 personal computer
15. 模拟计算机 analogue computer
16. 数字计算机 digital computer
17. 通用计算机 general-purpose computer
18. 处理器芯片 processor chip
19. 操作指令 operating instructions

20. 输入设备 input device

**III. Fill in each of the blanks with one of the words given in the following list, making changes if necessary:**

We can define a computer as a device that accepts input, processes data, stores data, and produces output. According to the *mode* of processing, computers are either analog or *digital*. They can also be classified as mainframes, *minicomputers*, workstations, or microcomputers. All else (for example, the age of the machine) being equal, this *categorization* provides some indication of the computer's *speed*, size, cost, and abilities.

Ever since the *advent* of computers, there have been constant changes. First-generation computers of historic *significance*, such as *UNIVAC* (通用自动计算机), introduced in the early 1950s, were *based* on vacuum tubes. Second-generation computers, *appearing* in the early 1960s, were those in which *transistors* replaced vacuum tubes. In third-generation computers, dating from the 1960s, integrated *circuits* replaced transistors. In fourth-generation computers such as *microcomputers*, which first appeared in the mid-1970s, large-scale *integration* enabled thousands of circuits to be incorporated on one *chip*. Fifth-generation computers are expected to *combine* very-large-scale integration with sophisticated approaches to *computing*, including artificial intelligence and true distributed processing.

**IV. Translate the following passage from English into Chinese:**

计算机将变得更加先进，也将变得更加容易使用。语音识别的改进将使计算机的操作更加容易。虚拟现实，即使用所有人类官能与计算机进行交互的技术，也将有助于创建更好的人机接口。人们正在开发其他的奇异计算模型，包括使用生物机体的生物计算、使用具有特定属性的分子的分子计算，以及使用遗传基本单位 DNA（脱氧核糖核酸）存储数据和执行运算的计算。这些都是可能的未来计算平台的例子，而它们迄今还能力有限或完全属于理论范畴。科学家们研究它们，是因为嵌入硅中的电路的微小型化受到物理限制。还有一些限制与即使最微小的晶体管也会产生热量有关。

# Unit 1/Section B

**I. Fill in the blanks with the information given in the text:**

1. experimentation
2. interfacing
3. interdisciplinary
4. microprocessor

**II. Translate the following terms or phrases from English into Chinese and vice versa:**

1. artificial neural network 人工神经网络
2. computer architecture 计算机体系结构
3. robust computer program 健壮的计算机程序
4. human-computer interface 人机接口
5. knowledge representation 知识表示
6. 数值分析 numerical analysis
7. 程序设计环境 programming environment
8. 数据结构 data structure
9. 存储和检索信息 store and retrieve information
10. 虚拟现实 virtual reality

## Unit 1/Section C

**I. Fill in the blanks with the information given in the text:**

1. supercomputer; petaflops
2. desktop
3. tablets
4. wearable

**II. Translate the following terms or phrases from English into Chinese and vice versa:**

1. niche device 小众设备，细分设备
2. fitness tracker 健身跟踪器，健身追踪器
3. touch-sensitive screen 触摸屏
4. mainframe computer 主机，大型机
5. 膝上型计算机 laptop computer
6. 平板电脑 tablet computer
7. 便携式媒体播放器 portable media player
8. 硬盘 hard disk

# Unit 2: Computer Architecture

## Unit 2/Section A

**I. Fill in the blanks with the information given in the text:**

1. input; output; storage
2. Basic Input/Output System

3. flatbed; hand-held
4. LCD (liquid crystal display); LED (light-emitting diode)
5. additive; FDM (fused deposition modeling)
6. RAM (random access memory)
7. magnetic; solid
8. serial; parallel

**II. Translate the following terms or phrases from English into Chinese and vice versa:**

1. function key 功能键，操作键，函数键
2. voice recognition module 语音识别模块
3. multifunction printer 多功能打印机
4. address bus 地址总线
5. additive manufacturing 增材制造
6. memory card 存储卡
7. parallel connection 并行连接
8. solid state drive 固态驱动器
9. boot loader 引导装入程序，引导加载程序
10. ink cartridge 墨盒
11. 只读存储器 ROM (read-only memory)
12. 液晶显示（器） LCD (liquid crystal display)
13. 喷墨打印机 ink jet printer
14. 数据总线 data bus
15. 串行连接 serial connection
16. 闪存 flash memory
17. 激光打印机 laser printer
18. 蓝光光盘 Blu-ray disc
19. 发光二极管 LED (light-emitting diode)
20. 随机存取存储器 RAM (random access memory)

**III. Fill in each of the blanks with one of the words given in the following list, making changes if necessary:**

Today's digital devices may use local storage and remote storage. Local storage refers to storage *devices* and media that can be directly attached to a *computer*, smartphone, or appliance. Local storage options include hard *drives*, CDs, DVDs, flash drives, solid state drives, and *memory* cards. Most digital devices have some type of *local* storage that is permanently available as you use the device. Built-in *storage* can be supplemented by removable storage, such as *flash* drives and memory cards.

In contrast, remote storage is housed on an *external* device that can be accessed from a *network*. Remote storage may be available on a home, *school*, or work network. It can also be available as an *Internet* service, in which case it is called *cloud* storage. The basic concept is that files can be *stored* in a ***subscriber's*** (用户) cloud-based storage area and *accessed* by logging in from any device. Some cloud implementations *offer* a ***synchronization*** (同步化) feature that automatically ***duplicates*** (复制) *files* stored on a local device by also saving them in the cloud.

**IV. Translate the following passage from English into Chinese:**

调制解调器是在模拟与数字信号之间进行转换的设备。计算机使用的是数字信号，这种信号由离散单元组成，通常用一系列 1 和 0 表示。模拟信号是连续变化的；声波就是模拟信号的一个例子。调制解调器经常用于实现计算机之间通过电话线的互相通信。调制解调器将发送端计算机的数字信号转换成可通过电话线传输的模拟信号。信号到达目的地后，另外一个调制解调器重构原来的数字信号，供接收端计算机处理。如果两个调制解调器可同时互相发送数据，那么它们采用的就是全双工工作方式；如果一次只有一个调制解调器可以发送数据，那么它们采用的则是半双工工作方式。

# Unit 2/Section B

**I. Fill in the blanks with the information given in the text:**

1. graphical
2. file; scheduler
3. virtual
4. slice

**II. Translate the following terms or phrases from English into Chinese and vice versa:**

1. interrupt handler 中断处理程序
2. virtual memory 虚拟存储（器），虚存，虚拟内存
3. context switch 上下文转换，语境转换

4. main memory 主存（储器）
5. bit pattern 位模式
6. 外围设备 peripheral device
7. 进程表 process table
8. 时间片 time slice
9. 图形用户界面 graphical user interface
10. 海量存储器 mass storage

## Unit 2/Section C

**I. Fill in the blanks with the information given in the text:**

1. repository
2. central; sub-systems
3. network
4. layered 或 abstract machine

**II. Translate the following terms or phrases from English into Chinese and vice versa:**

1. code generator 代码生成程序，代码发生器
2. abstract machine 抽象机
3. program editor 程序编辑程序，程序编辑器
4. configuration item 配置项
5. 计算机辅助设计 CAD (computer-aided design)
6. 数据冗余 data redundancy
7. 指挥与控制系统 command and control system
8. 视频压缩与解压缩 video compression and decompression

# Unit 3: Computer Language and Programming

## Unit 3/Section A

**I. Fill in the blanks with the information given in the text:**

1. artificial; instructions
2. low-level; high-level
3. machine
4. machine
5. functional; logic
6. statement
7. module
8. digital

**II. Translate the following terms or phrases from English into Chinese and vice versa:**

1. storage register 存储寄存器
2. function statement 函数语句
3. program statement 程序语句
4. object-oriented language 面向对象语言
5. assembly language 汇编语言
6. intermediate language 中间语言，中级语言
7. declarative language 声明式语言
8. artificial language 人工语言
9. data declaration 数据声明
10. functional language 函数式语言
11. 可执行程序 executable program
12. 程序模块 program module
13. 条件语句 conditional statement
14. 赋值语句 assignment statement
15. 逻辑语言 logic language
16. 机器语言 machine language
17. 结构化查询语言 SQL (Structured Query Language)
18. 程序设计语言 programming language
19. 运行计算机程序 run a computer program
20. 计算机程序员 computer programmer

**III. Fill in each of the blanks with one of the words given in the following list, making changes if necessary:**

A programming language is a language used to write instructions for the computer. It lets the programmer express data *processing* in a symbolic manner without regard to machine-specific details.

The difficulty of writing programs in the *machine* language of 0s and 1s led first to the development of assembly language, which allows *programmers* to use *mnemonics* (助记符) for instructions and symbols for variables. Such programs are then *translated* by a program known as an *assembler* (汇编程序) into the binary encoding used by the *computer*. Other pieces of system software known as *linking loaders* (链接装入程序) *combine* pieces of assembled code and load them into the machine's main *memory* unit, where they are then ready for execution. The concept of linking *separate* pieces of code was important, since it allowed "libraries" of *programs* to be built up to carry out common tasks—a first *step* toward the increasingly emphasized notion of software *reuse*. Assembly language was found to be sufficiently *inconvenient* that higher-level languages (closer to natural languages) were invented in the 1950s for easier, faster *programming*; along with them came the need for compilers, programs that translate *high-level* language programs into machine code. As programming languages became more *powerful* and abstract, building efficient compilers that create high-quality code in terms of *execution* speed and storage consumption became an interesting computer science problem in itself.

**IV. Translate the following passage from English into Chinese:**

面向对象程序设计语言，如C++和Java，基于传统的高级语言，但它们使程序设计员能够从合作对象集而非命令列表的角度进行思考。诸如圆之类的对象具有像圆的半径一类的属性，以及在计算机屏幕上绘制该对象的命令。对象类可以从其他的对象类继承特征。例如，定义正方形的类可以从定义长方形的类那里继承直角等特征。这一套程序设计类简化了程序设计员的工作，带来了更多"可复用的"计算机代码。可复用代码使程序设计员可以使用已经设计、编写和测试的代码。这使得程序设计员的工作变得比较容易，并带来更加可靠和高效的程序。

# Unit 3/Section B

**I. Fill in the blanks with the information given in the text:**

1. objects
2. platform-independent
3. multithreading
4. runtime

**II. Translate the following terms or phrases from English into Chinese and vice versa:**

1. native code　本机（代）码
2. header file　头标文件，页眉文件
3. multithreaded program　多线程程序
4. Java-enabled browser　支持Java的浏览器

5. malicious code 恶意代码
6. 机器码 machine code
7. 汇编码 assembly code
8. 特洛伊木马程序 Trojan horse
9. 软件包 software package
10. 类层次 class hierarchy

## Unit 3/Section C

**I. Fill in the blanks with the information given in the text:**

1. subscript; index
2. fixed
3. histogram
4. two-dimensional

**II. Translate the following terms or phrases from English into Chinese and vice versa:**

1. bar chart 条形图
2. frequency array 频率数组
3. graphical representation 图形表示
4. multidimensional array 多维数组
5. 用户视图 user('s) view
6. 下标形式 subscript form
7. 一维数组 one-dimensional array
8. 编程结构 programming construct

# Unit 4: Software Development

## Unit 4/Section A

**I. Fill in the blanks with the information given in the text:**

1. application; operating
2. assemblers
3. compiler
4. interpreter
5. debugger
6. loop
7. driver
8. John von Neumann

**II. Translate the following terms or phrases from English into Chinese and vice versa:**

1. inference engine 推理机
2. mathematical theorem 数学定理
3. compiled language 编译执行的语言
4. parallel computing 并行计算
5. pattern matching 模式匹配
6. memory location 存储单元
7. interpreter program 解释程序
8. library routine 库例程，库存程序
9. intermediate program 中间程序，过渡程序
10. source file 源文件
11. 解释执行的语言 interpreted language
12. 设备驱动程序 device driver
13. 分布式计算 distributed computing
14. 调试程序 debugging program
15. 目标代码 object code
16. 应用程序 application program
17. 实用程序 utility program
18. 逻辑程序 logic program
19. 系统调用 system call
20. 程序的存储与执行 program storage and execution

**III. Fill in each of the blanks with one of the words given in the following list, making changes if necessary:**

A compiler, in computer science, is a computer program that translates source code into object code. Software *engineers* write source code using high-level programming *languages* that people can understand. Computers cannot *directly* execute source code, but need a *compiler* to translate these instructions into a *low-level* language called machine code.

Compilers collect and reorganize (compile) all the *instructions* in a given set of source code to *produce* object code. Object code is often the same as or *similar* to a computer's machine code. If the object code is the same as the *machine* language, the computer can run the *program* immediately after the compiler produces its *translation*. If the object code is not in machine language, other programs—such as *assemblers*, *binders* (联编程序), linkers, and *loaders* (装入程序)—finish the translation.

Most computer languages use different *versions* of compilers for different types of *computers* or operating systems, so one language may have *different* compilers for personal computers (PC) and Apple Macintosh computers. Many different *manufacturers* often produce versions of the same programming language, so compilers for a language may vary between manufacturers.

**IV. Translate the following passage from English into Chinese:**

在软件中，错误是指导致程序发生故障或产生不正确结果的编码或逻辑错误。较轻微的错误，如光标表现异常，会造成不便或带来挫折，但不会对信息产生破坏性影响。较严重的错误会导致程序“中止”（对命令停止反应），可能使用户别无选择，只能重新启动程序，结果致使任何前面已经做好但尚未保存的工作丢失。两种情况无论是哪一种，程序员都必须凭借称为调试的过程，发现并改正错误。由于错误对重要数据的潜在危险，商用应用程序在发布前要经过尽可能全面的测试与调试。程序发布后发现的较轻微错误可在下一次更新时改正；较严重的错误有时可用称为补丁的特殊软件加以修补，以规避问题或减轻其影响。

# Unit 4/Section B

**I. Fill in the blanks with the information given in the text:**

1. entity
2. duration
3. data; process 或 process; data
4. implemented

**II. Translate the following terms or phrases from English into Chinese and vice versa:**

1. check box 复选框
2. structured design 结构化设计
3. building block 积木块，构建模块，构件
4. database schema 数据库模式
5. radio button 单选按钮
6. 系统建模技术 system modeling technique
7. 模型驱动开发 MDD (model-driven development)

8. 数据流程图 data flow diagram
9. 下拉菜单 drop-down (或 pull-down) menu
10. 滚动条 scroll bar

## Unit 4/Section C

**I. Fill in the blanks with the information given in the text:**

1. off-the-shelf
2. requirements; integration
3. throwaway
4. reuse-oriented; framework

**II. Translate the following terms or phrases from English into Chinese and vice versa:**

1. software life cycle 软件生命周期
2. evolutionary development process 演化开发过程
3. software reuse 软件复用
4. system design paradigm 系统设计范例
5. 瀑布模型 waterfall model
6. 系统集成 system integration
7. 商用现成软件 COTS (commercial off-the-shelf) software
8. 基于组件的软件工程 CBSE (component-based software engineering)

# Unit 5: Software Engineering

## Unit 5/Section A

**I. Fill in the blanks with the information given in the text:**

1. business; task-oriented
2. configuring; composite
3. standardized
4. XML; communication
5. operations
6. POST; DELETE
7. candidate; implementation
8. RESTful

**II. Translate the following terms or phrases from English into Chinese and vice versa:**

1. computationally intensive processing 计算密集型的处理
2. reusable component 可复用构件

3. data element 数据元（素）
4. service candidate identification 可选服务识别
5. application-specific feature 应用所特有的特征
6. high-throughput system 高吞吐量系统
7. utility service 实用服务
8. Representational State Transfer 表示层状态转化
9. service binding 服务绑定
10. task-oriented service 面向任务的服务
11. 面向服务的体系结构 service-oriented architecture
12. 面向实体的服务 entity-oriented service
13. 简单对象访问协议 SOAP (Simple Object Access Protocol)
14. 服务实现与部署 service implementation and deployment
15. 数字签名 digital signature
16. 工作流语言 workflow language
17. 业务服务 business service
18. 模式定义 schema definition
19. 可扩展标记语言 XML (Extensible Markup Language)
20. 移动设备 mobile device

**III. Fill in each of the blanks with one of the words given in the following list, making changes if necessary:**

There are three different types of software maintenance. Firstly, there is maintenance to repair software *faults*. Coding errors are usually relatively cheap to correct; design errors are more *expensive* as they may involve rewriting several program components. Requirements errors are the *most* expensive to repair because of the extensive system redesign that may be *necessary*. Secondly, there is maintenance to adapt the software to a *different* operating environment. This type of maintenance is required when some *aspect* of the system's environment such as the hardware, the platform operating *system* or other support software changes. The application system must be *modified* to adapt it to cope with these environmental changes. And thirdly, there is *maintenance* to add to or modify the system's functionality. This *type* of maintenance is necessary when the system requirements change in *response* to organizational or business change. The scale of the changes required to the *software* is often much greater than for the other types of maintenance. In practice, there isn't a clear-cut distinction between these types of maintenance. When you adapt the system to a new environment, you may add functionality to take advantage of new environmental features. Software faults are often exposed because users use the system in unanticipated ways. Changing the system to accommodate their way of working is the best way to fix these faults.

**IV. Translate the following passage from English into Chinese:**

软件工程是计算机科学的一个分支，致力于寻求指导大型复杂软件系统开发的原则。开发这类系统时所面临的问题并非只是编写小程序时所面临问题的放大。例如，开发这类系统需要不止一个人长时间努力，而在这期间，目标系统的需求可能会改变，分配给项目的人员可能会变动。因此，软件工程也包括诸如人员管理和项目管理之类的主题。

软件工程研究目前在两个层面上进行：一些有时称为实践者的研究者致力于开发直接应用的技术，而另一些称为理论家的研究者致力于寻找基本原则和理论，为有朝一日构建更稳定的技术提供基础。实践者过去开发和促进的许多方法已经被其他方法所代替，而这些替代方法本身也可能随着时间的推移而被废弃。与此同时，理论家的进展仍然很缓慢。

# Unit 5/Section B

**I. Fill in the blanks with the information given in the text:**

1. acceptance
2. code
3. automation; coverage
4. regression

**II. Translate the following terms or phrases from English into Chinese and vice versa:**

1. black box testing　黑盒测试（法）
2. acceptance testing　验收测试
3. code execution path　代码执行路径
4. test harness　测试框架
5. equivalence partitioning　等价（类）划分
6. 捕获 / 回放工具　capture/playback tool
7. 视频分辨率　video resolution
8. 白盒测试　white box testing

9. 测试脚本 test script
10. 用例 use case

## Unit 5/Section C

**I. Fill in the blanks with the information given in the text:**

1. problem
2. consequences
3. design
4. specific; general

**II. Translate the following terms or phrases from English into Chinese and vice versa:**

1. procedural language 过程语言
2. common design structure 通用设计结构
3. class and object interaction 类与对象交互
4. design constraint 设计约束
5. 设计模式 design pattern
6. 可复用软件 reusable software
7. 面向对象的系统 object-oriented system
8. 继承层次 inheritance hierarchy

# Unit 6: Database

## Unit 6/Section A

**I. Fill in the blanks with the information given in the text:**

1. flat
2. data
3. application; administrators
4. conceptual
5. tables
6. fragmented; replicated
7. structured
8. entity-relationship; attributes

**II. Translate the following terms or phrases from English into Chinese and vice versa:**

1. end user 最终用户，终端用户
2. atomic operation 原子操作
3. database administrator 数据库管理员

4. relational database model 关系数据库模型
5. local data 本地数据
6. object-oriented database 面向对象数据库
7. database management system (DBMS) 数据库管理系统
8. entity-relationship model (ERM) 实体关系模型
9. distributed database 分布式数据库
10. flat file 平面文件
11. 二维表 two-dimensional table
12. 数据属性 data attribute
13. 数据库对象 database object
14. 存储设备 storage device
15. 数据类型 data type
16. 数据插入与删除 data insertion and deletion
17. 层次数据库模型 hierarchical database model
18. 数据库体系结构 database architecture
19. 关系数据库管理系统 RDBMS (relational database management system)
20. 全局控制总线 global control bus

**III. Fill in each of the blanks with one of the words given in the following list, making changes if necessary:**

A database is any collection of data organized for storage in a computer memory and designed for easy _access_ by authorized users. The data may be in the form of _text_, numbers, or encoded graphics. Small databases were first _developed_ or funded by the U.S. government for agency or professional _use_. In the 1960s, some databases became commercially _available_, but their use was *funnelled* (传送) through a few so-called research _centers_ that collected information inquiries and handled them in *batches* (批，批量). _Online_ databases—that is, databases available to anyone who could _link_ up to them by computer—first appeared in the 1970s. Since their first, _experimental_ appearance in the 1950s, databases have become so _important_ that they can be found in almost _every_ field of information. Government, military, and industrial _databases_ are often highly restricted, and professional databases are usually of _limited_ interest. A wide range of commercial, governmental, and _nonprofit_ databases are available to the general *public*, however, and may be used by anyone who owns or has access to the *equipment* that they require.

**IV. Translate the following passage from English into Chinese:**

在关系数据库中，表的行表示记录（关于不同项的信息集），列表示字段（一个记录的特定属性）。在进行搜索时，关系数据库将一个表中的一个字段的信息与另一个表的一个相应字段的信息进行匹配，以生成将来自这两个表的所要求数据结合起来的另一个表。例如，如果一个表包含 EMPLOYEE-ID、LAST-NAME、FIRST-NAME 和 HIRE-DATE 字段，另一个表包含 DEPT、EMPLOYEE-ID 和 SALARY 字段，关系数据库可匹配这两个表中的 EMPLOYEE-ID 字段，以找到特定的信息，如所有挣到一定薪水的雇员的姓名或所有在某个日期之后受雇的雇员所属的部门。换言之，关系数据库使用两个表中的匹配值，将一个表中的信息与另一个表中的信息联系起来。微型计算机数据库产品一般是关系数据库。

# Unit 6/Section B

**I. Fill in the blanks with the information given in the text:**

1. commit
2. scheduler
3. shared; exclusive
4. wound-wait

**II. Translate the following terms or phrases from English into Chinese and vice versa:**

1. nonvolatile storage system 非易失性存储系统
2. equipment malfunction 设备故障
3. wound-wait protocol 受伤–等待协议
4. multiprogramming operating system 多道程序设计操作系统
5. database integrity 数据库完整性
6. 共享锁 shared lock
7. 数据库实现 database implementation
8. 级联回滚 cascading rollback
9. 数据项 data item
10. 排他锁 exclusive lock

## Unit 6/Section C

**I. Fill in the blanks with the information given in the text:**

1. dimensionality; non-traditional
2. intermediate; postprocessing
3. descriptive
4. association; anomaly

**II. Translate the following terms or phrases from English into Chinese and vice versa:**

1. data set 数据集
2. independent variable 自变量
3. predictive modeling 预测建模
4. cluster analysis 聚类分析
5. 数据挖掘 data mining
6. 因变量 dependent variable
7. 大数据 big data
8. 数据预处理 data preprocessing

# Unit 7: Computer Network

## Unit 7/Section A

**I. Fill in the blanks with the information given in the text:**

1. PAN; MAN
2. open; closed
3. bus; star
4. WiFi
5. Ethernet
6. client/server
7. messaging
8. network

**II. Translate the following terms or phrases from English into Chinese and vice versa:**

1. personal area network 个人域网
2. carrier sense 载波侦听，载波监听
3. protocol suite 协议簇，协议组
4. peer-to-peer model 对等模型
5. star topology 星型拓扑
6. communication channel 通信信道

7. access point （访问）接入点
8. proprietary network 专有网络
9. utility package 实用软件包，公用程序包
10. wireless headset 无线耳机
11. 局域网 LAN (local area network)
12. 无线鼠标 wireless mouse
13. 城域网 MAN (metropolitan area network)
14. 封闭式网络 closed network
15. 总线拓扑 bus topology
16. 客户机 / 服务器模型 client/server model
17. 以太网标准 Ethernet standards
18. 进程间通信 interprocess communication
19. 打印服务器 print server
20. 广域网 WAN (wide area network)

**III. Fill in each of the blanks with one of the words given in the following list, making changes if necessary:**

Computers can communicate with other computers through a series of connections and associated hardware called a network. The _advantage_ of a network is that data can be _exchanged_ rapidly, and software and hardware resources, such as hard-disk _space_ or printers, can be shared. Networks also allow _remote_ use of a computer by a user who cannot physically _access_ the computer.

One type of network, a local area network (LAN), consists of several PCs or _workstations_ connected to a special computer called a server, often _within_ the same building or office complex. The server stores and _manages_ programs and data. A server often contains all of a _networked_ group's data and enables LAN workstations or PCs to be set up _without_ large storage capabilities. In this *scenario* (方案), each PC may have "local" _memory_ (for example, a hard drive) specific to itself, but the _bulk_ of storage resides on the server. This _reduces_ the cost of the workstation or PC because less _expensive_ computers can be purchased, and it simplifies the _maintenance_ of software because the software resides only on the _server_ rather than on each individual workstation or PC.

**IV. Translate the following passage from English into Chinese:**

在计算机科学中，网络是指由通信设备连接在一起的一组计算机及其相关设备。网络可采用电缆等永久性连接，或者采用通过电话或其他通信链路而实现的临时性连接。网络可以是由少数计算机、打印机以及其他设备构成的小的局域网，也可以由分布在广大地理区域的许多小型与大型计算机组成。计算机网络无论大小，其存在的目的均在于为计算机用户提供以电子方式进行信息通信与传送的方法。有些类型的通信是简单的用户到用户的消息；有些称为分布式过程，可能涉及数台计算机，以及在执行任务中分担工作量或进行协作。

# Unit 7/Section B

**I. Fill in the blanks with the information given in the text:**

1. terminator
2. sub-central
3. tree; central
4. mesh

**II. Translate the following terms or phrases from English into Chinese and vice versa:**

1. routing path 路由选择通路
2. dual-ring topology 双环形拓扑（结构）
3. extended star topology 扩展星型拓扑（结构）
4. backbone network 基干网，骨干网
5. mesh topology 网格拓扑（结构）
6. 同轴电缆 coaxial cable
7. 逻辑拓扑结构 logical topology
8. 无冲突联网环境 collision-free networking environment
9. 树型拓扑结构 tree topology
10. 目的地节点 destination node

# Unit 7/Section C

**I. Fill in the blanks with the information given in the text:**

1. repeater
2. data-link
3. gateway
4. router

**II. Translate the following terms or phrases from English into Chinese and vice versa:**

1. destination address 目的地址
2. performance degradation 性能退化（或降级）
3. four-interface bridge 4 接口网桥
4. common bus 公共总线，公用总线
5. 数据链路层 data-link layer
6. 协议转换器 protocol converter
7. 开放式系统互联 OSI (Open Systems Interconnection)
8. 物理地址 physical address

# Unit 8: The Internet

## Unit 8/Section A

**I. Fill in the blanks with the information given in the text:**

1. research; DARPA (the Defense Advanced Research Projects Agency)
2. WANs
3. internet; Internet
4. ICANN (the Internet Corporation for Assigned Names and Numbers)
5. registrars
6. 128-bit
7. mnemonic; IP
8. servers

**II. Translate the following terms or phrases from English into Chinese and vice versa:**

1. cellular telephone 蜂窝电话，移动电话，手机
2. IP address IP 地址，网际协议地址
3. analog communications 模拟通信
4. access ISP 接入 ISP，因特网接入服务提供商
5. fiber-optic cable 光纤电缆，光缆
6. binary notation 二进制记数法
7. mnemonic name 助记名，缩写名
8. Internet-wide directory system 因特网范围的目录系统

9. name server 名称服务器，名字服务器
10. broadband Internet access 宽带因特网访问，宽带因特网接入
11. 助记标识符 mnemonic identifier
12. 电缆调制解调器 cable modem
13. 网络基础设施 network infrastructure
14. 顶级域名 TLD (top-level domain)
15. 因特网编址 Internet addressing
16. 点分十进制记数法 dotted decimal notation
17. 摄像机 video camera
18. 域名系统 domain name system
19. 卫星上行链路 satellite uplink
20. 有线电视 cable television

**III. Fill in each of the blanks with one of the words given in the following list, making changes if necessary:**

Early computer networks used leased telephone company lines for their connections. Telephone company systems of that time established a single connection between *sender* and receiver for each telephone call, and that connection carried all *data* along a single path. When a company wanted to connect computers it *owned* at two different locations, the company placed a telephone call to *establish* the connection, and then connected one computer to each end of that single *connection*.

The U.S. Defense Department was concerned about the inherent *risk* of this single-channel method for connecting computers, and its researchers *developed* a different method of sending information through multiple *channels*. In this method, files and messages are broken into *packets* that are labeled electronically with codes for their *origins*, sequences, and destinations. In 1969, Defense Department *researchers* in the Advanced Research Projects Agency (ARPA) used this network model to connect four *computers* into a network called the ARPANET. The ARPANET was the earliest of the *networks* that eventually combined to become what we now call the *Internet*. Throughout the 1970s and 1980s, many researchers in the *academic* community connected to the ARPANET and contributed to the *technological* developments that increased its speed and efficiency.

**IV. Translate the following passage from English into Chinese:**

因特网只是提供了将许许多多的计算机连接在一起的物理与逻辑基础结构。不少人认为，是万维网（WWW 或简称为 Web）为这个全球网络提供了“制胜法宝”。万维网被视为因特网的内容，它通过使用一套丰富的工具来提供各种信息，这套工具管理与链接文本、图形、声音和视频。在万维网上提供和查看信息是使用服务器应用程序和客户端应用程序完成的。

如果你已经探索过万维网，你就会看出客户端应用程序就是万维网浏览器。万维网浏览器接收、解释和显示来自万维网的网页信息。用户可以在网页之内浏览，可以通过点击超文本链接跳到其他网页，也可以指向万维网上的几乎任何网页。

# Unit 8/Section B

**I. Fill in the blanks with the information given in the text:**

1. application; transport; network; link
2. transport
3. different
4. link; network 或 network; link

**II. Translate the following terms or phrases from English into Chinese and vice versa:**

1. incoming message　来报，到来的报文
2. application layer　应用层
3. utility software　实用软件
4. sequence number　序列号，序号
5. remote login capabilities　远程登录能力
6. 端口号　port number
7. 软件例程　software routine
8. 转发表　forwarding table
9. 文件传送协议　FTP（File Transfer Protocol）
10. 万维网浏览器　Web browser

# Unit 8/Section C

**I. Fill in the blanks with the information given in the text:**

1. hyperlinks
2. home
3. tags
4. keywords; operators

**II. Translate the following terms or phrases from English into Chinese and vice versa:**

1. wildcard character 通配符
2. Copy command 复制命令
3. search operator 搜索算符
4. home page 主页
5. 回车键 Enter key
6. 搜索引擎 search engine
7. 嵌入代码 embedded code
8. 超文本标记语言 HTML 或 html (Hypertext Markup Language)

# Unit 9: Mobile and Cloud Computing

## Unit 9/Section A

**I. Fill in the blanks with the information given in the text:**

1. cluster; utility
2. Internet
3. intranet
4. hybrid
5. private; public
6. multi-core; disk
7. desktops
8. physical

**II. Translate the following terms or phrases from English into Chinese and vice versa:**

1. server farm 服务器农场
2. access protocol 存取协议，访问协议
3. storage area network 存储区域网（络）
4. high-throughput computing 高吞吐（量）计算
5. server cluster 服务器集群
6. public cloud 公共云
7. grid computing 网格计算
8. security-aware cloud architecture 具有安全意识的云体系结构
9. social networking 社交网络
10. utility computing 效用计算
11. 云计算提供商 cloud computing provider
12. 存储芯片 memory chip

13. 基于内部网的私有云 intranet-based private cloud
14. 网络带宽 network bandwidth
15. 混合云 hybrid cloud
16. 磁盘阵列 disk array
17. 软件即服务 SaaS (Software as a Service)
18. 集群计算 cluster computing
19. 虚拟化计算机资源 virtualized computer resources
20. 多核处理器 multi-core processor

**III. Fill in each of the blanks with one of the words given in the following list, making changes if necessary:**

The architecture of a cloud is developed at three layers: infrastructure, platform, and application. The infrastructure *layer* is deployed first to support IaaS (Infrastructure as a Service) services. This layer serves as the *foundation* for building the platform layer of the cloud for supporting PaaS (Platform as a Service) *services*. In turn, the platform layer is a foundation for implementing the *application* layer for SaaS applications.

The infrastructure layer is built with virtualized compute, *storage*, and network resources. The abstraction of these hardware *resources* is meant to provide the flexibility demanded by users. Internally, virtualization realizes *automated* provisioning of resources and optimizes the infrastructure management process. The *platform* layer is for general-purpose and repeated usage of the collection of *software* resources. This layer provides users with an environment to *develop* their applications, to test operation flows, and to monitor execution *results* and performance. In a way, the virtualized cloud platform *serves* as a "system *middleware* (中间件)" between the infrastructure and application layers of the *cloud*. The application layer is formed with a collection of all needed software modules for *SaaS* applications. Service applications in this layer include daily office *management* work. The application layer is also heavily used by enterprises in business *marketing* and sales, consumer relationship management, financial transactions, and supply chain management.

**IV. Translate the following passage from English into Chinese:**

云计算动态供应硬件、软件和数据集，以此方式按需应用一个拥有弹性资源的虚拟化平台。其目的是通过在数据中心使用服务器集群和巨型数据库，而将桌面计算转移到一个面向服务的平台。云计算充分利用其低成本与简单的特性，既惠及了用户又惠及了提供商。机器虚拟化使这样的成本效益成为可能。云计算意在同时满足许多用户应用程序的需要。在设计上，云生态系统必须具有安全性、可信性和可靠性。有些计算机用户将云视为一个集中式资源池，而另外一些计算机用户则认为云是一个在所用的各个服务器上实行分布式计算的服务器集群。

传统上，一个分布式计算系统往往为满足内部计算需要由一个自主管理域（如一个研究实验室或公司）所拥有和使用。然而，这些传统系统遭遇了数个性能瓶颈：需要不断进行系统维护；利用率差；与硬件 / 软件升级有关的费用日益增长。作为一种按需计算范式，云计算可解决这些问题或可使我们摆脱这些问题。

# Unit 9/Section B

**I. Fill in the blanks with the information given in the text:**

1. cloud; inner-edge
2. support
3. cognition; efficiency
4. deployment; mediation

**II. Translate the following terms or phrases from English into Chinese and vice versa:**

1. campus area network 校园区域网络，校园网
2. fog and edge computing 雾与边缘计算
3. X as a service X 即服务，一切皆服务
4. virtual machine 虚拟机
5. cybersecurity attack 网络安全攻击
6. 存储或缓存即服务 S/CaaS (storage or caching as a service)
7. 按需数据处理 ODP (on-demand data processing)
8. 软件定义网络 SDN (software-defined network)
9. 图形处理单元 GPU (graphics processing unit)
10. 软件开发工具包 SDK (Software Development Kit)

# Unit 9/Section C

**I. Fill in the blanks with the information given in the text:**

1. desktop
2. wireless
3. nodes
4. phones; computers

**II. Translate the following terms or phrases from English into Chinese and vice versa:**

1. notebook computer 笔记本
2. wireless hotspot 无线热点
3. Short Message Service 短信（服务）
4. wearable computer 可穿戴计算机，穿戴式计算机
5. 移动电话 mobile phone
6. 条形码阅读器 barcode reader
7. 网站 Web site
8. 智能手机 smart phone

# Unit 10: Computer Security

## Unit 10/Section A

**I. Fill in the blanks with the information given in the text:**

1. confidentiality; availability
2. data
3. authenticity
4. integrity
5. Passive; active
6. availability
7. ease; security
8. assurance

**II. Translate the following terms or phrases from English into Chinese and vice versa:**

1. backup system 备份系统
2. encryption key （加密）密钥
3. data confidentiality 数据机密性
4. system vulnerability 系统漏洞，系统脆弱之处
5. unauthorized access 未经授权的访问，越权存取
6. intrusion detection system 入侵检测系统
7. after-action recovery 事后恢复
8. software piracy 软件侵权
9. authorized user 特许用户
10. data unit 数据单元，数据单位
11. 软件版本 software version
12. 数据完整性 data integrity

13. 系统崩溃 system crash
14. 病毒检查软件 virus-checking software
15. 综合安全策略 comprehensive security strategy
16. 软件配置管理 software configuration management
17. 故障隔离 fault isolation
18. 统计数据库 statistical database
19. 保密的加密算法 secure encryption algorithm
20. 数据流 data stream

**III. Fill in each of the blanks with one of the words given in the following list, making changes if necessary:**

An access control mechanism *mediates* (调解) between a user (or a process executing on behalf of a user) and system _resources_, such as applications, operating systems, firewalls, routers, files, and databases. The _system_ must first *authenticate* (验证) a user seeking access. Typically the _authentication_ function determines whether the user is permitted to _access_ the system at all. Then the access control _function_ determines if the specific requested access by this _user_ is permitted. A security administrator maintains an authorization _database_ that specifies what type of access to which resources is _allowed_ for this user. The access control function consults this database to _determine_ whether to grant access. An *auditing* (审计) function monitors and keeps a _record_ of user accesses to system resources.

In practice, a number of _components_ may cooperatively share the access control function. All _operating_ systems have at least a *rudimentary* (基本的), and in many cases a quite _robust_, access control component. Add-on security packages can add to the _native_ access control capabilities of the OS. Particular applications or _utilities_, such as a database management system, also incorporate access _control_ functions. External devices, such as firewalls, can also provide access control services.

**IV. Translate the following passage from English into Chinese:**

入侵者攻击从温和的到严重的形形色色。在这一系列攻击的温和端，有许多人只是希望探查互联网，看看那里有什么。在严重端入侵者则试图阅读特许数据、对数据进行未经授权的修改或扰乱系统。

入侵者的目标是获得进入一个系统的机会，或者增加在一个系统中可以使用的特权范围。初始攻击大多利用系统或软件的漏洞，这些漏洞允许用户执行相应的代码，打开进入系统的后门。入侵者可对具有某些运行特权的程序实施像缓存溢出这样的攻击，利用此类攻击获得进入一个系统的机会。

另外，入侵者也试图获取应予保护的信息。在某些情况下，这种信息是以用户口令的形式存在的。知道了某个用户的口令之后，入侵者就可以登录系统并行使赋予合法用户的所有特权。

# Unit 10/Section B

**I. Fill in the blanks with the information given in the text:**

1. heuristic
2. quarantine
3. positive
4. hoax

**II. Translate the following terms or phrases from English into Chinese and vice versa:**

1. virus scan 病毒扫描
2. infected file 被感染的文件
3. quarantined file 被隔离的文件
4. virus hoax 病毒恶作剧
5. viral exploit 病毒漏洞利用
6. 启发式分析 heuristic analysis
7. 群发邮件蠕虫 mass-mailing worm
8. 病毒特征码 virus signature
9. 防病毒软件 antivirus software
10. 压缩文件 zipped file 或 compressed file

# Unit 10/Section C

**I. Fill in the blanks with the information given in the text:**

1. infection; payload
2. trapdoor
3. botnet
4. administrator; root

**II. Translate the following terms or phrases from English into Chinese and vice versa:**

1. maintenance hook 维护陷阱

2. multipartite virus 多成分病毒，跨领域病毒
3. authentication procedure 验证过程，认证过程
4. instant messaging 即时通信，即时消息
5. 系统登录程序 system login program
6. 逻辑炸弹 logic bomb
7. 多威胁恶意软件 multiple-threat malware
8. 源代码 source code

# Unit 11: Cyberculture

## Unit 11/Section A

**I. Fill in the blanks with the information given in the text:**

1. system
2. copyright
3. rights
4. deleted
5. formatting 或 format
6. signature
7. screens
8. emoticons

**II. Translate the following terms or phrases from English into Chinese and vice versa:**

1. mailing list 邮件发送清单，邮件列表
2. proprietary software 专有软件
3. cc line 抄送行
4. bcc line 密送行
5. forwarded e-mail message 转发的电子邮件
6. e-mail convention 电子邮件常规
7. click on an icon 点击图标
8. confidential document 密件，秘密文件
9. classified information 密级信息
10. recovered e-mail message 恢复的电子邮件
11. 常用情感符 commonly used emoticon
12. 已删除电子邮件 deleted e-mail
13. 电子系统 electronic system
14. 附件行 Attachments line

15. 版权法 copyright law
16. 电子邮件网规 e-mail netiquette
17. 信息高速公路 information superhighway
18. 签名文件 signature file
19. 电子数据表程序 spreadsheet program
20. 文字处理软件 word processor

**III. Fill in each of the blanks with one of the words given in the following list, making changes if necessary:**

Instant messaging has always been a good way to stay in *touch* with friends. As companies look to cut *costs* and improve efficiency, instant messaging is a good way to replace *in-person* meetings, while allowing people to *collaborate* (合作) through their *computers*. Instant messaging is a valuable new tool that is becoming as common as *e-mail* in the workplace.

There are companies which give you free *access* to instant messaging. To use their systems, you will need to *download* their software to your computer from their *websites*. Some instant messaging programs include the ability to send live *audio* or video. With a microphone and/or digital *camera* mounted on your computer, you can let other people *see* and hear you talking.

Most instant messaging *programs* ask you to sign in with a password. Then, you can *create* a list of other people with whom you want to *converse* (交谈). To begin a conversation, send a *message* to the others to see if they are at their computers. Their computers will open a *window* or make a sound to tell them you want to talk to them. If they *respond*, you can start writing back and forth.

**IV. Translate the following passage from English into Chinese:**

新技术带来的变化速度对于全世界人们的生活、工作和娱乐方式产生了重要影响。新兴技术对传统的教学过程以及教育的管理方式提出了挑战。人们经常断言，推动当今社会发生变化的最强有力因素将被证明是信息技术。信息技术虽然本身就是一个重要的学习领域，但同时也在所有的课程领域产生着重大影响。方便的全球通信使我们可以即时访问巨量的数据。快速的通信加上在家里、工作场所和教育机构使用信息技术的机会的增多，可能意味着学习成为一项真正持续终生的活动；在这项活动中，技术变化的速度迫使我们要对学习过程本身不断做出评估。

# Unit 11/Section B

**I. Fill in the blanks with the information given in the text:**

1. intentional; practices
2. safety; disaster
3. limitations; expert
4. default

**II. Translate the following terms or phrases from English into Chinese and vice versa:**

1. safety-critical system 安全攸关的系统，安全苛求系统
2. professional competence 职业能力，专业能力
3. code of professional conduct 职业行为准则，专业行为规范
4. code of ethics 道德规范，道德准则
5. data manipulation 数据操作
6. 专家系统 expert system
7. 默认设置 default setting
8. 产权 property rights
9. 项目经理 project manager
10. 计算机专业人员 computer professional

# Unit 11/Section C

**I. Fill in the blanks with the information given in the text:**

1. neutrality
2. networks; content
3. recognition
4. Phishing

**II. Translate the following terms or phrases from English into Chinese and vice versa:**

1. message board 留言板，消息板
2. software vendor 软件供应商，软件厂商
3. anonymous message 匿名消息
4. video clip 视频剪辑，视频片段
5. 过滤软件 filtering software

6. 版权侵犯 copyright infringement
7. 网络中立性 network neutrality
8. 网络运营商 network operator

# Unit 12: Smart World

## Unit 12/Section A

**I. Fill in the blanks with the information given in the text:**

1. autonomous
2. neurology; mechanical
3. sensors; actuators
4. procedural; declarative
5. engineering; theoretical
6. performance-oriented; simulation-oriented
7. Turing
8. interactive; analyst

**II. Translate the following terms or phrases from English into Chinese and vice versa:**

1. autonomous machine 自主机器
2. intelligent agent 智能代理，智能主体
3. performance-oriented methodology 面向性能的方法论
4. declarative knowledge 陈述性知识
5. reflex action 反射作用，本能反应，本能动作
6. soil sampling device 土壤采样设备，土壤取样装置
7. physical attribute 物理属性
8. goal-directed behavior 目标导向行为
9. common sense 常识
10. audio pattern 音频模式
11. 图灵测试 Turing test
12. 语音合成器 speech synthesizer
13. 测试对象 test subject
14. 文档翻译程序 document translator
15. 交互程序 interactive program
16. 距离传感器 range sensor
17. 面向模拟的方法论 simulation-oriented methodology
18. 图像处理与分析 image processing and analysis

19. 过程性知识 procedural knowledge
20. 试错过程 trial-and-error process

**III. Fill in each of the blanks with one of the words given in the following list, making changes if necessary:**

The conjecture that machines can be programmed to exhibit intelligent behavior is known as *weak* AI and is accepted, to varying degrees, by a wide audience today. However, the *conjecture* that machines can be programmed to possess intelligence and, in fact, *consciousness*, which is known as strong AI, is widely *debated*. Opponents of strong AI argue that a machine is inherently *different* from a human and thus can never feel love, tell *right* from wrong, and think about itself in the *same* way that a human does. However, *proponents* (支持者) of *strong* AI argue that the human mind is constructed from small *components* that individually are not human and are not conscious but, when *combined*, are. Why, they argue, would the same phenomenon not be *possible* with machines? The problem in resolving the strong AI debate is that such *attributes* as intelligence and consciousness are internal characteristics that cannot be identified *directly*. As Alan Turing pointed out, we credit other humans with *intelligence* because they behave intelligently—even though we cannot *observe* their internal mental states. Are we, then, prepared to grant the same *latitude* (自由度) to a *machine* if it exhibits the external characteristics of consciousness?

**IV. Translate the following passage from English into Chinese:**

人工智能领域的早期工作是在显式编写程序来模拟智能的背景下探索这一主题。然而，如今许多人认为，人类智能的基础不是复杂程序的执行，而是经过世代进化的简单的刺激–反应机能。这种“智能”理论称为“基于行为的智能”，因为“智能的”刺激–反应机能似乎是某些行为的结果，这些行为在有的个体未能生存和繁衍的同时却使有的个体得以存活和繁衍。

基于行为的智能似乎能回答人工智能界的几个问题，例如，为什么基于冯·诺依曼体系结构的机器在计算技能上能轻易胜过人类，却很难展现常识性的东西？因此，基于行为的智能可能会对人工智能研究产生重要影响。基于行为的技术已经应用于以下领域：人工神经网络领域，教神经元如何以期望的方式表现；遗传算法领域，为较传统的程序设计过程提供替代方法；机器人技术领域，通过反应策略改进机器性能。

# Unit 12/Section B

**I. Fill in the blanks with the information given in the text:**

1. virtual; registered
2. tracking; visualization
3. conferencing; interface
4. online

**II. Translate the following terms or phrases from English into Chinese and vice versa:**

1. situated visualization 情境可视化，位置可视化
2. spatial annotation 空间注释
3. video conferencing 视频会议（技术）
4. live video feed 实况视频馈送
5. 3D graphics rendering 三维图形渲染
6. 头戴式显示器 HMD (head-mounted display)
7. 增强现实 augmented reality
8. 实时 real time
9. 视点控制 viewpoint control
10. 紧密耦合的反馈回路 tightly coupled feedback loop

# Unit 12/Section C

**I. Fill in the blanks with the information given in the text:**

1. enabling
2. tablet; smart
3. sensing; clouds
4. autonomous

**II. Translate the following terms or phrases from English into Chinese and vice versa:**

1. computer vision 计算机视觉
2. ubiquitous computing 普适计算
3. command line 命令行
4. smart power grid 智能电网
5. 移动网络 mobile network
6. 供应链 supply chain
7. 物联网 Internet of Things
8. 传感器融合 sensor fusion

# Glossary

# （词汇表）

## A

| | |
|---|---|
| abbreviate /ə'bri:vieit/ v. | 缩写（1C） |
| abstract /æb'strækt/ v. | 把…抽象出来；提取，抽取（5C） |
| abstract machine | 抽象机（2C） |
| abstraction /æb'strækʃən/ n. | 抽象；提取（3A） |
| abundance /ə'bʌndəns/ n. | 大量，丰富，充足（6C） |
| abusive /ə'bju:siv/ a. | 谩骂的；毁谤的（11A） |
| academia /ˌækə'di:miə/ n. | 学术界（6C） |
| accommodate /ə'kɔmədeit/ v. | 容纳；使适应（8A） |
| accountability /əˌkauntə'biliti/ n. | 负有责任（10A） |
| accusation /ˌækju:'zeiʃən/ n. | 指控，控告（11C） |
| achiever /ə'tʃi:və/ n. | 事业成功的人（11A） |
| activate /'æktiveit/ v. | 激活，启动（2C） |
| actuation /ˌæktju'eiʃən, -tʃu-/ n. | 驱动；启动（9B） |
| actuator /'æktjueitə/ n. | 执行器，执行机构；致动器（12A） |
| additive /'æditiv/ a. & n. | 添加的 / 添加物；添加剂（2A） |
| additive manufacturing | 增材制造（2A） |
| address /ə'dres/ v. | 编址；寻址（3C） |
| Address box | 地址框（8C） |
| address bus | 地址总线（2A） |
| addressee /ˌædre'si:/ n. | 收信人；收件人（8B） |
| addressing /ə'dresiŋ/ n. | 编址；寻址（5A） |
| adjacent /ə'dʒeisənt/ a. | 相邻的，毗连的（2C） |
| administrator /əd'ministreitə/ n. | （系统、程序等的）管理员（6A） |
| adorable /ə'dɔ:rəbəl/ a. | 值得崇拜的；可爱的（12A） |
| advent /'ædvent/ n. | 出现，到来（1A） |
| adverse /'ædvə:s, æd'və:s/ a. | 不利的，有害的（2C） |
| aesthetical /i:s'θetikəl/ a. | 美学的；审美的（12A） |
| aforementioned /ə'fɔ:ˌmenʃənd/ a. | 前面提到的（9A） |
| aggregate /'ægrigət/ a.& n. | 聚集的；合计的 / 总数，合计；聚集体（10A） |
| agile /'ædʒail/ a. | 敏捷的；灵活的（9A） |
| agility /ə'dʒiliti/ n. | 敏捷；机敏（9B） |
| albeit /ɔ:l'bi:it/ conj. | 尽管，即使（12B） |
| algorithm /'ælgəriðəm/ n. | 算法（1A） |
| alignment /ə'lainmənt/ n. | 对准，对齐（12B） |
| allocate /'æləkeit/ v. | 分配；分派（2B） |
| allot /ə'lɔt/ v. | 分配；分派（2B） |
| alternate /ɔ:l'tə:nit, 'ɔ:ltə-/ a. | 供选择的，供替换的；备用的（7B） |
| alternatively /ɔ:l'tə:nətivli/ ad. | 或者，非此即彼（7B） |

ambiguous /æm'bigjuəs/ a. 含糊不清的，模棱两可的（3A）
amendment /ə'mendmənt/ n. 修正案，修正条款（11C）
amplify /'æmplifai/ v. 放大；增强；扩大（12B）
analog(ue) /'ænəlɔg/ a. 模拟的（1A）
analogous /ə'næləgəs/ a. 相似的；可比拟的（*to/with*）（8B）
analogy /ə'nælədʒi/ n. 比拟，类推，类比（5C）
analyst /'ænəlist/ n. 分析员，分析师（4B）
analytic(al) /ˌænə'litik(əl)/ a. 分析的（1A）
Analytical Engine 分析机，解析机（1A）
analytics /ˌænə'litiks/ n. 分析方法；分析学（9B）
analyzer /'ænəlaizə/ n. 分析程序，分析器（2C）
animated /'ænimeitid/ a. 动画（片）的；活生生的（12B）
animation /ˌæni'meiʃən/ n. 动画（制作）（8C）
annotate /'ænəuteit/ v. 给…做注解（或注释、评注）（9C）
annotation /ˌænəu'teiʃən/ n. 注解，注释，评注（9C）
anomalous /ə'nɔmələs/ a. 异常的（6C）
anomaly /ə'nɔməli/ n. 异常（现象）（6A）
anonymous /ə'nɔniməs/ a. 匿名的（11C）
antenna /æn'tenə/ n. 天线（12C）
antitrust /ˌænti'trʌst/ a. 反托拉斯的，反垄断的（11A）
antivirus software 防病毒软件（10B）
app /æp/ n. 应用程序（*app*lication 的缩略）（1C）
appealing /ə'pi:liŋ/ a. 吸引人的；有感染力的（3B）
append /ə'pend/ v. 附加（8B）
applet /'æplət/ n. 小应用程序（3B）
application /ˌæpli'keiʃən/ n. 应用程序，应用软件（1B）
approximation /əˌprɔksi'meiʃən/ n. 近似（值）（1A）
arcane /ɑ:'kein/ a. 神秘的，晦涩难解的（3B）
architecture /'ɑ:kitektʃə/ n. 体系结构（1B）
archive /'ɑ:kaiv/ n. [常作～s] 档案；档案馆，档案室（9A）
arms race 军备竞赛（11C）
array /ə'rei/ n. 数组；阵列；一系列（3A）
artery /'ɑ:təri/ n. 动脉；干线（8A）
artifact /'ɑ:tifækt/ n. 人工制品，制造物（5B）
artificial intelligence 人工智能（1A）
aspirational /ˌæspə'reiʃənəl/ a. 有志向的；有抱负的（11B）
assembler /ə'semblə/ n. 汇编程序，汇编器（4A）
assembly code 汇编代码（3B）
assembly language 汇编语言（3A）
assignment statement 赋值语句（3A）
assuming /ə'sju:miŋ; ə'su:-/ conj. 假定，假如（5A）
assurance /ə'ʃuərəns/ n. 保证；把握，信心（12C）
asterisk /'æstərisk/ n. 星号（3C）
astounding /ə'staundiŋ/ a. 令人震惊的；使人惊骇的（1C）
attachment /ə'tætʃmənt/ n. 附件（10B）
audit /'ɔ:dit/ v. 审计，查账（12C）
augment /ɔ:g'ment/ v. 增强；增加；扩大（12B）
augmentation /ˌɔ:gmen'teiʃən/ n. 增强；增加；扩大（12B）
augmented reality 增强现实（12B）
authenticate /ɔ:'θentikeit/ v. 验证，鉴别（11C）

authentication /ɔːˌθentiˈkeiʃən/ n. 验证，鉴别（9B）
authenticity /ˌɔːθenˈtisiti/ n. 可靠性，真实性（10A）
authorization /ˌɔːθəraiˈzeiʃən; -riˈz-/ n. 授权；委托（9B）
authorized /ˈɔːθəraizd/ a. 经授权的，特许的（5A）
autocorrelation /ˈɔːtəuˌkɔriˈleiʃən/ n. 自相关（性）（6C）
automated /ˈɔːtəmeitid/ a. 自动化的（1B）
automation /ˌɔːtəˈmeiʃən/ n. 自动化（1A）
autonomous /ɔːˈtɔnəməs/ a. （独立）自主的；自治的（9B）
autonomy /ɔːˈtɔnəmi/ n. 自治（权），自主性（12C）
avatar /ˈævətɑː/ n. 化身（印度教和佛教中化作人形或兽形的神）（12B）
avoidance /əˈvɔidəns/ n. 避免；避开（7A）

# B

backbone /ˈbækbəun/ n. 骨干（网），主干（网），基干（网）（7B）
backcountry /ˈbækˌkʌntri/ a. 偏僻乡村的（8C）
backdoor /ˌbækˈdɔː/ n. 后门（10C）
backend /ˈbækend/ n. & a. 后端（的）（9A）
backlight /ˈbæklait/ v. （-lighted 或-lit /-lit/）从背后照亮（2A）
backlighting /ˈbæklaitiŋ/ n. 背后照明；背光（2A）
backup /ˈbækʌp/ n. & a. 备份，后备（2C）/ 备份的，后备的（6B）
balance /ˈbæləns/ n. 余额；差额；结算（6B）
bandwidth /ˈbændwidθ/ n. 带宽（9A）
banking /ˈbæŋkiŋ/ n. 银行业（务）（1A）
bar /bɑː/ n. 条；条形图（3C）
bar chart 条形图（3C）
barcode /ˈbɑːkəud/ n. 条形码（9C）
base 10 notation 以10为底的记数法（8A）
baseline /ˈbeislain/ n. 基线；基准；基础（5B）
batch /bætʃ/ n. 批，批量，成批（9A）
battlefield /ˈbætlfiːld/ n. 战场（12C）
beep /biːp/ n. 短促而尖厉的声音，嘟（9C）
benchmark /ˈbentʃmɑːk/ n. 基准（12A）
beneficiary /ˌbeniˈfiʃiəri/ n. 受益人，受惠者（12C）
big data 大数据（6C）
billing /ˈbiliŋ/ n. 开（账）单；记账（9A）
binary /ˈbainəri/ a. & n. 二进制的；二元的（1A）/ 二进制（数）（8A）
binary notation 二进制记数法（8A）
bioinformatics /ˌbaiəuˌinfəˈmætiks/ n. 生物信息学（6C）
biometrics /ˌbaiəuˈmetriks/ n. 生物统计学（12C）
biotechnology /ˌbaiəutekˈnɔlədʒi/ n. 生物技术（6C）
bit /bit/ n. 位，比特（1A）
bit pattern 位模式（2B）
black box testing 黑盒测试（法）（5B）
blackboard model 黑板法模型（2C）
blackjack /ˈblækdʒæk/ n. 21点（一种牌戏）（11C）
blanket /ˈblæŋkit/ v. 用毯子（或毯状物）盖（或裹）；覆在…的上面（9C）
blooper /ˈbluːpə/ n. 过失，失礼（11A）
blueprint /ˈbluːprint/ n. 蓝图（4B）

Bluetooth /ˈbluːtuːθ/ n. 蓝牙（技术）（9B）
blurry /ˈbləːri/ a. 模糊的，难辨认的（1C）
bombsight /ˈbɔmsait/ n. 轰炸瞄准器（1A）
boot /buːt/ v. 启动，引导（*up*）（2A）
boot loader 引导装入程序，引导加载程序（2A）
bot /bɔt/ n. 机器人；僵尸程序（= robot）（10C）
botnet /ˈbɔtnet/ n. 僵尸网络（10C）
boundary value 边界值（5B）
bounds checking 边界检查（3B）
boxer /ˈbɔksə/ n. 拳击运动员，拳师（8C）
bracket /ˈbrækit/ n. 括号（3C）
branding /ˈbrændiŋ/ n. 品牌推广，品牌宣传（12B）
breach /briːtʃ/ n. 破坏；违反（10A）
brick-and-mortar /ˌbrikəndˈmɔːtə/ a. 实体的，具体的（6C）
bridge /bridʒ/ n. 网桥，桥接器（7C）
broadband /ˈbrɔːdbænd/ n. & a. 宽带（的）（8A）
broker /ˈbrəukə/ n. 代理者；代理程序（3B）
browse /brauz/ v. 浏览（3B）
brute /bruːt/ a. 粗暴的；野蛮的（12A）
bug /bʌg/ n. （程序）错误，故障（4A）
buggy /ˈbʌgi/ a. 有（程序）错误的，有故障的（11C）
building block 积木块，构建模块，构件（4B）
built-in /ˈbiltˈin/ a. 内置的，内部的（1A）
bureaucratic /ˌbjuərə*u*ˈkrætik/ a. 官僚（政治）的；官僚主义的（4B）
bus /bʌs/ n. 总线（1A）
bus topology 总线拓扑（结构）（7A）
byte /bait/ n. 字节（1A）
bytecode /ˈbaitkəud/ n. 字节码（3B）

# C

cable television 有线电视（8A）
cabling /ˈkeibliŋ/ n. 电缆（9C）
caching /ˈkæʃiŋ/ n. 缓存，高速缓存（9B）
calculation /ˌkælkjuˈleiʃən/ n. 计算（1A）
calculator program 计算（器）程序（10C）
callout /ˈkɔːlaut/ n. 标注（9A）
callout box 标注框（9A）
calorie /ˈkæləri/ n. 卡（路里）；热量（1C）
campus area network 校园区域网络，校园网（9B）
cancellation /ˌkænsəˈleiʃən/ n. 取消；删去（6B）
capitalization /ˌkæpitəlaiˈzeiʃən; -liˈz/ n. 大写字母的使用（8C）
captivating /ˈkæptiveitiŋ/ a. 迷人的，可爱的（8C）
cardinality /ˌkɑːdiˈnæliti/ n. 基数性；（集的）势，（集的）基数（6A）
carnivore /ˈkɑːnivɔː/ n. 食肉动物（11C）
carrier /ˈkæriə/ n. 载波（7A）/ 运输公司；运输工具；通信公司（9C）
carrier sense 载波侦听，载波监听（7A）
cartridge /ˈkɑːtridʒ/ n. 盒，匣（2A）
carve /kɑːv/ v. 雕刻，刻（9A）

| | |
|---|---|
| cascade /kæ'skeid/ n. | 小瀑布；级联；层叠（4C） |
| cascading /kæ'skeidiŋ/ a. | 级联的；层叠的（6B） |
| cascading rollback | 级联回滚（6B） |
| cash register | 现金出纳机，收银机（9C） |
| casino /kə'si:nəu/ n. | （有表演、舞池等的）卡西诺赌场（11C） |
| cassette tape | 盒式磁带（2A） |
| categorical /ˌkæti'gɔrikəl/ a. | 类的，分类的（6C） |
| categorize /'kætigəraiz/ v. | 将…分类（7B） |
| cater /'keitə/ v. | 满足需要；迎合；考虑（*for, to*）（4C） |
| cell phone | 蜂窝电话，移动电话，手机（1C） |
| cellular /'seljulə/ a. | 蜂窝状的，多孔的（8A） |
| cellular (tele)phone | 蜂窝电话，移动电话，手机（8A） |
| censor /'sensə/ v. | 审查，检查（11C） |
| census /'sensəs/ n. | 人口普查（1A） |
| central processing unit | 中央处理器（1A） |
| chain-letter /'tʃeinˌletə/ v. | 向…发送连锁信（或连锁邮件）（11A） |
| challenging /'tʃælindʒiŋ/ a. | 挑战性的（12A） |
| chatty /'tʃæti/ a. | 聊天式的，轻松而亲切的；爱闲聊的（11A） |
| check box | 复选框（4B） |
| checkers /'tʃekəz/ n. | 〈美〉西洋跳棋（12A） |
| checkout /'tʃekaut/ n. | 结账（离去）；付款台（12C） |
| chip /tʃip/ n. | 芯片（1A） |
| circuitry /'sə:kitri/ n. | 电路（1A） |
| circumvent /ˌsə:kəm'vent/ v. | 绕过；规避（4C） |
| clamshell /'klæmʃel/ n. | 蛤壳（1C） |
| clarity /'klæriti/ n. | 清晰，明晰（2A） |
| class hierarchy | 类层次（3B） |
| classified /'klæsifaid/ a. | 归入密级的，保密的（11A） |
| click /klik/ v. | （鼠标）单击（2B） |
| client /'klaiənt/ n. | 客户机；客户程序（2C） |
| clip /klip/ n. | 剪下来的东西；电影（或电视）片段（2C） |
| cloak /kləuk/ v. | 掩盖；掩饰；伪装（10B） |
| cloud computing | 云计算（9A） |
| cluster /'klʌstə/ n. | 一群，一串，一簇；聚类（6C） |
| clustering /'klʌstəriŋ/ n. | 聚类（6C） |
| clutter /'klʌtə/ n. | 凌乱，杂乱；杂乱的东西（3B） |
| coaxial /ˌkəu'æksiəl/ a. | 同轴的（7B） |
| coaxial cable | 同轴电缆（7B） |
| cobble /'kɔbəl/ v. | 修补（鞋子）；匆忙草率地拼凑（11B） |
| cognition /kɔg'niʃən/ n. | 认识，认知（1B） |
| cognitive /'kɔgnitiv/ a. | 认识的，认知的（12B） |
| coherent /kəu'hiərənt/ a. | 相干的；一致的；协调的（6A） |
| cold chain | 冷链（12C） |
| collaboration /kəˌlæbə'reiʃən/ n. | 合作；协作（1B） |
| collaborative /kə'læbəreitiv/ a. | 合作的，协作的（12B） |
| collide /kə'laid/ v. | 冲突；碰撞（7A） |
| colon /'kəulən/ n. | 冒号（8C） |
| comic strip | （报刊上的）连环漫画（9C） |
| command line | 命令行（12C） |
| commit /kə'mit/ v. | 提交，委托；承诺（6B） |

commit point 提交点（6B）
compatible /kəm'pætəbl/ a. 兼容的（2C）
compelling /kəm'peliŋ/ a. 令人信服的；有强烈吸引力的（12B）
competence /'kɔmpitəns/ n. 能力；胜任（11B）
compilation /ˌkɔmpi'leiʃən/ n. 编译；汇编（2C）
compile /kəm'pail/ v. 汇编；编译（1A）
compiled code 编译执行的代码（3B）
compiled language 编译执行的语言，编译型语言（4A）
compiler /kəm'pailə/ n. 编译程序，编译器（3A）
complexity /kəm'pleksiti/ n. 复杂性，错综性（1B）
composite /'kɔmpəzit, kəm'pɔz-/ a. 综合成的；合成的；复合的（5A）
compositing /'kɔmpəzitiŋ, kəm'pɔz-/ n. 混合；合成（12B）
comprehend /ˌkɔmpri'hend/ v. 理解，领会（1B）
compression /kəm'preʃən/ n. 压缩（2C）
computational /ˌkɔmpju:'teiʃənəl/ a. 计算（机）的（1B）
computer vision 计算机视觉（1B）
conceive /kən'si:v/ v. （构）想出（1A）
conceptual /kən'septjuəl/ a. 概念的（6A）
conceptually /kən'septjuəli/ ad. 概念上（1A）
concise /kən'sais/ a. 简明的，简要的（11A）
concurrent /kən'kʌrənt/ a. 同时发生的，并发的（2B）
conditional statement 条件语句（3A）
conducive /kən'dju:siv/ a. 有助的，有益的（*to*）（8A）
conferencing /'kɔnfərənsiŋ/ n. 召开会议；会议技术（12B）
confidential /ˌkɔnfi'denʃəl/ a. 秘密的，机密的（10A）
confidentiality /'kɔnfiˌdenʃi'æliti/ n. 机密性（10A）
configuration /kənˌfigju'reiʃən/ n. 配置（1C）
configuration item 配置项（2C）
configure /kən'figə/ v. 配置（1C）
confluence /'kɔnfluəns/ n. 汇合；聚集（6C）
conform /kən'fɔ:m/ v. 遵照；一致，符合（*to, with*）（5A）
congest /kən'dʒest/ v. 拥挤；拥塞（9C）
conjecture /kən'dʒektʃə/ n. 推测，猜想（1A）
connectivity /kənek'tiviti/ n. 连通（性），连接（性）（6C）
connotation /ˌkɔnəu'teiʃən/ n. 内涵（意义），含义（12A）
consecutive /kən'sekjutiv/ a. 连续的（5B）
console /'kɔnsəul/ n. 控制台，操纵台（1C）
consolidate /kən'sɔlideit/ v. （把…）联为一体，合并（4B）
constrain /kən'strein/ v. 约束，限制（3B）
constraint /kən'streint/ n. 约束，限制（4C）
consumer electronics 消费电子产品（9C）
contagion /kən'teidʒən/ n. （接）触（传）染（10C）
context switch 上下文转换，语境转换（2B）
controller /kən'trəulə/ n. 控制器（2B）
convergence /kən'və:dʒəns/ n. 会聚；结合；收敛（9C）
conversion /kən'və:ʃən/ n. 转换；转变（8A）
converter /kən'və:tə/ n. 转换器，转换程序（7C）
convertible /kən'və:tibəl/ a. 可转变的；可转换的；可折叠的（1C）
convoluted /'kɔnvəlu:tid/ a. 盘绕的；错综复杂的（3B）
cookie /'kuki/ n. 小甜饼（指一种临时保存网络用户信息的结构）（11C）

cooperative /kəu'ɔpərətiv/ a. 合作的，协作的（12C）
copier /'kɔpiə/ n. 复印机（2A）
Copy command 复制命令（8C）
corporate /'kɔ:pərit/ a. 公司的；社团的（8C）
correlation /ˌkɔri'leiʃən/ n. 相关（性），相互关联（6C）
corrupt /kə'rʌpt/ v. 破坏，损坏；腐蚀（2A）
corruption /kə'rʌpʃən/ n. 破坏；腐化；讹误（6B）
cosmetic /kɔz'metik/ a. 化妆用的；装饰性的；非实质性的（8C）
cost-effective /'kɔsti'fektiv/ a. 有成本效益的；合算的（4C）
counteract /ˌkauntə'rækt/ v. 对…起反作用；对抗；抵消（5B）
countermeasure /'kauntəˌmeʒə/ n. 对策，对抗手段（10A）
counterpart /'kauntəpɑ:t/ n. 对应的物（或人）（4A）
country code 国家代码（8A）
courier /'kuriə/ n. 信使（11A）
coverage /'kʌvəridʒ/ n. 覆盖（范围）（5B）
coworker /kəu'wə:kə/ n. 同事（11A）
crack /kræk/ v. 破译（1A）
credit card 信用卡（9C）
criterion /krai'tiəriən/ n. （[复] -ria/-riə/或 -rions）标准，准则（5B）
critter /'kritə/ n. 〈美口〉生物；动物（11A）
cryptic /'kriptik/ a. 隐秘的；令人困惑的（3B）
culprit /'kʌlprit/ n. 罪犯；被控犯罪的人（11C）
cursor /'kə:sə/ n. 光标（2A）
customary /'kʌstəməri/ a. 习惯的；按惯例的（8A）
customizable /'kʌstəmaizəbəl/ a. 可定制的（9B）
customization /kʌstəmai'zeiʃən/ n. 定制，用户化（9A）
customize /'kʌstəmaiz/ v. 定制，使用户化（2B）
customized /'kʌstəmaizd/ a. 定制的，用户化的（9B）
cyber- /'saibə/ comb. form 表示“计算机（网络）的”（12C）
cyber-physical /ˌsaibə'fizikəl/ a. 信息物理的（12C）
cybersecurity /ˌsaibəsi'kjuəriti/ n. 网络安全（9B）
cyberspace /'saibəspeis/ n. 计算机空间，网络空间（8C）

# D

data bus 数据总线（2A）
data coloring 数据着色（9A）
data declaration 数据声明（3A）
data element 数据元（素）（5A）
data flow diagram 数据流程图（4B）
data item 数据项（6B）
data link 数据链路（7C）
data logging 数据（事件）记录；数据登录（12C）
data mining 数据挖掘（6C）
data set 数据集（6C）
data stream 数据流（9B）
datum /'deitəm/ n. （[复] data）数据（1A）
de facto /di:'fæktəu/ a. 〈拉〉实际的，事实上的（3B）
deadlock /'dedlɔk/ n. 死锁；僵局（6B）

| | |
|---|---|
| dealer /'di:lə/ n. | 发牌者；商人（11C） |
| debit /'debit/ n. | 借方；借记，借入（9C） |
| debit card | 借记卡，借方卡（9C） |
| debug /di:'bʌg/ v. | 调试，排除（程序）中的错误（4A） |
| debugger /di:'bʌgə/ n. | 调试程序，排错程序（4A） |
| decimal /'desiməl/ a. | 十进制的（1A） |
| decimal notation | 十进制记数法（8A） |
| deck /dek/ n. | 卡片叠，卡片组（1A） |
| declarative language | 声明式语言（3A） |
| decode /ˌdi:'kəud/ v. | 译（码），解（码）（2B） |
| decompression /ˌdi:kəm'preʃən/ n. | 解压缩（2C） |
| decrement /'dekrimənt/ v. | 减少，减缩（6B） |
| dedicate /'dedikeit/ v. | 把…献给；把…用于（*to*）（1C） |
| dedicated /'dedikeitid/ a. | 专用的（1A） |
| deduce /di'dju:s/ v. | 推论，推断（3A） |
| deem /'di:m/ v. | 认为，视为（5B） |
| default /di'fɔ:lt/ n. | 默认，缺省，系统设定值（4A） |
| defective /di'fektiv/ a. | 有缺点的，有缺陷的，有毛病的（11C） |
| definitive /di'finitiv/ a. | 决定性的；确定的；规定的（4C） |
| degradation /ˌdegrə'deiʃən/ n. | 降级，退化（7C） |
| degrade /di'greid/ v. | 使降级，使退化（10A） |
| degraded /di'greidid/ a. | 降级的，退化的（7B） |
| déjà vu /ˌdeiʒɑ:'vju:/ n. | 〈法〉似曾经历的错觉（5C） |
| delineate /di'linieit/ v. | 勾画出…的轮廓；画出；描绘（1C） |
| deliverable /di'livərəbəl/ n. | ［常作复］可交付使用的产品（4B） |
| delve /delv/ v. | 搜索，翻查（3B） |
| demodulate /di:'mɔdjuleit/ v. | 解调（12C） |
| demodulator /di:'mɔdjuleitə/ n. | 解调器（2A） |
| denote /di'nəut/ v. | 表示；意思是（6A） |
| dependency /di'pendənsi/ n. | 依靠，依赖（2C） |
| dependent variable | 因变量（6C） |
| deploy /di'plɔi/ v. | 部署；展开（5A） |
| deployment /di'plɔimənt/ n. | 部署；展开（5A） |
| derivation /ˌderi'veiʃən/ n. | 派生（物）（8C） |
| designate /'dezigneit/ v. | 指定；命名；指派（3C） |
| desk /desk/ n. | 服务台；部门（8C） |
| deskside /'desksaid/ a. | 桌边（型）的（9A） |
| desktop /'desktɔp/ a. & n. | 桌面的；台式（计算机）的（1A）/ 台式（计算）机；桌面（1C） |
| detach /di'tætʃ/ v. | 拆卸；使分开；使分离（12B） |
| deterministic /diˌtə:mi'nistik/ a. | 决定论的；确定性的（12C） |
| deterrence /di'terəns/ n. | 威慑（10A） |
| devastating /'devəsteitiŋ/ a. | 破坏性极大的，毁灭性的（6B） |
| device driver | 设备驱动程序（2B） |
| dexterity /dek'steriti/ n. | 灵巧，敏捷（1B） |
| diagnostic /ˌdaiəg'nɔstik/ a. | 诊断的（9C） |
| dialog box | 对话框（3B） |
| dichotomy /dai'kɔtəmi/ n. | 一分成二；二分法（12A） |
| Difference Engine | 差分机（1A） |
| differentiate /ˌdifə'renʃieit/ v. | 区分，区别（1C） |

diffusion /di'fju:ʒən/ n. 扩散（12C）
digit /'didʒit/ n. 数字（1A）
digitize /'didʒitaiz/ v. 使数字化（2C）
dimensionality /di,menʃə'næliti, dai-/ n. 维度，维数（6C）
diminish /di'miniʃ/ v. 减少；降低（11B）
dinosaur /'dainəsɔ:/ n. 恐龙；（尤指废弃过时的）庞然大物（2B）
diode /'daiəud/ n. 二极管（2A）
dire /'daiə/ a. 可怕的；极紧迫的；极端的（10B）
directory /di'rektəri/ n. 目录（2B）
directory path 目录路径（2B）
disability /,disə'biliti/ n. 无能力；残疾（1B）
disastrous /di'zɑ:strəs/ a. 灾难性的（6B）
disclose /dis'kləuz/ v. 使显露；泄露，透露（10A）
disclosure /dis'kləuʒə/ n. 泄露，透露（10A）
discrete /'diskri:t/ a. 分离的；离散的（6C）
discretion /dis'kreʃən/ n. 斟酌决定（或处理）的自由（2B）
discrimination /dis,krimi'neiʃən/ n. 区别；歧视（11A）
disk array 磁盘阵列（9A）
dismay /dis'mei/ n. 失望；气馁；惊愕（12A）
dispatcher /dis'pætʃə/ n. 分派程序（2B）/（车辆）调度员（9C）
disrupt /dis'rʌpt/ v. 扰乱；使中断（7B）
disruption /dis'rʌpʃən/ n. 扰乱；中断（10A）
disruptive /dis'rʌptiv/ a. 扰乱性的，制造混乱的（10C）
disseminate /di'semineit/ v. 散布；传播（6C）
distorted /dis'tɔ:tid/ a. 扭曲的，变形的（11C）
distributed /di'stributid/ a. 分布（式）的（2C）
disturbance /di'stə:bəns/ n. 打扰；干扰；扰动；骚乱（6C）
diversity /dai'və:siti/ n. 多样性；差异（性）（1B）
document file 文档文件（10B）
documentation /,dɔkjumen'teiʃən/ n. 文档编制；［总称］文件证据，文献资料（4B）
domain /dəu'mein/ n. 领域，域（7A）
domain name 域名（8A）
doorway /'dɔ:wei/ n. 出入口，门口（8C）
dormant /'dɔ:mənt/ a. 休眠的；暂停活动（或作用）的（10C）
dotted decimal notation 点分十进制记数法（8A）
down /daun/ v. 击倒；击落；倒下（7B）
downright /'daunrait/ ad. 彻底地，完全地（11C）
dressing room 试衣室，试衣间（12B）
drive /draiv/ n. 驱动器（2A）
driver /'draivə/ n. 驱动程序，驱动器（2B）
drone /drəun/ n. 游手好闲者，寄生虫（10C）
drop-down menu 下拉（式）菜单（4B）
drought /draut/ n. 干旱；旱灾（6C）
drum /drʌm/ n. 磁鼓（2A）
dual /'dju:əl/ a. 双的；双重的（5A）
dub /dʌb/ v. 给…起绰号；把…称为（1C）
duct /dʌkt/ n. 管道；导管（9C）
duke /dju:k/ n. 公爵（3B）
duplicate /'dju:plikeit/ v. 复制；重复（8C）

# E

eavesdrop /ˈiːvzdrɔp/ v. 偷听，窃听（*on*）（10A）
ecommerce /iːˈkɔməːs/ n. 电子商务（1C）
ecosystem /ˈiːkəuˌsistəm/ n. 生态系统（6C）
edge computing 边缘计算（9B）
editor /ˈeditə/ n. 编辑程序，编辑器（2C）
eerie /ˈiəri; ˈiri/ a. （因怪诞或阴森而）引起恐惧的；怪异的（12A）
e-label /ˈiːleibəl/ n. 电子标签（12C）
elastic /iˈlæstik/ a. 有弹性的；可伸缩的（9A）
eldercare /ˈeldəkɛə/ n. 老年保健，老年护理（9B）
electrical contact 电触点（1A）
electronic book reader 电子（图）书阅读器（9C）
electronic picture frame 电子相框（9C）
electrostatic /iˌlektrəˈstætik/ a. 静电的；静电学的（2A）
embed /imˈbed/ v. 把…嵌入（1A）
embedded /imˈbedid/ a. 嵌入（式）的（8C）
emission /iˈmiʃən/ n. 发出；射出；发射（2A）
emoticon /iˈməutikɔn/ n. 情感符（*emoti*on *icon* 的缩合）（11A）
empower /imˈpauə/ v. 授权给；使能够（12A）
encapsulation /inˌkæpsjuˈleiʃən; -sə-/ n. 封装（3B）
encipher /inˈsaifə/ v. 把…译成密码（1A）
encode /enˈkəud/ v. 把…编码；把…译成电码（或密码）（4A）
encompass /inˈkʌmpəs/ v. 包含，包括（8C）
encrypt /inˈkript/ v. 把…加密（10B）
encryption /inˈkripʃən/ n. 加密（1A）
encryption key （加密）密钥（10A）
end user 最终用户，终端用户（6A）
endpoint /ˈendpɔint/ n. 端点（7B）
energy harvesting 能量收集（12C）
enforcement /inˈfɔːsmənt/ n. 实施；强制执行（9C）
enormity /iˈnɔːmiti/ n. 巨大，广大（6C）
entail /inˈteil/ v. 使成为必要，需要（12C）
Enter key 回车键（8C）
entirety /inˈtaiəti/ n. 全部；整体（6B）
entity /ˈentiti/ n. 实体（4B）
entity relationship diagram 实体关系图，实体联系图，E-R 图（4B）
entity-relationship model 实体关系模型（6A）
envision /inˈviʒən/ v. 想象；设想（9A）
equivalence /iˈkwivələns/ n. 等价；相等（5B）
equivalence partitioning 等价（类）划分（5B）
eradicate /iˈrædikeit/ v. 根除；消灭（10B）
eradication /iˌrædiˈkeiʃən/ n. 根除；消灭（10C）
erasable /iˈreizəbəl/ a. 可擦（除）的；可消除的（2A）
erroneous /iˈrəuniəs/ a. 错误的，不正确的（6B）
essence /ˈesəns/ n. 本质，实质（3A）
etch /etʃ/ v. 蚀刻（1A）
Ethernet /ˈiːθəˌnet/ n. 以太网（标准）（亦作 ethernet）（7A）
ethical /ˈeθikəl/ a. 道德的；伦理的（11B）

ethics /ˈeθiks/ n. [用作单] 伦理学；[用作单或复] 道德准则（11B）
etiquette /ˈetiket/ n. 礼节；（行业中的）道德规范；规矩（11A）
exabyte /ˈeksəbait/ n. 艾字节，百亿亿字节（6C）
excerpt /ekˈsəːpt/ v. 摘录；引用（11A）
exclusive lock 排他锁，互斥（型）锁（6B）
execution /ˌeksiˈkjuːʃən/ n. 执行，运行（1A）
execution path 执行路径（5B）
exercising test 压力测试（5B）
exhaustive /igˈzɔːstiv/ a. 全面而彻底的，详尽无遗的（5B）
expedite /ˈekspidait/ v. 迅速执行；加速（12C）
expertise /ˌekspəːˈtiːz/ n. 专门知识（或技能），专长（11B）
expiration /ˌekspiˈreiʃən/ n. 满期，届期（9C）
expire /ikˈspaiə/ v. 满期，届期（9C）
explanatory /ikˈsplænətəri/ a. 解释的，说明的（6C）
exploit /ˈeksplɔit, ikˈs-/ n. 漏洞利用（10B）
exploratory /ikˈsplɔrətəri/ a. 探索的；勘探的（4C）
exponential /ˌekspəˈnenʃəl/ a. 指数的；迅速增长的（6C）
expression /ikˈspreʃən/ n. 表达式（3A）
extensibility /ikˌstensəˈbiliti/ n. 可扩展性，可扩充性（5C）
extension /ikˈstenʃən/ n. 扩展；扩展名（2B）
extract /ikˈstrækt/ v. 提取，抽取，析取（6C）

# F

fabricate /ˈfæbrikeit/ v. 制作（1A）
factor /ˈfæktə/ v. 把…分解成（*into*）（5C）
factoring /ˈfæktəriŋ/ n. 因子分解，因式分解（1A）
familiarize /fəˈmiliəraiz/ v. 使熟悉；使通晓（7B）
fault isolation 故障隔离（10A）
femtocell /femtəuˈsel/ n. 毫微微蜂窝（9B）
ferry /ˈferi/ n. & v. 渡船；摆渡；渡口 / 渡运；运送（8C）
fiber optics 光学纤维，光纤；纤维光学（8A）
fiber-optic /ˈfaibəˈɔptik/ a. 光学纤维的，光纤的；纤维光学的（8A）
fictional /ˈfikʃənəl/ a. 小说的；虚构的（2B）
field /fiːld/ n. & v. 字段；域；信息组（6A）/ 派…上场；实施（3B）
filament /ˈfiləmənt/ n. 细丝；灯丝（2A）
file server 文件服务器（2C）
filestore /ˈfailstɔː/ n. 文件存储（器）（2C）
film clip 剪片（2C）
fine-grained /ˈfainˈgreind/ a. 细粒度的（2B）
firewall /ˈfaiəwɔːl/ n. 防火墙（9A）
firmware /ˈfəːmwɛə/ n. [总称] 固件（2A）
fitness tracker 健身跟踪器，健身追踪器（1C）
flag /flæg/ v. 用标志表明，做标记（10B）
flame /fleim/ v. （向…）发送争论（或争辩）邮件（11A）
flaming /ˈfleimiŋ/ n. 争论（特指在邮件讨论组或网络论坛中争论）（11A）
flash drive 闪存驱动器（2A）
flash memory 闪存，快闪存储器（2A）
flat file 平面文件，展开文件（6A）

| | |
|---|---|
| flatbed scanner | 平板扫描仪（2A） |
| flaw　/flɔ:/　n. | 缺点，瑕疵（3B） |
| flawed　/flɔ:d/　a. | 有缺点的，有瑕疵的（5C） |
| floppy　/ˈflɔpi/　a. | （松）软的（2A） |
| floppy disk | 软（磁）盘（2A） |
| flowchart　/ˈfləutʃɑ:t/　n. | 流程图（3C） |
| flux　/flʌks/　n. | （不断的）变动；波动（6B） |
| flyer　/ˈflaiə/　n. | （广告）传单（12B） |
| focal　/ˈfəukəl/　a. | 焦点的（7A） |
| focal point | 焦点，活动（或注意、兴趣等）的中心（7A） |
| fog computing | 雾计算（9B） |
| folder　/ˈfəuldə/　n. | 文件夹（2B） |
| folklore　/ˈfəuklɔ:/　n. | 民间传说；民俗（学）（12A） |
| font　/fɔnt/　n. | 字体（1A） |
| footage　/ˈfutidʒ/　n. | （影片的）连续镜头，片段（12B） |
| forensic　/fəˈrensik/　a. | （用于）法庭的；法医的（10A） |
| forfeit　/ˈfɔ:fit/　v. | 丧失，失去（5B） |
| form factor | （电子产品等的）物理尺寸和形状；规格（1C） |
| formulate　/ˈfɔ:mjuleit/　v. | 构想出；系统地阐述（8C） |
| forward slash | 正斜杠（8C） |
| foul-up　/ˈfaulʌp/　n. | 混乱，一团糟（11B） |
| frame　/freim/　n. | 帧，画面（2C）/ 图文框（4B） |
| frequency array | 频率数组（3C） |
| frequent　/friˈkwent/　v. | 常到，常去，时常出入于（11C） |
| frivolous　/ˈfrivələs/　a. | 轻薄的；琐屑的（11A） |
| frontend　/ˈfrʌntˈend/　n. & a. | 前端（的）（9A） |
| front-end　/ˈfrʌntˈend/　n. | 前端（9B） |
| function key | 功能键，操作键，函数键（2A） |
| function statement | 函数语句（3A） |
| functional language | 函数式语言（3A） |
| functionality　/ˌfʌŋkʃəˈnæliti/　n. | 功能性（2B） |
| fuse　/fju:z/　v. | 熔合；熔凝（2A） |
| fusion　/ˈfju:ʒən/　n. | 熔合；融合（12C） |

# G

| | |
|---|---|
| gaggle　/ˈgægl/　n. | （紊乱而有联系的）一堆（3B） |
| garment　/ˈgɑ:mənt/　n. | （一件）衣服（12B） |
| gas guzzler | 耗油量大的汽车，油老虎（9C） |
| gateway　/ˈgeitwei/　n. | 网关（7C） |
| gauge　/geidʒ/　v. | 估计，判断；计量（4B） |
| generalization　/ˌdʒenərəlaiˈzeiʃən; -liˈz-/　n. | 概括，归纳（5A） |
| generalize　/ˈdʒenərəlaiz/　v. | 对…进行概括，归纳出（5A） |
| generator　/ˈdʒenəreitə/　n. | 生成程序，生成器（2C） |
| generic　/dʒiˈnerik/　a. | 类属的；一般的；通用的（2B） |
| geocaching　/ˌdʒi:əuˈkæʃiŋ/　n. | 地理藏宝（12C） |
| geo-coordinate　/ˌdʒi:əukəuˈɔ:dinət/　n. | 地理坐标（12B） |
| geolocation　/ˌdʒi:əuləuˈkeiʃən/　n. | 地理位置，地理定位（12C） |
| geotagging　/ˌdʒi:əuˈtægiŋ/　n. | 地理标记，地理标签（12C） |

| | |
|---|---|
| good-sized /ˈgudˈsaizd/ a. | （相当）大的；宽大的（1C） |
| grammatical /grəˈmætikəl/ a. | （符合）语法的（3A） |
| granularity /ˌgrænjuˈlæriti/ n. | （颗）粒度（5C） |
| graph /græf, grɑːf/ v. | 用图（或图表、曲线图等）表示（1C） |
| graphic(al) /ˈgræfik(əl)/ a. | 图形的，图示的（1B） |
| graphical user interface | 图形用户界面（2B） |
| graphics /ˈgræfiks/ n. | 图形，图形显示（1A） |
| green /griːn/ n. | （高尔夫）球穴区；高尔夫球场（8C） |
| greeting card | 贺卡（8C） |
| grid /grid/ n. | 网格，格网（9A） |
| grim /grim/ a. | 严厉的；阴森的（11A） |
| gripper /ˈgripə/ n. | 夹子；抓爪（器）（12A） |
| groundbreaking /ˈgraundˌbreikiŋ/ a. | 开辟新天地的，开拓性的（10C） |
| gullible /ˈgʌlibəl/ a. | 易受骗的，易上当的（11A） |
| gustatory /ˈgʌstətəri/ a. | 味觉的（12B） |
| guzzler /ˈgʌzlə/ n. | 狂饮者；滥吃者；大量消耗者（9C） |
| gyroscope /ˈdʒaiərəskəup/ n. | 陀螺仪，回转仪（12A） |

# H

| | |
|---|---|
| habitat /ˈhæbitæt/ n. | （动植物的）自然生存环境，栖息地（12C） |
| hack /hæk/ v. | 非法闯入（计算机网络），黑客攻击（9C） |
| hand-held scanner | 手持式扫描仪（2A） |
| handle /ˈhændl/ n. | 句柄；称号（5C） |
| handler /ˈhændlə/ n. | 处理程序，处理器（2B） |
| handshaking /ˈhændˌʃeikiŋ/ n. | 握手，信号交换（2B） |
| haptic /ˈhæptik/ a. | 触觉的（12B） |
| harass /ˈhærəs/ v. | 骚扰；烦扰（11A） |
| harassment /ˈhærəsmənt, həˈræs-/ n. | 骚扰；烦扰（11A） |
| hard disk | 硬（磁）盘（1C） |
| hard disk drive | 硬（磁）盘驱动器（2A） |
| hard drive | 硬盘驱动器（4A） |
| header /ˈhedə/ n. | 标题；头标；页眉（3B） |
| header file | 头文件（3B） |
| headset /ˈhedset/ n. | （头戴式）耳机（7A） |
| head-up display | 前导显示器，平视显示器（1B） |
| healthcare /ˈhelθkɛə/ n. | 医疗保健（9B） |
| heterogeneous /ˌhetərəuˈdʒiːniəs/ a. | 各种各样的；异构的（6C） |
| heuristic /hjuˈristik/ a. | 启发（式）的；探索的（10B） |
| heuristics /hjuˈristiks/ n. | 启发法；探索法（10B） |
| hierarchical /ˌhaiəˈrɑːkikəl/ a. | 分级的，分层的，层次的（2B） |
| hierarchy /ˈhaiəˌrɑːki/ n. | 层次，分层（结构），分级（结构）（3A） |
| high-rise /ˈhaiˈraiz/ a. | （建筑）高层的（12C） |
| hinge /hindʒ/ v. | 依…而定，以…为转移（*on, upon*）（11A） |
| hinged /hindʒd/ a. | 有铰链的（1C） |
| hire /haiə/ n. | 新雇员；新员工（6C） |
| histogram /ˈhistəugræm/ n. | 直方图，矩形图；频率分布图（3C） |
| hoax /həuks/ n. | 恶作剧；骗局（10B） |
| hoaxbuster /ˈhəuksˌbʌstə/ n. | 恶作剧揭穿者，辟谣者（10B） |

hockey /ˈhɔki/ n. 曲棍球；冰上曲棍球，冰球（12B）
home appliance 家用电器（8A）
home page 主页（8C）
homogeneous /ˌhɔməuˈdʒiːniəs/ a. 同种类的；同性质的；同构的（5B）
hop /hɔp/ v. 跳跃（8B）
host /həust/ n. & v. 主机（8A）/ 作…的主机（8C）
host program 主机程序，宿主程序，主程序（10C）
hot spot （网络）热点（8A）
howl /haul/ v. 吼叫；怒吼（9C）
HTML tag HTML 标记（8C）
hub /hʌb/ n. （网络）集线器；（轮）毂；（兴趣、活动等的）中心（7B）
Hungarian /hʌŋˈgɛəriən/ a. 匈牙利的（1A）
hurricane /ˈhʌrikən/ n. 飓风（6C）
hybrid /ˈhaibrid/ n. & a. 混合物（7B）/ 混合的（9A）
hydroelectric /ˌhaidrəuiˈlektrik/ a. 水力发电的；（电）由水力发的（9A）
hyperlink /ˈhaipəliŋk/ n. 超（级）链接（6C）
hypertext /ˈhaipətekst/ n. 超（级）文本（2C）
Hypertext Markup Language 超文本标记语言（8C）
Hypertext Transfer Protocol 超文本传送协议，超文本传输协议（2C）
hypervisor /ˈhaipəvaizə/ n. （系统）管理程序（9B）
hypothesis /haiˈpɔθisis/ n. （[复] -ses /-siːz/）假设，假说（6C）
hypothesize /haiˈpɔθisaiz/ v. 假设，假定（6C）

# I

ice hockey 冰上曲棍球，冰球（12B）
icon /ˈaikɔn/ n. 图标，图符（2B）
identifier /aiˈdentifaiə/ n. 标识符（5A）
identity theft 身份（信息）盗取（11C）
immigration /ˌimiˈgreiʃən/ n. 移民，移居（8C）
impersonate /imˈpəːsəneit/ v. 扮演；模仿；假冒（10A）
implant /imˈplɑːnt; -ˈplænt/ v. 植入，埋置（9C）
implementation /ˌimplimenˈteiʃən/ n. 实现，执行（1B）
inadvertent /ˌinədˈvəːtənt/ a. 因疏忽造成的；粗心大意的（6B）
incarnation /ˌinkɑːˈneiʃən/ n. （某种品质、概念、思想等的）典型体现，化身（12B）
incoming /ˈinˌkʌmiŋ/ a. 进来的；输入的（8B）
incompatibility /ˈinkəmˌpætəˈbiliti/ n. 不兼容性（5A）
incompetence /inˈkɔmpitəns/ n. 无能力；不胜任（11B）
inconsistency /ˌinkənˈsistənsi/ n. 不一致（2C）
incorporate /inˈkɔːpəreit/ v. 包含；把…合并；使并入（1A）
increment /ˈinkrimənt/ v. 增加，增长（6B）
incremental /ˌinkriˈmentəl/ a. 增量的；递增的（2C）
indefinite /inˈdefinit/ a. 不确定的，不定的（5B）
independent variable 自变量（6C）
indexing /ˈindeksiŋ/ n. 编索引；标引；加下标；变址（3C）
indicative /inˈdikətiv/ a. 标示的；指示的（12C）
indiscreet /ˌindiˈskriːt/ a. 不慎重的，轻率的（11A）

indistinguishable /ˌindisˈtiŋgwiʃəbəl/ a. 难以区别的，难以分辨的（12B）
infallible /inˈfæləbəl/ a. 不可能出错的；绝对可靠的（10B）
inference engine 推理机（4A）
infiltrate /ˈinfiltreit, inˈfil-/ v. 渗入，渗透（10B）
info /ˈinfəu/ n. 〈口〉信息（= information）（2C）
information superhighway 信息高速公路（11A）
informative /inˈfɔːmətiv/ a. 提供有用信息的；增长知识的（12B）
infrastructure /ˈinfrəˌstrʌktʃə/ n. 基础设施；基础结构（5A）
infringe /inˈfrindʒ/ v. 违犯，侵犯（权利等）（11C）
infringement /inˈfrindʒmənt/ n. （对他人权利等的）侵犯，侵害（11C）
inheritance /inˈheritəns/ n. 继承（性）（3B）
inheritance hierarchy 继承层次（5C）
inhibit /inˈhibit/ v. 抑制，约束（10A）
initiate /iˈniʃieit/ v. 开始；发起（2B）
inject /inˈdʒekt/ v. 注射；注入；引入（10C）
ink cartridge 墨盒（2A）
ink jet printer 喷墨打印机（2A）
innovation /ˌinəuˈveiʃən/ n. 革新，创新（5A）
innovative /ˈinəuveitiv/ a. 革新的，新颖的（5A）
input stream 输入（信息）流（1A）
instant message 即时消息（10B）
instant messaging 即时通信，即时消息（7A）
instantaneous /ˌinstənˈteiniəs/ a. 瞬间的，即刻的（3B）
institutional /ˌinstiˈtjuːʃənəl/ a. （公共）机构的（11B）
instruction set 指令集（1A）
instrumental /ˌinstruˈmentəl, ˌinstrə-/ a. 起作用的，有帮助的（*in, to*）（8C）
instrumentation /ˌinstrumenˈteiʃən, ˌinstrə-/ n. 仪表化；装设仪器；（一套）仪器（12B）
insulin /ˈinsjulin; ˈinsə-/ n. 胰岛素（制剂）（9C）
insulin pump 胰岛素泵（9C）
intact /inˈtækt/ a. 完整无缺的；未受损伤的（12C）
integer /ˈintidʒə/ n. 整数（5B）
integral /ˈintigrəl/ a. 不可缺少的，必要的（1B）
integrated /ˈintigreitid/ a. 集成的，综合的，一体化的（1A）
integrated circuit 集成电路（1A）
integrator /ˈintigreitə/ n. 积分器（1A）
intelligent agent 智能代理，智能主体（12A）
intent /inˈtent/ n. 意图，目的（5C）
interactive /ˌintərˈæktiv/ a. 交互（式）的（7A）
interactivity /ˌintərækˈtiviti/ n. 交互性（12B）
interchange /ˌintəˈtʃeindʒ/ v. 交换，互换（2C）
interchangeable /ˌintəˈtʃeindʒəbəl/ a. 可交换的，可互换的（7C）
interdisciplinary /ˌintəˈdisiplinəri/ a. 学科间的，跨学科的（1B）
interface /ˈintəfeis/ n. 界面；接口（1B）
interfacing /ˈintəˌfeisiŋ/ n. 接口技术（1B）
interleave /ˈintəliːv/ v. 交错，交叉，交替（4C）
intermediary /ˌintəˈmiːdiəri/ n. 媒介（物）；调解人；中间人（2B）
intermediate language 中间语言，中级语言（3A）
intermix /ˌintəˈmiks/ v. （使）混合（*with*）（5B）
internet /ˈintənet/ n. 互联网，互连网（7A）
Internet Service Provider 因特网服务提供商（7B）

| | |
|---|---|
| internetwork /ˌintəˈnetwəːk/ n. | 互联网，互连网（7C） |
| interoperability /ˈintərˌɔpərəˈbiliti/ n. | 互操作性，互用性（9B） |
| interoperable /ˌintərˈɔpərəbl/ a. | 可互操作的，能互用的（9A） |
| interpreted code | 解释执行的代码（3B） |
| interpreted language | 解释执行的语言，解释型语言（4A） |
| interpreter /inˈtəːpritə/ n. | 解释程序，解释器（3B） |
| interpreter program | 解释程序（4A） |
| interprocess communication | 进程间通信（7A） |
| interrogator /inˈterəugeitə/ n. | 讯问者；审问者（12A） |
| interrupt /ˌintəˈrʌpt/ n. | 中断（2B） |
| interrupt handler | 中断处理程序（2B） |
| interrupt signal | 中断信号（2B） |
| intersection /ˌintəˈsekʃən/ n. | 交，相交，交集；交点（6A） |
| intertwine /ˌintəˈtwain/ v. | （使）缠结，（使）缠绕在一起（3B） |
| intervention /ˌintəˈvenʃən/ n. | 干涉，干预（10B） |
| interweave /ˌintəˈwiːv/ v. | （使）交织（6B） |
| intranet /ˈintrənet/ n. | 内联网，内部网络（8A） |
| intrusion /inˈtruːʒən/ n. | 侵入；打扰（6C） |
| inundate /ˈinəndeit/ v. | 淹没；（似洪水般）布满（8C） |
| inventory /ˈinvəntəri; -tɔːri/ n. & v. | 存货（清单），库存（9C）/ 为…开列存货清单；编制…的目录；盘存（12C） |
| inverted /inˈvəːtid/ a. | 反向的；倒置的（6A） |
| invoke /inˈvəuk/ v. | 调用；激活（10C） |
| involved /inˈvɔlvd/ a. | （由于复杂或混乱而）难处理的，棘手的；复杂的（6A） |
| irony /ˈaiərəni/ n. | 反语；讽刺文体（11A） |
| irrespective /ˌiriˈspektiv/ a. | 不考虑的；不顾的（*of*）（4C） |
| iteration /ˌitəˈreiʃən/ n. | 迭代（法）；重复（4C） |

# J

| | |
|---|---|
| jack /dʒæk/ n. | 插座；插口（9C） |
| jewelry /ˈdʒuːəlri/ n. | 〈美〉[总称] 珠宝，首饰（8C） |
| joystick /ˈdʒɔistik/ n. | 控制杆，操纵杆，游戏杆（2A） |
| junk /dʒʌŋk/ n. | 废旧物品；破烂，垃圾（11C） |
| Jurassic /dʒuˈræsik/ n. & a. | 侏罗纪（的）（12B） |
| jurisdiction /ˌdʒuərisˈdikʃən/ n. | 司法权；管辖权；管辖范围（8A） |

# K

| | |
|---|---|
| kernel /ˈkəːnəl/ n. | 内核，内核程序，核心程序（2B） |
| keyword /ˈkiːwəːd/ n. | 关键词，关键字（6C） |
| kiosk /ˈkiːɔsk, kiˈɔsk/ n. | 亭子；信息亭（12B） |

# L

| | |
|---|---|
| laptop /ˈlæptɔp/ a. & n. | 膝上型的（1C）/ 膝上型计算机（1A） |

laser printer　激光打印机（2A）
latency /'leitənsi/ n.　潜伏；等待时间，时延，延迟（9B）
launching pad　发射台；跳板；起点（10C）
layered /'leiəd/ a.　分层的（2C）
legacy /'legəsi/ a.　遗留的，旧版本的（8A）
legality /li'gæliti/ n.　合法（性）（7A）
legitimate /li'dʒitimit/ a.　合法的；正当的（10A）
lever /'li:və; 'le-/ n.　（杠）杆；控制杆（2A）
leverage /'li:vəridʒ; 'le-/ v.　充分利用（9A）
library /'laibrəri/ n.　库，程序（或文件、对象）库（2C）
library routine　库例程，程序库例行程序（4A）
licensing /'laisənsiŋ/ n.　发给许可证；许可，特许（9A）
life cycle　生命周期（1B）
light pen　光笔（2A）
light-emitting diode　发光二极管（2A）
likelihood /'laiklihud/ n.　可能，可能性（10A）
limelight /'laimlait/ n.　公众注意的中心（12C）
linear /'liniə/ a.　线（性）的；直线的（3C）
linguist /'liŋgwist/ n.　语言学家（12A）
linguistic /liŋ'gwistik/ a.　语言（学）的（3A）
linguistics /liŋ'gwistiks/ n.　语言学（1B）
linkage /'liŋkidʒ/ n.　联系；连接；链接（8A）
linker /'liŋkə/ n.　连接程序，链接程序（3B）
liquid crystal display　液晶显示（器）（2A）
listserv /'listsə:v/ n.　邮件发送清单（或邮件列表）管理程序（11A）
live-and-let-live a.　自己活也让别人活的；互相宽容的；互不相扰的（11C）
local area network　局域网（7A）
localize /'ləukəlaiz/ v.　使局部化，使本地化（2C）
locator /ləu'keitə; 'ləukeitə/ n.　定位器，定位符（8C）
lock-in /'lɔkin/ n.　锁定；同步（9A）
log /lɔg/ n. & v.　（运行）记录；（系统）日志（6B）/登录，注册，进入系统（*on, in*）（10A）
logic bomb　逻辑炸弹（10C）
logic language　逻辑语言（3A）
logical element　逻辑元件（1A）
login /'lɔgin; 'lɔ:g-/ n.　注册，登录，进入系统（8B）
logistics /ləu'dʒistiks/ n.　后勤（学），物流（12C）
logo /'ləugəu/ n.　标识，标志，徽标（12B）
loom /lu:m/ n.　织机（1A）
looping /'lu:piŋ/ n.　循环；构成环形（3C）
lowercase /'ləuə'keis/ a.　（字母）小写的（8C）
lurk /lə:k/ v.　潜伏；暗藏；潜行（6B）

# M

machine vision　机器视觉（12C）
macro /'mækrəu/ n.　宏，宏指令（3A）
macrocell /mækrəu'sel/ n.　宏蜂窝（9B）

magnetic disk 磁盘（2A）
magnetize /'mægnitaiz/ v. （使）磁化（2A）
mail server 邮件服务器（8B）
mailing list 邮件发送清单，邮件列表（11A）
main memory 主存（储器）（2A）
main page 主页（8C）
mainframe /'meinfreim/ n. 主机，大型机（1A）
mainstream /'meinstri:m/ a. 主流的（5C）
maintenance hook 维护陷阱（10C）
malicious /mə'liʃəs/ a. 恶意的（3B）
mall /mɔ:l, mæl/ n. （非露天的）购物区，购物街（12C）
malware /'mælwεə/ n. 恶意软件（10B）
manager /'mænidʒə/ n. 管理程序，管理器（2B）
mandate /'mændeit/ n. 授权；命令；指令（12B）
mandatory /'mændətəri; -ˌtɔ:ri/ a. 命令的；强制的（11B）
maneuver /mə'nu:və/ v. 〈美〉调动；机动；移动（12A）
manifest /'mænifest/ v. 显示；使显现（5B）
map /mæp/ v. 映射；变换，变址（2C）
masquerade /ˌmæskə'reid, ˌmɑ:s-/ n. & v. 假面舞会；伪装；假扮（10A）/ 假冒；假扮（11C）
mass storage 海量存储器，大容量存储器（2A）
mass-mailing worm 群发邮件蠕虫（10B）
mass-produce /'mæsprəˌdju:s/ v. （使用机器）大量生产（1B）
maximize /'mæksimaiz/ v. 使增加到最大限度（4B）
mediation /ˌmi:di'eiʃən/ n. 调解（9B）
medic /'medik/ n. 〈口〉医生；医科学生（8C）
medication /ˌmedi'keiʃən/ n. 药物治疗；药物（8C）
medicinal /me'disinəl/ a. （医）药的；药用的；有疗效的（8C）
memo /'meməu, 'mi:-/ n. 〈口〉备忘录（= memorandum）（11A）
memorize /'meməraiz/ v. 记住；熟记（12B）
memory /'meməri/ n. 存储器，内存（1A）
memory card 存储卡（2A）
memory location 存储单元（4A）
merchandise /'mə:tʃəndaiz, -dais/ n. ［总称］商品（1C）
merge /mə:dʒ/ v. 合并；结合（4B）
mesh /meʃ/ n. 网；网格；网状结构；网状网络（7B）
message board 留言板，消息板（11C）
messaging /'mesidʒiŋ/ n. 消息接发，通信（7A）
methodology /ˌmeθə'dɔlədʒi/ n. （学科的）一套方法；方法论（4B）
metropolitan /ˌmetrə'pɔlitən/ a. 大城市的，大都会的（7A）
metropolitan area network 城域网（7A）
microarray /ˌmaikrəuə'rei/ n. 微阵列（6C）
microcell /maikrəu'sel/ n. 微蜂窝（9B）
microchip /'maikrəutʃip/ n. 微芯片（1A）
microcomputer /'maikrəukəmˌpju:tə/ n. 微型计算机（1A）
microcontroller /'maikrəukən'trəulə/ n. 微控制器（1C）
microprocessor /ˌmaikrəu'prəusesə/ n. 微处理器（1A）
microscopic /ˌmaikrə'skɔpik/ a. 非用显微镜不可见的；微小的（2A）
microsecond /'maikrəuˌsekənd/ n. 微秒（2B）
microwave oven 微波炉（1A）
millennium /mi'leniəm/ n. 一千年；千禧年（11C）

millisecond　/ˈmiliˌsekənd/　n.　毫秒（2B）
mimic　/ˈmimik/　v.　模仿（1B）
miniaturization　/ˌminitʃəraiˈzeiʃən; ˌminiətʃərə-/　n.　小型化，微型化（12C）
minicomputer　/ˈminikəmˌpjuːtə/　n.　小型计算机（1A）
minimal　/ˈminiməl/　a.　最小的；最低限度的（2B）
minimize　/ˈminimaiz/　v.　使减少到最低限度；使最小化（3A）
minuscule　/ˈminəskjuːl, miˈnʌs-/　a.　非常小的，微小的（12C）
mischief　/ˈmistʃif/　n.　恶作剧；淘气；损害（11C）
misconception　/ˌmiskənˈsepʃən/　n.　误解，错误想法（12B）
mistakenly　/miˈsteikənli/　ad.　错误地，误解地（10B）
mnemonic　/niːˈmɔnik/　a.　助记的；记忆的（8A）
mnemonic address　助记地址（8A）
mnemonic name　助记名，缩写名（8A）
mobile phone　移动电话（9C）
modeling　/ˈmɔdəliŋ/　n.　建模，模型化（4B）
moderator　/ˈmɔdəreitə/　n.　（电视或广播节目的）主持人（12B）
modification　/ˌmɔdifiˈkeiʃən/　n.　修改，更改（4C）
modulate　/ˈmɔdjuleit/　v.　调制（12C）
modulator　/ˈmɔdjuleitə/　n.　调制器（2A）
module　/ˈmɔdjuːl/　n.　模块（2A）
mom　/mɔm/　n.　〈主美口〉妈妈（3B）
monetary　/ˈmʌnitəri/　a.　货币的；金钱的（10A）
monopoly　/məˈnɔpəli/　n.　垄断；独有（11C）
mortar　/ˈmɔːtə/　n.　砂浆，灰浆（6C）
motion picture　〈美〉影片，电影（7A）
motivation　/ˌməutiˈveiʃən/　n.　动机，诱因（10C）
multidimensional　/ˌmʌltidiˈmenʃənəl, -daiˈm-/　a.　多维的（3C）
multidimensional array　多维数组（3C）
multimedia　/ˌmʌltiˈmiːdiə/　a.　（使用）多媒体的（9B）
multipartite　/ˌmʌltiˈpɑːtait/　a.　分成多部分的（10C）
multiprogramming　/ˈmʌltiˌprəugræmiŋ/　n.　多（道）程序设计（2B）
multitasking　/ˈmʌltiˌtɑːskiŋ/　n.　多任务（处理）（2B）
multitenant　/ˌmʌltiˈtenənt/　a.　多租户的（9A）
multithreaded　/ˈmʌltiˌθredid/　a.　多线程的（3B）
multithreading　/ˈmʌltiˌθrediŋ/　n.　多线程操作（3B）
multitude　/ˈmʌltitjuːd/　n.　大量，许多（8A）
mundane　/mʌnˈdein/　a.　世俗的；平凡的（9C）

# N

name server　名称服务器，名字服务器（8A）
native code　本机（代）码（3B）
navigate　/ˈneivigeit/　v.　航行（于）；（为…）领航；指引（2A）
navigation　/ˌnæviˈgeiʃən/　n.　导航；航行（1C）
navigator　/ˈnævigeitə/　n.　导航仪，导航系统；航海者（12C）
netiquette　/ˈnetiket/　n.　网络礼节，网规（*net*work et*iquette* 的缩合）（11A）
network　/ˈnetwəːk/　v.　连网，联网，建网（2C）
networking　/ˈnetˌwəːkiŋ/　n.　连网，联网，建网（7A）
neural　/ˈnjuərəl/　a.　神经的（1A）

| | |
|---|---|
| neural network | 神经网络（1B） |
| neurology /ˌnjuəˈrɔlədʒi/ n. | 神经病学；神经学（12A） |
| neuron /ˈnjuərɔn/ n. | 神经元，神经细胞（1B） |
| neurophysiology /ˈnjuərəuˌfiziˈɔlədʒi/ n. | 神经生理学（1B） |
| neutrality /njuːˈtræliti/ n. | 中立（11C） |
| niche /nitʃ, niːʃ/ n. | 缝隙市场，利基市场，小众市场（1C） |
| node /nəud/ n. | 节点（7B） |
| nonetheless /ˌnʌnðəˈles/ ad. | 尽管如此，然而（6C） |
| nonprofit /nɔnˈprɔfit/ a. | 非营利的（8A） |
| nonrepudiation /ˈnɔnriˌpjuːdiˈeiʃən/ n. | 不拒绝；不否认；不可抵赖（10A） |
| nonvolatile /nɔnˈvɔlətail/ a. | 非易失（性）的（6B） |
| normalize /ˈnɔːməlaiz/ v. | （使）正常化（6A） |
| notation /nəuˈteiʃən/ n. | 标记法；记号（3C） |
| notepad /ˈnəutpæd/ n. | 记事本；便笺簿；拍纸簿（1C） |
| novice /ˈnɔvis/ n. | 新手，初学者（5C） |
| nozzle /ˈnɔzəl/ n. | 喷嘴；管嘴（2A） |
| nugget /ˈnʌgit/ n. | 小块；天然金块；有价值的东西（11C） |
| numeric(al) /njuːˈmerik(əl)/ a. | 数字的；数值的（1A） |
| numerical analysis | 数值分析（1B） |

# O

| | |
|---|---|
| object code | 目标（代）码（4A） |
| object diagram | 对象图（4B） |
| object-oriented language | 面向对象语言（3A） |
| obsolete /ˈɔbsəliːt, ˌɔbsəˈliːt/ a. | 废弃的；淘汰的；过时的（6A） |
| obstruct /əbˈstrʌkt, ɔb-/ v. | 阻塞；阻碍（8B） |
| office tower | 办公大楼（12C） |
| offload /ˈɔfləud/ v. | 卸下，卸载（5A） |
| offshoot /ˈɔfʃuːt/ n. | 支族，旁系；衍生事物（3B） |
| off-the-shelf /ˈɔfðəˈʃelf/ a. | 现成的，非专门设计（或定制）的（4C） |
| olfactory /ɔlˈfæktəri/ a. | 嗅觉的（12B） |
| omission /əuˈmiʃən/ n. | 省略；遗漏；疏忽（4B） |
| on-board /ˈɔnbɔːd, ˈɔːn-/ a. | 车载的；机载的；舰载的（9C） |
| one-dimensional array | 一维数组（3C） |
| operator /ˈɔpəreitə/ n. | （运）算子，算符（8C） |
| opportunistic /ˌɔpətjuːˈnistik/ a. | 机会主义的（5A） |
| optics /ˈɔptiks/ n. | 光学（8A） |
| optimal /ˈɔptiməl/ a. | 最优的，最佳的（1B） |
| optimization /ˌɔptimaiˈzeiʃən/ n. | （最）优化，最佳化，（1B） |
| optimize /ˈɔptimaiz/ v. | 使优化，使最佳化（2C） |
| ordinal /ˈɔːdinəl/ a. | 顺序的（3C） |
| ordinal number | 序数（3C） |
| orientation /ˌɔːrienˈteiʃən/ n. | 方向；定向（11A） |
| -oriented /ˈɔːrientid/ comb. form | 表示“面向…的”（3A） |
| originate /əˈridʒineit/ v. | 发源；产生（7C） |
| originator /əˈridʒineitə/ n. | 创始人；起源（10A） |
| outgoing /ˈautˌgəuiŋ/ a. | 往外去的，离去的（10B） |
| outgrowth /ˈautgrəuθ/ n. | 发展结果；产物（3A） |

outlier /'aut,laiə/ n. 与主体分开的人(或物);离群值;离群点;异常值(6C)
out-of-core /'autəv'kɔ:/ a. 核外的,基于外存的(6C)
outsource /aut'sɔ:s/ v. (将…)外包(9A)
outweigh /,aut'wei/ v. 在价值(或重要性、影响等)方面超过(3B)
overhead /'əuvəhed/ n. 经常(或管理、间接)费用(3B)
overlap /,əuvə'læp/ v. 重叠,层叠(1B)
overlay /'əuvəlei/ n. 覆盖物(12B)
overlay /,əuvə'lei/ v. 置(一物)于他物之上;在…上覆盖(12B)
overload /,əuvə'ləud/ v. 使过载,使超载(10A)
overly /'əuvəli/ ad. 过度地(5A)
oversee /,əuvə'si:/ v. 监视;监督(2B)
overshadow /,əuvə'ʃædəu/ v. 使相形见绌(1A)
overturn /,əuvə'tə:n/ v. 推翻;废除(1A)
overview /'əuvəvju:/ n. 概述;概观(1A)

# P

pacemaker /'peis,meikə/ n. 起搏器(9C)
package /'pækidʒ/ n. 程序包,软件包(3B)
paging /'peidʒiŋ/ n. 页面调度(技术或方法),分页(2B)
palmtop /'pɑ:mtɔp/ n. 掌上型计算机(1A)
paper-tape reader 纸带阅读器(1A)
paradigm /'pærədaim/ n. 范例;范式(3A)
parameter /pə'ræmitə/ n. 参数(3A)
paraphrase /'pærəfreiz/ v. 将…释义(或意译)(11A)
parasitic /,pærə'sitik/ a. 寄生的(10C)
parking lot 〈美〉停车场(9C)
parking meter 汽车停放计时器,汽车停放收费计(9C)
participant /pɑ:'tisipənt/ n. 参加者,参与者(12C)
partition /pɑ:'tiʃən/ v. & n. 划分,分区,分割(4C)/分割;划分;(分割而成的)部分(5B)
passageway /'pæsidʒwei/ n. 走廊;通道(12A)
patchwork /'pætʃwə:k/ n. 拼缀物;拼凑的东西,杂烩(9C)
patent /'peitənt/ n. 专利(权)(1A)
path expression 路径表达式(2B)
pay-as-you-go /'peiəzju'gəu/ n. & a. 即付即用(的)(9A)
paycheck /'peitʃek/ n. 薪金支票;薪水,工资(6A)
payload /'peiləud/ n. 有效载荷,有效负载(10C)
payroll /'peirəul/ n. 工资表;在职人员名单;工薪总额(6A)
peer /piə/ n. 同级设备;同层,对等层(7A)
peer-to-peer /'piətə'piə/ a. 对等的(7A)
per se /,pə:'sei, -'si:/ ad. 〈拉〉自身,本身(5B)
perforated /'pə:fəreitid/ a. 穿孔的(1A)
periodic /,piəri'ɔdik/ a. 周期(性)的;定期的(10B)
peripheral /pə'rifərəl/ a. 外围的,外部的(2B)
peripheral device 外围设备,外部设备(2B)
perishable /'periʃəbəl/ a. 易腐的,易烂的(12C)
perplexing /pə'pleksiŋ/ a. 使人困惑的;令人费解的(10B)

| | |
|---|---|
| personal area network | 个人域网（7A） |
| personalize /ˈpəːsənəlaiz/ v. | 使个性化（9A） |
| pertinent /ˈpəːtinənt/ a. | 相关的（5C） |
| pervasive /pəˈveisiv/ a. | 遍布的；普遍的（12C） |
| petabyte /ˈpetəbait/ n. | 拍字节，千万亿字节（6C） |
| petaflops /ˈpetəflɔps/ n. | 每秒千万亿次浮点运算，每秒 $10^{15}$ 次浮点运算（1C） |
| pharmaceutic(al) /ˌfɑːməˈsjuːtik(əl)/ a. | 制药的；药（物）的；药用的（6A） |
| phishing /ˈfiʃiŋ/ n. | 网络钓鱼（11C） |
| photocopier /ˈfəutəuˌkɔpiə/ n. | 复印机（2A） |
| physiologist /ˌfiziˈɔlədʒist/ n. | 生理学家（1B） |
| pickup /ˈpikʌp/ n. | 搭车；接人；提货（9C） |
| picocell /piːkəuˈsel/ n. | 微微蜂窝（9B） |
| pictorial /pikˈtɔːriəl/ a. | 用图表示的，图示的；形象化的（2B） |
| pinpoint /ˈpinpɔint/ v. | 精确地确定…的位置；确定（5B） |
| piracy /ˈpaiərəsi/ n. | 侵犯版权，盗版，剽窃（10A） |
| pirate /ˈpaiərət/ v. | 剽窃；盗用；非法翻印（11C） |
| plagiarism /ˈpleidʒiərizəm/ n. | 剽窃，抄袭（12A） |
| player piano | 自动钢琴（4A） |
| playwright /ˈpleirait/ n. | 剧作家（5C） |
| pluck /plʌk/ v. | 拔；摘，采（11C） |
| plus /plʌs/ ad. | 〈口〉外加地；另外（11A） |
| pointer /ˈpɔintə/ n. | 指针（光标）；指示字，指示符（2A） |
| police /pəˈliːs/ v. | 维持…的治安；管理；监督（11C） |
| pollutant /pəˈluːtənt/ n. | 污染性物质（12C） |
| polymorphic /ˌpɔliˈmɔːfik/ a. | 多（种）形（式）的，多态的（5A） |
| polymorphism /ˌpɔliˈmɔːfizəm/ n. | 多态性，多形性（3B） |
| pornography /pɔːˈnɔgrəfi/ n. | ［总称］色情（或淫秽）作品（11C） |
| port /pɔːt/ n. | 端口，通信口（8B） |
| port number | 端口号（8B） |
| portability /ˌpɔːtəˈbiliti/ n. | 便于携带；可移植性（1C） |
| portable /ˈpɔːtəbəl/ a. & n. | 便携式的；可移植的（1C）/ 便携式计算机（1C） |
| potent /ˈpəutənt/ a. | 有（效）力的，效力大的（8A） |
| pothole /ˈpɔthəul/ n. | 路面凹坑，坑洞（9C） |
| power grid | 电网（12C） |
| practitioner /prækˈtiʃənə/ n. | 实践者；开业者（6C） |
| pragmatic /prægˈmætik/ a. | 注重实效的；实际的；实用主义的（5C） |
| precede /ˌpriːˈsiːd, pri-/ v. | 处在…之前，先于（4A） |
| precursor /ˌpriːˈkəːsə, pri-/ n. | 先驱（1A） |
| predate /priːˈdeit/ v. | 在日期上早于（或先于）（10C） |
| predecessor /ˈpriːdisesə; ˈpre-/ n. | 前任；前身；先驱（2C） |
| prediction /priˈdikʃən/ n. | 预言，预测（6C） |
| predictive /priˈdiktiv/ a. | 预言性的，预测性的（6C） |
| preempt /ˌpriːˈempt, pri-/ v. | 预先制止；抢先，先占（2B） |
| prehistoric /ˌpriːhiˈstɔrik/ a. | 史前的（2B） |
| premature /ˌpreməˈtjuə; ˌpriː-/ a. | 不成熟的；仓促的（4C）/ 提早的；早产的（11B） |
| premium /ˈpriːmiəm/ n. | 奖品；额外费用（4B） |
| preparatory /priˈpærətəri; -tɔːri/ a. | 准备性的，预备性的（5B） |
| preprocessor /priːˈprəusesə/ n. | 预处理程序，预处理器（3B） |
| presentation software | 演示软件（11A） |
| preview /ˈpriːvjuː/ n. & v. | 预映；试演；预看（12B） |

print server 打印服务器（7A）
printing press 印刷机（11C）
privacy /ˈprivəsi, ˈprai-; ˈprai-/ n. 隐私；秘密（1A）
privileged /ˈprivilidʒd/ a. 享有特权的；特许的（6B）
procedural language 过程语言（3A）
procedure call 过程调用（2C）
procedure statement 过程语句（3A）
process switch 进程转换（2B）
process table 进程表（2B）
procure /prəuˈkjuə/ v. 取得，获得；采办（4C）
profitable /ˈprɔfitəbl/ a. 有利可图的，有盈利的（9C）
profound /prəˈfaund/ a. 深邃的；深刻的（11B）
program module 程序模块（3A）
program unit 程序单元（4C）
programming construct 编程结构（3C）
proliferation /prəuˌlifəˈreiʃən/ n. 激增；扩散（10A）
prompt /prɔmpt/ v. 提示，提醒（4A）
prop /prɔp/ n. （戏剧、电影等中用的）道具（12B）
propagate /ˈprɔpəgeit/ v. 传播；繁殖；蔓延（10C）
propagation /ˌprɔpəˈgeiʃən/ n. 传播；繁殖；蔓延（10C）
prophet /ˈprɔfit/ n. 预言者，预言家（12B）
proposition /ˌprɔpəˈziʃən/ n. 提议，建议（12B）
proprietary /prəuˈpraiətəri/ a. 专有的，专用的；专利的（5A）
protocol /ˈprəutəkɔl/ n. 协议（2C）
protocol suite 协议簇，协议组（7A）
prototype /ˈprəutətaip/ n. 原型；样机（1A）
prototyping /ˈprəutəutaipiŋ/ n. 原型法；原型开发（或设计）；样机研究（4C）
prover /ˈpru:və/ n. 证明程序（3B）
provision /prəuˈviʒən/ v. 供应，提供（9A）
proxy /ˈprɔksi/ n. 代理（人）（9B）
psychotherapy /ˌsaikəuˈθerəpi/ n. 心理疗法，心理治疗（12A）
punch card 穿孔卡片（4A）
punched card 穿孔卡片（1A）
punctuation /ˌpʌŋktjuˈeiʃən, -tʃu-/ n. 标点符号（3A）
punctuation mark 标点符号（8C）
pursuit /pəˈsju:t/ n. 追求；追赶（12A）

# Q

quadrillion /kwɔˈdriljən/ n. 〈英〉$10^{24}$（100 万的 4 次幂）；〈美〉$10^{15}$（1000 的 5 次幂）（1C）
quantum /ˈkwɔntəm/ n. （[复] -ta /-tə/）量子（1A）
quarantine /ˈkwɔrənti:n/ v. & n. 隔离（10B）
query /ˈkwiəri/ n. 查询（2C）
quip /kwip/ n. 妙语，俏皮话（11A）
quotation marks [复]（一对）引号（8C）

# R

| | |
|---|---|
| radio button | 单选按钮（4B） |
| radio frequency | 射频；无线电频率（12C） |
| radio-frequency identification | 射频识别（12C） |
| rage /reidʒ/ v. | 发怒；激烈进行（11C） |
| railroad car | 〈美〉火车车厢（8C） |
| random /'rændəm/ a. | 随机的；任意的（2A） |
| random access memory | 随机（存取）存储器（2A） |
| range sensor | 距离传感器（12A） |
| read-only memory | 只读存储器（2A） |
| real time | 实时（12B） |
| realign /ˌriːə'lain/ v. | 重新调整（4A） |
| recipe /'resipi/ n. | 烹饪法，食谱；处方（8C） |
| recipient /ri'sipiənt/ n. | 接受者；收件人（11A） |
| recur /ri'kəː/ v. | 再发生；反复出现（5C） |
| redundancy /ri'dʌndənsi/ n. | 冗余（2C） |
| redundant /ri'dʌndənt/ a. | 冗余的（7B） |
| reference /'refərəns/ n. & v. | 引用，参考；基准（3C）/引用，参考（3B） |
| reflex /'riːfleks/ n. | 反射（作用）；（对刺激的）本能反应（12A） |
| reflex action | 反射作用；本能反应；本能动作（12A） |
| reformat /ˌriː'fɔːmæt/ v. | 重新格式化（4A） |
| refrigeration /riˌfridʒə'reiʃən/ n. | 冷藏；制冷（12C） |
| refund /'riːfʌnd/ n. | 退款；退还（11C） |
| register /'redʒistə/ n. | 寄存器（3A） |
| registrar /ˌredʒi'strɑː, 'redʒis-/ n. | 登记员；（域名）注册（服务）商（8A） |
| registration /ˌredʒi'streiʃən/ n. | 登记；注册（6B） |
| registry /'redʒistri/ n. | 注册表；登记簿（5A） |
| regression /ri'greʃən/ n. | 回归（5B） |
| regression test | 回归测试（5B） |
| relay /riː'lei/ v. | 接力传送；传递；转播（9C） |
| release /ri'liːs/ n. | （程序或软件的）版本；发布（3B） |
| relevance /'reləvəns/ n. | 相关（性），关联；重要性（6C） |
| render /'rendə/ v. | 对…做艺术处理（或进行渲染）（12B） |
| rendering /'rendəriŋ/ n. | 艺术处理；渲染（12B） |
| rental /'rentəl/ n. | 租赁，出租（9C） |
| repeater /ri'piːtə/ n. | 中继器，转发器（7C） |
| repetitive /ri'petitiv/ a. | 重复的（1B） |
| replica /'replikə/ n. | 复制品，拷贝（6A） |
| replicate /'replikeit/ v. | 复制，重复（6A） |
| repository /ri'pɔzitəri/ n. | 存储库，仓库（2C） |
| reprisal /ri'praizəl/ n. | 报复（11C） |
| reside /ri'zaid/ v. | 驻留，常住（2B） |
| residential /ˌrezi'denʃəl/ a. | 居住的；住宅的（12C） |
| resiliency /ri'ziliənsi/ n. | 弹性；复原力（9A） |
| resin /'rezin/ n. | 树脂；合成树脂（2A） |
| resize /riː'saiz/ v. | 调整大小（4B） |
| respondent /ri'spɔndənt/ n. | 调查对象，（调查表的）回答者（10A） |
| restructure /riː'strʌktʃə/ v. | 重建；改组；调整（12A） |

retailer /ˈriːteilə/ n. 零售商；零售店（12C）
retailing /ˈriːteiliŋ/ n. 零售业（12C）
retrieval /riˈtriːvəl/ n. 检索（6A）
retrieve /riˈtriːv/ v. 检索（1B）
revoke /riˈvəuk/ v. 撤销；废除（6A）
rigidity /riˈdʒiditi/ n. 严格；刻板（4B）
rigorous /ˈrigərəs/ a. 严密的；严格的（3A）
ring topology 环型拓扑（结构）（7B）
roam /rəum/ v. 漫步；漫游（9C）
roboticist /rəuˈbɔtisist/ n. 机器人专家（1B）
robotics /rəuˈbɔtiks/ n. 机器人学；机器人技术（1B）
robust /rəuˈbʌst, ˈrəubʌst/ a. 健壮的，坚固的（1B）
rodent /ˈrəudənt/ n. 啮齿目动物（如鼠、松鼠、河狸等）（9C）
rollback /ˈrəulbæk/ n. 回滚，回退（2C）
romance /rəuˈmæns, ˈrəumæns/ n. 浪漫文学，传奇文学（5C）
rootkit /ˈruːtkit/ n. 根工具包（10C）
roulette /ruːˈlet/ n. 轮盘赌（11C）
roulette wheel （轮盘赌台上的）轮盘（11C）
route /ruːt/ v. （按特定路径）发送，传递（7C）
router /ˈruːtə/ n. 路由器（7B）
routine /ruːˈtiːn/ n. 例程，例行程序（4A）
routing /ˈruːtiŋ/ n. 路由选择（7B）
routing path 路由路径，路由选择通路（7B）
runtime /ˈrʌntaim/ n. 运行时刻（3B）
runway /ˈrʌnwei/ n. （机场的）跑道（7C）

# S

safeguard /ˈseifgɑːd/ v. 保护，维护（9A）
sampling /ˈsɑːmpliŋ; ˈsæ-/ n. 取样，抽样，采样（6C）
sandbox /ˈsændbɔks/ n. （供儿童游戏的）沙池；沙箱，沙盒（9B）
sarcasm /ˈsɑːkæzəm/ n. 讽刺，挖苦（11A）
scalability /ˌskeiləˈbiliti/ n. 可缩放性，可伸缩性（6C）
scalable /ˈskeiləbl/ a. 可缩放的，可伸缩的（6C）
scale /skeil/ v. 攀登；（阶梯）逐步升高（9A）
scanner /ˈskænə/ n. 扫描仪；扫描程序（2A）
scenario /siˈnɑːriəu/ n. 方案；情况；脚本（9A）
scheduler /ˈʃedjuːələ; ˈskedʒuːələr/ n. 调度程序（2B）
schema /ˈskiːmə/ n. 模式；纲要（2C）
scour /skauə/ v. 四处搜索；细查（6C）
screencast /ˈskriːnkɑːst/ n. 截屏视频，屏幕录像，录屏（12B）
scroll /skrəul/ v. & n. 滚动（4B）
scroll bar 滚动条（4B）
search engine 搜索引擎（5A）
secrecy /ˈsiːkrəsi/ n. 秘密；保密（10A）
segregate /ˈsegrigeit/ v. 隔离；分开（3B）
self-contained /ˌselfkənˈteind/ a. 自给的，独立自足的（10C）
semantic /siˈmæntik/ a. 语义（学）的（12C）
Semantic Grid 语义网格（12C）

| | |
|---|---|
| Semantic Web | 语义网（12C） |
| semantically /si'mæntikəli/ ad. | 在语义上（3B） |
| semantics /si'mæntiks/ n. | 语义（学）（10C） |
| sensation /sen'seiʃən/ n. | 感觉；轰动；激动（12A） |
| sensor /'sensə/ n. | 传感器；敏感元件（1B） |
| sensor network | 传感器网络，传感网，感知网（9C） |
| sensory /'sensəri/ a. | 感觉的；感官的（1B） |
| sequence number | 序列号，序号（8B） |
| sequential /si'kwenʃəl/ a. | 顺序的，序列的（6C） |
| serial /'siəriəl/ a. | 串行的；连续的（2A） |
| server /'sə:və/ n. | 服务器；服务程序（1C） |
| server cluster | 服务器集群（9A） |
| server farm | 服务器农场（9A） |
| session /'seʃən/ n. | 对话（期），会话（7C） |
| severity /si'veriti/ n. | 剧烈；严重；严厉（10C） |
| sexual orientation | （对同性或异性的）性取向，性倾向（11A） |
| shaft /ʃɑ:ft/ n. | 轴（1A） |
| shared lock | 共享锁（6B） |
| shipping /'ʃipiŋ/ n. | 运送，运输；航运业（8B） |
| shirk /ʃə:k/ v. | 逃避（工作、责任等）（11B） |
| shoplifting /'ʃɔp,liftiŋ/ n. | （混在顾客群中进行的）商店货物扒窃（12C） |
| shopping mall | （同一建筑物里有许多商店的）购物中心（12C） |
| Short Message Service | 短信（服务）（9C） |
| shortcut /'ʃɔ:tkʌt/ n. | 近路，捷径（1A） |
| show room | （商品样品的）陈列室，展览室（12B） |
| showbiz /'ʃəubiz/ n. | 〈口〉娱乐性行业；娱乐界（= show business）（8C） |
| showcase /'ʃəukeis/ v. | 陈列；展示（12B） |
| shuffle /'ʃʌfl/ v. | 洗（牌）；混洗（2B） |
| signature /'signətʃə/ n. | （人或物的）识别标志；鲜明特征（10B） |
| signature file | 签名文件（11A） |
| simulate /'simjuleit/ v. | 模拟，仿真（1C） |
| simulation /,simju'leiʃən/ n. | 模拟，仿真（12A） |
| simultaneous /,siməl'teiniəs; ,sai-/ a. | 同时（发生）的（1C） |
| situated /'sitjueitid, -tʃu-/ a. | 位于…的；处于…境地的（12B） |
| situated visualization | 情境可视化，位置可视化（12B） |
| skip /skip/ v. | 略过，跳过（4B） |
| slack /slæk/ n. | 松弛（部分）（7B） |
| slash /slæʃ/ n. | 斜线，斜杠（2B） |
| slate /sleit/ n. & a. | 石板（的）（1C） |
| slate tablet | 平板电脑（1C） |
| sloppy /'slɔpi/ a. | 〈口〉马虎的；凌乱的（11A） |
| slot /slɔt/ n. | 狭长孔，狭缝（11C） |
| slot machine | 吃角子老虎（一种投硬币的赌具）（11C） |
| small-time /'smɔ:l'taim/ a. | 〈口〉次要的，无关紧要的（11C） |
| smart card | 智能卡（12C） |
| smartcard /'smɑ:tkɑ:d/ n. | 智能卡（9C） |
| smartphone /'smɑ:tfəun/ n. | 智能手机（1C） |
| smartwatch /'smɑ:twɔtʃ/ n. | 智能手表（1C） |
| smirk /smə:k/ v. | 假笑；得意地笑（11A） |
| snapshot /'snæpʃɔt/ n. | 快照；瞬象，瞬态图（6B） |

snoop /snu:p/ v. 窥探；打探（11C）
social network 社交网络（11C）
social networking 社交网络（9A）
social security number 社会安全（或保险、保障）号（码）(6A)
societal /sə'saiətəl/ a. 社会的（12C）
software agent 软件代理（12C）
software package 软件包（3B）
software routine 软件例程（8B）
solid state drive 固态驱动器（2A）
sophistication /səˌfisti'keiʃən/ n. 复杂性；尖端性（3A）
source code 源（代）码（3B）
source file 源文件（4A）
source program 源程序（4A）
space /speis/ n. 空格，空白（8C）
spacing /'speisiŋ/ n. 间隔（1A）
spam /spæm/ n. & v. 垃圾邮件（11A）/（向…）发送垃圾邮件（11A）
spammer /'spæmə/ n. 垃圾邮件发送者（11C）
sparing /'spɛəriŋ/ a. 节约的；有节制的（11A）
spatial /'speiʃəl/ a. 空间的（6C）
spec /spek/ n. 〈口〉明确说明；规格，规范（= specification）(5B)
-specific /spi'sifik/ comb. form 表示“限定的”“特有的”(5A)
specification /ˌspesifi'keiʃən/ n. 明确说明；[常作~s] 规格，规范（1B）
spectrum /'spektrəm/ n. 频谱；系列；范围（12C）
speculative /'spekjulətiv, -lei-/ a. 推测的；投机性的（1B）
speech synthesizer 语音合成器（12A）
spoof /spu:f/ v. 哄骗，欺骗（3B）
spreadsheet /'spredʃi:t/ n. 电子表格，电子数据表（11A）
spreadsheet program 电子数据表程序（11A）
sprout /spraut/ v. 发芽；长出（1B）
spyware /'spaiwɛə/ n. 间谍软件（10B）
square bracket 方括号（3C）
standalone /'stændəˌləun/ a. 独立的（2A）
star topology 星型拓扑（结构）(7A)
state of the art （学科、技术等当前的或某一时期的）发展水平，最新水平（10A）
statement /'steitmənt/ n. 语句（3A）
statistician /ˌstæti'stiʃən/ n. 统计学家；统计员（1A）
status word 状态字（2B）
stealthy /'stelθi/ a. 偷偷摸摸的，暗中进行的，秘密的（10C）
stimulus /'stimjuləs/ n. ([复] -li /-lai/ 或 -luses）刺激（物），激励（物）(12A)
storage register 存储寄存器（3A）
storage volume 存储卷（10B）
store /stɔ:/ n. 存储（器）(2C)
straightforward /ˌstreit'fɔ:wəd/ a. 径直的；简单的（2C）
stream /stri:m/ v. 流播（1C）
strive /straiv/ v. 努力；力争（1B）
structure chart 结构图（4B）
stylus /'stailəs/ n. ([复] -luses 或 -li /-lai/）输入笔，光笔（2A）
subdirectory /ˌsʌbdi'rektəri/ n. 子目录（2B）

subdomain /ˌsʌbdəʊ'mein/ n. 子域（8A）
submarine /ˌsʌbmə'riːn/ n. 潜艇（1A）
subnetting /'sʌbˌnetiŋ/ n. 子网划分，子网组建（7B）
subroutine /'sʌbruːˌtiːn/ n. 子例程，子（例行）程序（4A）
subscribe /səb'skraib/ v. 订阅；订购（9B）
subscriber /səb'skraibə/ n. 用户，订户（9C）
subscript /'sʌbskript/ n. 下标，脚注（3C）
subscription /səb'skripʃən/ n. 订购；订阅（9A）
subservient /səb'səːviənt/ a. 屈从的；恭顺的；顺从的（12A）
subset /'sʌbset/ n. 子集（6C）
substrate /'sʌbstreit/ n. 衬底，基底（1A）
subvert /səb'vəːt, sʌb-/ v. 颠覆；（暗中）破坏（2C）
sue /sjuː; suː/ v. 控告，起诉（11C）
suffix /'sʌfiks/ n. 后缀（8A）
suite /swiːt/ n. （同类物的）系列，组，套（1C）
supercomputer /'sjuːpəkəmˌpjuːtə/ n. 超级计算机，巨型计算机（1A）
superfluous /sjuː'pəːfluəs/ a. 多余的；过剩的（3B）
superimpose /ˌsjuːpərim'pəuz/ v. 添加；附加；叠加（12B）
supervisor /'sjuːpəvaizə/ n. 监督人；管理人；指导者（11A）
supply chain 供应链（12C）
supposedly /sə'pəuzidli/ ad. 据推测；据称；大概，可能（10B）
suppress /sə'pres/ v. 压制；抑制；阻止（10A）
suppression /sə'preʃən/ n. 压制；抑制；阻止（12C）
surf /səːf/ v. （在…）冲浪（9C）
surround sound 环绕立体声（2B）
surveillance /səː'veiləns, sə-/ n. 监视（10B）
susceptible /sə'septəbl/ a. 易受影响的（*to*）（10A）
suspicious /sə'spiʃəs/ a. 可疑的（10B）
swap /swɔp/ v. 交换（10C）
swarm /swɔːm/ n. 一大群；蜂群（7A）
switch /'switʃ/ n. 交换（设备）；开关；转换（1A）
symptomatic /ˌsimptə'mætik/ a. 作为征候（或征兆）的；表明的（*of*）（5C）
synchronization /ˌsiŋkrənai'zeiʃən/ n. 同步（化）（3B）
synchronize /'siŋkrənaiz/ v. （使）同步；（使）协调（4B）
synchrony /'siŋkrəni/ n. 同时（性）；同步（性）（2C）
synergistic /ˌsinə'dʒistik/ a. 增效的；协同作用的；协作的（12C）
syntactic /sin'tæktik/ a. （按照）句法的（3B）
syntax /'sintæks/ n. 句法（3A）
synthesizer /'sinθisaizə/ n. 合成器（12A）
system call 系统调用（4A）
system crash 系统崩溃（10A）
system integration 系统集成（4C）
system routine 系统例程（4A）

# T

tab /tæb/ n. 标记，标签；制表键；工作表选项卡（10B）
tablet /'tæblit/ n. 平板电脑（1C）
tack /tæk/ v. 附加，追加（*on*）（9C）

| | |
|---|---|
| tactic /ˈtæktik/ n. | 战术；策略，手段（8A） |
| tactile /ˈtæktail; -til/ a. | （有）触觉的（2B） |
| tad /tæd/ n. | 〈口〉微量（1C） |
| tangible /ˈtændʒəbl/ a. | 可触摸的；有形的；明确的（12C） |
| tapestry /ˈtæpistri/ n. | 花毯；挂毯（4A） |
| targeted /ˈtɑːgitid/ a. | 有目标的；指向目标的（8C） |
| taskbar /ˈtɑːskbɑː; ˈtæsk-/ n. | 任务条，任务栏（10B） |
| telecom /ˈtelikɔm/ n. | 电信（= telecommunication）（12C） |
| telemedicine /ˈteliˌmedisin/ n. | 远程医学（9A） |
| teleoperation /ˌteliɔpəˈreiʃən/ n. | 远程操作，遥操作（12C） |
| telepresence /ˈteliˌprezəns/ n. | 远程呈现（12B） |
| televise /ˈtelivaiz/ v. | 用电视播放（12B） |
| template /ˈtempleit, -plit/ n. | 模板，样板（5C） |
| temporal /ˈtempərəl/ a. | 时间的（6C） |
| tenant /ˈtenənt/ n. | 租户，承租人（9A） |
| terabyte /ˈterəbait/ n. | 太字节，万亿字节（6C） |
| terminate /ˈtəːmineit/ v. | （使）终止（2B） |
| terminator /ˈtəːmineitə/ n. | 终结器；端子；终结符（7B） |
| terminology /ˌtəːmiˈnɔlədʒi/ n. | ［总称］术语（7A） |
| test harness | 测试框架（5B） |
| test script | 测试脚本（5B） |
| testbed /ˈtestbed/ n. | 试验台，测试台（5B） |
| text editor | 文本编辑程序，文本编辑器（4A） |
| textbox /ˈtekstbɔks/ n. | 文本（或正文、文字）框（11A） |
| texting /ˈtekstiŋ/ n. | 发短信（9C） |
| textual /ˈtekstjuəl, -tʃuəl/ a. | 文本的，正文的（2B） |
| theorem /ˈθiərəm/ n. | 定理（3B） |
| theorem prover | 定理证明程序，定理证明器（3B） |
| theoretician /ˌθiəriˈtiʃən/ n. | 理论家（12A） |
| therapeutic /ˌθerəˈpjuːtik/ a. | 治疗的；有疗效的（12A） |
| therapist /ˈθerəpist/ n. | （特定治疗法的）治疗专家（12A） |
| thermostat /ˈθəːməustæt/ n. | 温度自动调节器，恒温器（12C） |
| thesis /ˈθiːsis/ n. | （［复］-ses /-siːz/）命题；论点；（学位）论文（12A） |
| thread /θred/ n. | 线程，线索（3B） |
| three-dimensional /ˈθriːdiˈmenʃənəl/ a. | 三维的，立体的（1B） |
| throughput /ˈθruːput/ n. | 吞吐量；吞吐率（5A） |
| throwaway /ˈθrəuəˌwei/ a. | 使用后抛弃的，一次性使用的（4C） |
| thrust /θrʌst/ n. | 要点，要旨；目标（10A） |
| tier /tiə/ n. | （一）层（2C） |
| tilt /tilt/ v. | （使）倾斜，（使）倾侧（12A） |
| time slice | 时间片（2B） |
| timer /ˈtaimə/ n. | 计时器，定时器；定时程序（2B） |
| time-sharing /ˈtaimˌʃɛəriŋ/ n. & a. | 分时（的）（2B） |
| toehold /ˈtəuhəuld/ n. | （攀登悬崖等时脚趾大小的）立足点，支点（12A） |
| toolset /ˈtuːlset/ n. | 成套工具，工具箱（2C） |
| top-level domain | 顶级域名（8A） |
| topology /təuˈpɔlədʒi, tɔ-/ n. | 拓扑（结构），布局（7A） |
| torpedo /tɔːˈpiːdəu/ n. | 鱼雷（1A） |

| | |
|---|---|
| touch screen | 触摸屏，触屏（2B） |
| touchscreen /ˈtʌtʃskriːn/ n. | 触摸屏，触屏（1C） |
| touch-sensitive /ˈtʌtʃˌsensitiv/ a. | 触敏的（1C） |
| touch-sensitive screen | 触摸屏（1C） |
| trackball /ˈtrækbɔːl/ n. | 跟踪球，轨迹球（2A） |
| trade-off /ˈtreidɔf/ n. | 平衡，权衡（亦作 tradeoff）（5C） |
| trafficwise /ˈtræfikwaiz/ ad. | 在交通方面（7C） |
| tragic(al) /ˈtrædʒik(əl)/ a. | 悲剧（性）的；悲惨的（5C） |
| trajectory /trəˈdʒektəri/ n. | 轨迹，轨道（6C） |
| transistor /trænˈzistə, -ˈsis-/ n. | 晶体管（1A） |
| translator /trænsˈleitə/ n. | 翻译程序，翻译器（2C） |
| trapdoor /ˈtræpdɔː/ n. | 陷阱门，天窗（10C） |
| triad /ˈtraiəd, -æd/ n. | 三件一套（或组）；三位一体（10A） |
| trial-and-error /ˈtraiələndˈerə/ a. | 试错法的；反复试验的（12A） |
| trigger /ˈtrigə/ v. & n. | 触发（2C）/ 引爆器；触发器（10C） |
| trojan /ˈtrəudʒən/ n. | 特洛伊木马（程序或病毒），木马（10B） |
| Trojan /ˈtrəudʒən/ (horse) n. | 特洛伊木马（程序或病毒），木马（3B） |
| trust delegation | 信任委托（9A） |
| trustworthiness /ˈtrʌstˌwəːðinis/ n. | 值得信任；可信；可靠（9B） |
| trustworthy /ˈtrʌstˌwəːði/ a. | 值得信任的；可信的；可靠的（3B） |
| tuple /ˈtʌpl, ˈtjuːpl/ n. | 元组，字节组（6A） |
| Turing test | 图灵测试（12A） |
| tussle /ˈtʌsəl/ n. | 争执，争辩（11C） |
| tweet /twiːt/ v. | 上推特，发微博（9C） |
| two-dimensional array | 二维数组（3C） |
| two-dimensional table | 二维表（6A） |
| typically /ˈtipikəli/ ad. | 一般地，通常（1C） |
| typo /ˈtaipəu/ n. | 打字（或排印）错误（11A） |

# U

| | |
|---|---|
| ubiquitous /juːˈbikwitəs/ a. | 普遍存在的，无所不在的（9B） |
| ubiquitous computing | 普适计算（12C） |
| ultra- /ˈʌltrə/ pref. | 表示“极端”“超”（9B） |
| unattended /ˌʌnəˈtendid/ a. | 没人看管的；没人照料的（12C） |
| unauthorized /ˌʌnˈɔːθəraizd/ a. | 未经授权的（3B） |
| underlie /ˌʌndəˈlai/ v. | 构成…的基础（或起因）；支持（1B） |
| underline /ˈʌndəˌlain/ n. | 下划线（8C） |
| underlying /ˌʌndəˈlaiiŋ/ a. | 基本的，根本的（2C） |
| undertaking /ˌʌndəˈteikiŋ, ˈʌndət-/ n. | 任务；事业（8A） |
| underway /ˈʌndəˈwei/ a. | 在进行中的（3B） |
| unforeseen /ˌʌnfɔːˈsiːn/ a. | 未预见到的；意料之外的（1B） |
| Uniform Resource Locator | 统一资源定位符，统一资源定位器（8C） |
| unify /ˈjuːnifai/ v. | 使联合，统一；使一致（1B） |
| unimpaired /ˌʌnimˈpɛəd/ a. | 未受损的；未削弱的（10A） |
| union /ˈjuːnjən/ n. | 共用体（定义关键字），共用（数据类型）（3B） |
| unit testing | 单元测试（4C） |
| unscrupulous /ˌʌnˈskruːpjuləs/ a. | 肆无忌惮的；无耻的（10C） |

untrustworthy /ˌʌn'trʌstˌwə:ði/ a. 不可信赖的；靠不住的（10B）
update /'ʌpdeit/ n. 更新；修改（6B）
update /ʌp'deit/ v. 更新；修改（4B）
upgrade /ʌp'greid, 'ʌpg-/ v. 使升级；改善（2C）
uphold /ʌp'həuld/ v. 举起；支持；维护（11B）
uplink /'ʌpliŋk/ n. 上行链路（8A）
upload /ˌʌp'ləud/ v. 上传，上载（9C）
uppercase /'ʌpə'keis/ a. （字母）大写的（8A）
use case 用例（5B）
user ID 用户标识（符）（10C）
user interface 用户界面（2B）
utility /ju:'tiliti/ n. 实用程序，公用程序（4A）
utility package 实用软件包，公用程序包（7A）
utility program 实用程序，公用程序（4A）
utility software 实用软件（8B）
utilization /ˌju:tilai'zeiʃən; -li'z-/ n. 利用（9A）

# V

vacuum tube 真空管（1A）
validate /'vælideit/ v. 确认（…有效），证实，验证（4B）
validation /ˌvæli'deiʃən/ n. 确认，验证（4C）
validity /və'liditi/ n. 有效（性）（10A）
vantage /'vɑ:ntidʒ; 'væn-/ n. 优势；有利地位（12B）
vantage point 有利位置；观点，看法（12B）
vector /'vektə/ n. 传病媒介；矢量（10C）
velocity /vi'lɔsiti/ n. 速度；速率（6C）
vendor /'vendə/ n. 卖主；厂家，厂商（3B）
verbal /'və:bəl/ a. 用言辞的；文字上的；口头的（11C）
versatile /'və:sətail/ a. 多用途的，通用的（1A）
versatility /ˌvə:sə'tiliti/ n. 多用途；通用性（3A）
versus /'və:səs/ prep. 对；与…相对（或相比）（10A）
viable /'vaiəbəl/ a. 切实可行的；可实施的（1B）
video camera 摄像机（8A）
video conferencing （召开）视频会议；视频会议技术（12B）
video game 电子游戏（2A）
view /vju:/ n. 视图（3C）
violate /'vaiəleit/ v. 违反，违犯（11A）
viral /'vaiərəl/ a. 病毒（性）的；病毒引起的（10B）
virtual /'və:tʃuəl/ a. 虚拟的（1B）
virtual machine 虚拟机（9B）
virtual memory 虚拟内存，虚拟存储器（2B）
virtual reality 虚拟现实（1B）
virtualization /və:tʃuəlai'zeiʃən/ n. 虚拟化（9A）
virtualize /'və:tʃuəlaiz/ v. 虚拟化（9A）
virus signature 病毒特征码（10B）
visualization /ˌvizjuəlai'zeiʃən/ n. 可视化，直观化（1B）
visualize /'vizjuəlaiz/ v. 使可视化；使直观化（4B）

volatile /ˈvɔlətail/ a. 易失（性）的（2A）
vulnerability /ˌvʌlnərəˈbiliti/ n. 易遭攻击的地方；脆弱（性）；漏洞（10A）
vulnerable /ˈvʌlnərəbl/ a. 脆弱的；易受攻击的（3B）

# W

walkthrough /ˈwɔːkθruː/ n. 走查（5B）
warrant /ˈwɔrənt; ˈwɔː-/ n. 授权（令）（11C）
waterfall /ˈwɔːtəfɔːl/ n. 瀑布（4B）
waterfall model 瀑布模型（4C）
wearable computer 可穿戴计算机，穿戴式计算机（1C）
Web browser 网络浏览器（2C）
Web page （万维）网页（5A）
Web server 万维网服务器（8C）
Web site 网站，站点（1C）
webpage /ˈwebpeidʒ/ n. 网页（11A）
website /ˈwebsait/ n. 网站，站点（5A）
whereby /hwɛəˈbai/ ad. 靠那个；借以（9A）
white box testing 白盒测试（法）（5B）
wide area network 广域网（6A）
wildcard /ˈwaildkɑːd/ n. 通配符，万能符（8C）
wildcard character 通配符（8C）
wink /wiŋk/ v. 眨眼；眨眼示意，使眼色（11A）
witty /ˈwiti/ a. 诙谐的；说话风趣的（11A）
word processor 文字处理软件（11A）
workflow /ˈwəːkfləu/ n. 工作流（5A）
workload /ˈwəːkləud/ n. 工作量；工作负荷（9A）
workstation /ˈwəːkˌsteiʃən/ n. 工作站（1A）
worm /wəːm/ n. 蠕虫（病毒）（10B）
wristwatch /ˈristwɔtʃ/ n. 手表（1A）

# Y

yardage /ˈjɑːdidʒ/ n. 码数；以码计量的长度（12B）

# Z

zebra /ˈziːbrə, ˈze-/ n. 斑马（9C）
zip /zip/ v. 压缩（文件）（10B）
zombie /ˈzɔmbi/ n. 还魂尸，僵尸；僵（进程）（10C）

# Abbreviations

# （缩略语表）

| | | |
|---|---|---|
| 3D | three-dimensional | 三维的，立体的（2A） |
| 3D | three dimensions | 三维，立体（12B） |
| ABC | American Broadcasting Company | 美国广播公司（8C） |
| ACM | Association for Computing Machinery | （国际）计算机协会（11B） |
| AI | artificial intelligence | 人工智能（1B） |
| ANSI/SPARC | American National Standards Institute Standards Planning and Requirements Committee | 美国国家标准协会标准计划与需求委员会（6A） |
| AP | access point | （访问）接入点（7A） |
| API | Application Program Interface | 应用程序接口（3B） |
| APSE | Ada Programming Support Environment | Ada 程序设计支持环境（2C） |
| AR | augmented reality | 增强现实（12B） |
| AWS | Amazon Web Services | 亚马逊网络服务（9A） |
| BASIC | Beginner's All-purpose Symbolic Instruction Code | BASIC 语言，初学者通用符号指令码（3A） |
| bcc | blind carbon copy | 密送（11A） |
| BD | Blu-ray disc | 蓝光光盘（2A） |
| BIOS | Basic Input/Output System | 基本输入 / 输出系统（2A） |
| BLURS | bandwidth, latency, uninterrupted, resource-constraint, and security | 带宽、时延、不间断、资源约束和安全性（9B） |
| CaaS | context as a service | 上下文即服务（9B） |
| CAD | computer-aided design | 计算机辅助设计（2C） |
| CAN | campus area network | 校园区域网络，校园网（9B） |
| CAPTCHA | Completely Automated Public Turing Test to Tell Computers and Humans Apart | 全自动区分计算机和人类的图灵测试，验证码（11C） |
| CASE | computer-aided software engineering | 计算机辅助软件工程（2C） |
| CBSE | component-based software engineering | 基于组件的软件工程（4C） |
| cc | carbon copy | 抄送（11A） |
| CD | compact disc | 光盘（2A） |
| CD-ROM | compact disc read-only memory | 只读光盘（存储器）（10A） |
| CE | containers engine | 容器引擎（9B） |
| CIA | confidentiality, integrity, and availability | 机密性、完整性与可用性（10A） |
| CIoT | Cloud-centric Internet of Things | 以云为中心的物联网（9B） |

| | | |
|---|---|---|
| CLOS | *Common Lisp Object System* | 公共 Lisp 对象系统（5C） |
| CNN | *Cable News Network* | （美国）有线电视新闻网（8C） |
| COBOL | *Common Business-Oriented Language* | COBOL 语言，面向商业的通用语言（3A） |
| CORBA | *Common Object Request Broker Architecture* | 公用对象请求代理（程序）体系结构（3B） |
| COTS | *commercial off-the-shelf* | 现成商用的，现成民用的（4C） |
| CPU | *central processing unit* | 中央处理器（1C） |
| CSMA/CA | *Carrier Sense, Multiple Access with Collision Avoidance* | 带冲突避免的载波侦听多址（或多路）访问（协议）（7A） |
| CSMA-CD | *Carrier Sense, Multiple Access with Collision Detection* | 带冲突检测的载波侦听多址（或多路）访问（协议）（7A） |
| DARPA | *Defense Advanced Research Projects Agency* | （美国）国防部高级研究计划局（8A） |
| DBA | *database administrator* | 数据库管理员（6A） |
| DBMS | *database management system* | 数据库管理系统（6A） |
| DCS | *Digital Collection System* | 数字收集系统（11C） |
| DMCA | *Digital Millennium Copyright Act* | 《数字千年版权法》《千禧年数字版权法》（11C） |
| DNA | *deoxyribonucleic acid* | 脱氧核糖核酸（1A） |
| DNS | *domain name system* | 域名系统（8A） |
| DSL | *Digital Subscriber Line* | 数字用户线路（8A） |
| DVD | *digital versatile disc* | 数字多功能光盘（2A） |
| ENIAC | *Electronic Numerical Integrator And Computer* | 电子数字积分计算机，ENIAC 计算机（1A） |
| EPC | *evolved packet core* | 演进分组核心（9B） |
| ERM | *entity-relationship model* | 实体关系模型（6A） |
| FAST | *Framework for the Application of Systems Techniques* | 系统技术应用框架（4B） |
| FBI | *Federal Bureau of Investigation* | （美国）联邦调查局（11C） |
| FDM | *fused deposition modeling* | 熔融沉积成型（2A） |
| FEC | *fog and edge computing* | 雾与边缘计算（9B） |
| flops | *floating-point operations per second* | 每秒浮点运算次数（1C） |
| Fortran | *Formula Translation* | Fortran 语言，公式翻译程序语言（3A） |
| FPGA | *field programmable gate array* | 现场可编程门阵列（9B） |
| FTP | *File Transfer Protocol* | 文件传送协议，文件传输协议（8B） |
| GAE | *Google App Engine* | 谷歌应用引擎（9A） |
| GB | *gigabyte* | 吉字节，千兆字节（2B） |
| GPS | *Global Positioning System* | 全球（卫星）定位系统（1C） |
| GPU | *graphics processing unit* | 图形处理单元，图形处理器（9B） |
| GUI | *graphical user interface* | 图形用户界面（2B） |
| HDD | *hard disk drive* | 硬（磁）盘驱动器（2A） |

| | | |
|---|---|---|
| HMD | head-mounted display | 头戴式显示器（12B） |
| HTC | high-throughput computing | 高吞吐（量）计算（9A） |
| HTML, html | Hypertext Markup Language | 超文本标记语言（8B） |
| HTTP, http | Hypertext Transfer Protocol | 超文本传送协议，超文本传输协议（2C） |
| HTTPS, https | Hypertext Transfer Protocol over Secure Sockets Layer | 使用安全套接层的超文本传输协议，安全超文本传输协议（5A） |
| I/PaaS | infrastructure or platform as a service | 基础设施或平台即服务（9B） |
| IaaS | infrastructure as a service | 基础设施即服务（9B） |
| IBM | International Business Machines | （美国）国际商用机器公司（1C） |
| IC | integrated circuit | 集成电路（1A） |
| ICANN | Internet Corporation for Assigned Names and Numbers | 因特网名称与数字地址分配机构（8A） |
| ICL | International Computers Limited | 国际计算机有限公司（2C） |
| ID | identification | 身份证明，身份识别，标识（10C） |
| IE | information engineering | 信息工程（4B） |
| IE | Internet Explorer | IE 网页浏览器，IE 浏览器（8C） |
| IEEE | Institute of Electrical and Electronics Engineers | （美国）电气和电子工程师协会（7A） |
| IEEE CS | IEEE Computer Society | IEEE 计算机学会（11B） |
| IoT | Internet of Things | 物联网（9B） |
| IP | Internet Protocol | IP 协议，网际协议，网间协议（8A） |
| IPv6 | Internet Protocol version 6 | IP 协议第 6 版（12C） |
| ISP | Internet Service Provider | 因特网服务提供商（8A） |
| IT | information technology | 信息技术（9A） |
| KB | kilobyte | 千字节（2B） |
| KDD | knowledge discovery in databases | 数据库知识发现（6C） |
| LAN | local area network | 局域网（7A） |
| LCD | liquid crystal display | 液晶显示（器）（2A） |
| LED | light-emitting diode | 发光二极管（2A） |
| LISP | List Processing | LISP 语言，表处理语言（5C） |
| LSI | large-scale integrated | 大规模集成的（1A） |
| | large-scale integration | 大规模集成（电路）（1A） |
| MAN | metropolitan area network | 城域网（7A） |
| MDD | model-driven development | 模型驱动开发（4B） |
| NFC | Near Field Communication | 近场通信，近距离无线通信（9C） |
| NIST | National Institute of Standards and Technology | （美国）国家标准与技术研究院（10A） |
| NRCPC | National Research Center of Parallel Computer Engineering and Technology | （中国）国家并行计算机工程技术研究中心（1C） |

| | | |
|---|---|---|
| ODP | on-demand data processing | 按需数据处理（9B） |
| OOAD | object-oriented analysis and design | 面向对象分析与设计（4B） |
| OS | operating system | 操作系统（1C） |
| OSI | Open Systems Interconnection | 开放式系统互联（2C） |
| P2P | peer-to-peer | 对等的（7A） |
| PaaS | platform as a service | 平台即服务（9B） |
| PAN | personal area network | 个人域网（7A） |
| PC | personal computer | 个人计算机，个人电脑（1C） |
| PDA | personal digital assistant | 个人数字助理（1C） |
| PFLOPS | petaflops | 每秒千万亿次浮点运算（1C） |
| PMI | Project Management Institute | 项目管理协会（11B） |
| Prolog | Programming in Logic | Prolog 语言，逻辑程序设计语言（3A） |
| PUA | potentially unwanted application | 可能有害的应用程序（10B） |
| QoS | Quality of Service | 服务质量（9A） |
| R | recordable | 可写的（2A） |
| RAM | random access memory | 随机（存取）存储器（2A） |
| RC2 | Research Compute Cloud | 研究计算云（9A） |
| RDBMS | relational database management system | 关系数据库管理系统（6A） |
| REST | Representational State Transfer | 表示层状态转化（5A） |
| RF | radio frequency | 射频；无线电频率（12C） |
| RFID | radio-frequency identification | 射频识别（9C） |
| ROM | read-only memory | 只读存储器（1A） |
| RPG | Report Program Generator | RPG 语言，报表程序生成器（3A） |
| RW | rewritable | 可重写的（2A） |
| S/CaaS | storage or caching as a service | 存储或缓存即服务（9B） |
| SaaS | Software as a Service | 软件即服务（9A） |
| SAN | storage area network | 存储区域网（络）（9A） |
| SCALE | security, cognition, agility, latency, and efficiency | 安全、认知、敏捷、低延迟和高效率（9B） |
| SCANC | storage, compute, acceleration, networking, and control | 存储、计算、加速、组网和控制（9B） |
| SDK | Software Development Kit | 软件开发工具包（9B） |
| SDN | software-defined network | 软件定义网络（9B） |
| SF | science fiction | 科学幻想小说，科幻小说（10C） |
| SLA | service-level agreement | 服务水平（或级别、等级）协议（9A） |
| SMTP | Simple Mail Transfer Protocol | 简单邮件传送协议，简单邮件传输协议（8B） |
| SOA | service-oriented architecture | 面向服务的体系结构（5A） |
| SOAP | Simple Object Access Protocol | 简单对象访问协议（5A） |

| | | |
|---|---|---|
| SQL | *S*tructured *Q*uery *L*anguage | 结构化查询语言（3A） |
| SSD | *s*olid *s*tate *d*rive | 固态驱动器（2A） |
| SSH | *S*ecure *Sh*ell | 安全外壳（8B） |
| TCP/IP | *T*ransmission *C*ontrol *P*rotocol/*I*nternet *P*rotocol | TCP/IP 协议，传输控制（协议）/ 网际（或网间）协议（7A） |
| TLD | *t*op-*l*evel *d*omain | 顶级域名（8A） |
| UDDI | *U*niversal *D*escription, *D*iscovery, and *I*ntegration | 通用描述、发现与集成（5A） |
| UK | *U*nited *K*ingdom | 联合王国，英国（2C） |
| URL | *U*niform *R*esource *L*ocator, *U*niversal *R*esource *L*ocator | 统一资源定位符，统一资源定位器（5A） |
| USB | *u*niversal *s*erial *b*us | 通用串行总线（2A） |
| VLSI | *v*ery *l*arge-*s*cale *i*ntegrated | 超大规模集成的（1A） |
| | *v*ery *l*arge-*s*cale *i*ntegration | 超大规模集成（电路）（1A） |
| VM | *v*irtual *m*achine | 虚拟机（9A） |
| VR | *v*irtual *r*eality | 虚拟现实（12B） |
| WAN | *w*ide *a*rea *n*etwork | 广域网（7A） |
| WiFi, Wi-Fi | *Wi*reless *Fi*delity | 无线保真（7A） |
| WLAN | *w*ireless *l*ocal *a*rea *n*etwork | 无线局域网（9B） |
| WS | *W*eb *S*ervice | Web 服务（5A） |
| WS-BPEL | *W*eb *S*ervice *B*usiness *P*rocess *E*xecution *L*anguage | Web 服务业务流程执行语言（5A） |
| WSDL | *W*eb *S*ervice *D*escription *L*anguage | Web 服务描述语言（5A） |
| WSN | *w*ireless *s*ensor *n*etwork | 无线传感器网络（12C） |
| WWW | *W*orld *W*ide *W*eb | 万维网（2C） |
| XaaS | *X* *a*s *a* *s*ervice | X 即服务，一切皆服务（9B） |
| XHTML | e*X*tensible *H*yper*t*ext *M*arkup *L*anguage | 可扩展超文本标记语言（5A） |
| XML | e*X*tensible *M*arkup *L*anguage | 可扩展标记语言（2C） |
| XSD | *X*ML *S*chema *D*efinition | XML 模式定义（5A） |
| XSLT | e*X*tensible *S*tylesheet *L*anguage *T*ransformations | 可扩展样式表转换语言（5A） |